Advancements in Wireless Communications, Networks and Signal Processing

Advancements in Wireless Communications, Networks and Signal Processing

Editors

Runzhou Zhang
Lin Zhang
Yang Yue
Hao Feng
Zheda Li
Dawei Ying

MDPI

Basel • Beijing • Wuhan • Barcelona • Belgrade • Novi Sad • Cluj • Manchester

Editors
Runzhou Zhang
University of Southern California
Los Angeles
USA

Lin Zhang
Tianjin University
Tianjin
China

Yang Yue
Xi'an Jiaotong University
Xi'an
China

Hao Feng
Intel Corporation
Santa Clara
USA

Zheda Li
Amazon Lab126
Sunnyvale
USA

Dawei Ying
Intel Corporation
Santa Clara
USA

Editorial Office
MDPI AG
Grosspeteranlage 5
4052 Basel, Switzerland

This is a reprint of articles from the Special Issue published online in the open access journal *Applied Sciences* (ISSN 2076-3417) (available at: https://www.mdpi.com/journal/applsci/special_issues/Wireless_Communications_Networks_Signal_Processing).

For citation purposes, cite each article independently as indicated on the article page online and as indicated below:

Lastname, A.A.; Lastname, B.B. Article Title. *Journal Name* **Year**, *Volume Number*, Page Range.

ISBN 978-3-7258-1733-7 (Hbk)
ISBN 978-3-7258-1734-4 (PDF)
doi.org/10.3390/books978-3-7258-1734-4

Contents

Runzhou Zhang, Lin Zhang, Yang Yue, Hao Feng, Zheda Li and Dawei Ying
Editorial for Special Issue "Advancements in Wireless Communications, Networks, and Signal Processing"
Reprinted from: *Appl. Sci.* **2024**, *14*, 5725, doi:10.3390/app14135725 1

Mirko Stojčić, Milorad K. Banjanin, Milan Vasiljević, Dragana Nedić, Aleksandar Stjepanović, Dejan Danilović and Goran Puzić
Predictive Modeling of Delay in an LTE Network by Optimizing the Number of Predictors Using Dimensionality Reduction Techniques
Reprinted from: *Appl. Sci.* **2023**, *13*, 8511, doi:10.3390/app13148511 3

Hanbing Shi and Juan Wang
Intelligent TCP Congestion Control Policy Optimization
Reprinted from: *Appl. Sci.* **2023**, *13*, 6644, doi:10.3390/app13116644 27

Yumei Cao, Peng Li, Tianmian Liang, Xiaojun Wu, Xiaoming Wang and Yuanru Cui
A Novel Opportunistic Network Routing Method on Campus Based on the Improved Markov Model
Reprinted from: *Appl. Sci.* **2023**, *13*, 5217, doi:10.3390/app13085217 40

Yilan Wang, Linbo Yang, Zhiqun Yang, Yaping Liu, Zhanhua Huang and Lin Zhang
High-Performance Microwave Photonic Transmission Enabled by an Adapter for Fundamental Mode in MMFs
Reprinted from: *Appl. Sci.* **2023**, *13*, 1794, doi:10.3390/app13031794 59

Xutao Wang, Honglin Sun, Huihui Wang, Zhiqun Yang, Yaping Liu, Zhanhua Huang and Lin Zhang
Heterogeneously Integrated Multicore Fibers for Smart Oilfield Applications
Reprinted from: *Appl. Sci.* **2023**, *13*, 1579, doi:10.3390/app13031579 65

Batzorig Bazargur, Otgonbayar Bataa and Uuganbayar Budjav
Reliability Study for Communication System: A Case Study of an Underground Mine
Reprinted from: *Appl. Sci.* **2023**, *13*, 821, doi:10.3390/app13020821 75

MariaCarmen de Toro, Carlos Borrego and Sergi Robles
A Controller-Driven Approach for Opportunistic Networking
Reprinted from: *Appl. Sci.* **2022**, *12*, 12479, doi:10.3390/app122312479 95

Diego Freire, Carlos Borrego and Sergi Robles
Corpus for Development of Routing Algorithms in Opportunistic Networks
Reprinted from: *Appl. Sci.* **2022**, *12*, 9240, doi:10.3390/app12189240 131

Jie Shen, Yijun Hao, Yuqian Yang and Cong Zhao
User-BS Selection Strategy Optimization with RSSI-Based Reliability in 5G Wireless Networks
Reprinted from: *Appl. Sci.* **2022**, *12*, 6082, doi:10.3390/app12126082 150

Fei Duan, Yuhao Guo, Zenghui Gu, Yanlong Yin, Yixin Wu and Teyan Chen
Optical Beamforming Networks for Millimeter-Wave Wireless Communications
Reprinted from: *Appl. Sci.* **2023**, *13*, 8346, doi:10.3390/app13148346 169

applied sciences

MDPI

Editorial

Editorial for Special Issue "Advancements in Wireless Communications, Networks, and Signal Processing"

Runzhou Zhang [1,*], Lin Zhang [2], Yang Yue [3], Hao Feng [4], Zheda Li [5] and Dawei Ying [4]

1 Department of Electrical and Computer Engineering, University of Southern California, Los Angeles, CA 90089, USA
2 School of Precision Instrument and Optoelectronics Engineering, Tianjin University, Tianjin 300072, China
3 School of Information and Communications Engineering, Xi'an Jiaotong University, Xi'an 710049, China
4 Intel Corporation, Santa Clara, CA 95054, USA; haofeng.fh666@gmail.com (H.F.)
5 Amazon Lab126, Sunnyvale, CA 94089, USA
* Correspondence: runzhou@usc.edu

1. Introduction

Due to dramatic increases in data traffic over the last decade, there has been growing interest in enhancing system performance levels in data communications and networking. Moreover, data are represented in various digital forms, such as text, voice, or video. As data capacity continues to increase, it imposes ever-higher requirements on data-processing capabilities and networking system architectures. The desired system characteristics include, but are not limited to, the following: (i) lower power consumption, (ii) minimized networking latency, (iii) higher spectral efficiency/throughput, and (iv) enhanced data security.

These technical advancements are required across the entire data infrastructure industry, in wireless networks, data-center interconnects, long haul telecommunications, and other communication platforms. To achieve these advances, extensive research have been conducted in different layers of system architectures, including (i) physical layer schemes to enhance the bandwidth utilization of physical media; (ii) digital signal processing algorithms to optimize signal-over-noise ratio and recovered data quality; (iii) networking control and policies to minimize system latency and increase system throughput; and (iv) full-stack software design to enhance system resilience and data security.

In this Editorial, we are proud to introduce the Special Issue "Advancements in Wireless Communications, Networks, and Signal Processing". This Special Issue highlights research efforts dedicated to addressing the technical challenges faced in the broader data communication field, focusing on original system and algorithm approaches, which can enhance state-of-the-art techniques for communications and networking.

2. Contributions

This Special Issue features original research and review articles addressing recent advancements in wireless communications, networks, and digital signal processing. The research studies in this publication cover a variety of communication applications, including (i) LTE/5G/millimeter-wave wireless network modeling and optimization; (ii) ethernet network control and optimization; and (iii) transmission studies for optical fiber communications.

This Special Issue showcases research articles that aim to predict and optimize LTE/5G/ethernet network performance using quantitative modeling approaches, including (i) dimensionality reduction [1]; (ii) user-BS selection strategy [2]; and (iii) TCP-congestion control policy optimization [3]. Furthermore, this Special Issue highlights the latest research on opportunistic networks, including improved routing approaches based on improved corpus and Markov modeling [4,5] and controller-driven approaches to opportunistic networking [6].

Citation: Zhang, R.; Zhang, L.; Yue, Y.; Feng, H.; Li, Z.; Ying, D. Editorial for Special Issue "Advancements in Wireless Communications, Networks, and Signal Processing". *Appl. Sci.* **2024**, *14*, 5725. https://doi.org/10.3390/app14135725

Received: 15 June 2024
Accepted: 18 June 2024
Published: 30 June 2024

In addition to wireless networks, this Special Issue also presents optical fiber transmission studies based on space-division multiplexing techniques (either via multi-mode or multi-core fibers), including microwave transmission in multi-mode fiber [7] and heterogeneously integrated multi-core fiber [8]. Finally, this Special Issue features a communication reliability case study [9] and a review article on an optical beam forming technique for millimeter-wave wireless systems [10].

3. Conclusions

This Special Issue showcases research articles sharing advancements in communication system performance, achieved by optimizing physical-layer bandwidth/spectrum utilization, digital signal-processing efficiency, or networking protocols and latency. As data capacity is increasing at an accelerating speed, the Editors believe that there are significant opportunities for researchers to move beyond state-of-the-art communication technologies in this fast-expanding field. Future research could focus on enhancing different aspects of system performance, including broadening system bandwidths, increasing spectral efficiency, reducing networking latency, enhancing system security, and optimizing system resilience.

Acknowledgments: The authors thank the anonymous reviewers for contributing to this Special Issue by reviewing the submitted manuscripts.

Conflicts of Interest: The authors declare no conflicts of interest.

References

1. Stojčić, M.; Banjanin, M.K.; Vasiljević, M.; Nedić, D.; Stjepanović, A.; Danilović, D.; Puzić, G. Predictive Modeling of Delay in an LTE Network by Optimizing the Number of Predictors Using Dimensionality Reduction Techniques. *Appl. Sci.* **2023**, *13*, 8511. [CrossRef]
2. Shen, J.; Hao, Y.; Yang, Y.; Zhao, C. User-BS Selection Strategy Optimization with RSSI-Based Reliability in 5G Wireless Networks. *Appl. Sci.* **2022**, *12*, 6082. [CrossRef]
3. Shi, H.; Wang, J. Intelligent TCP Congestion Control Policy Optimization. *Appl. Sci.* **2023**, *13*, 6644. [CrossRef]
4. Cao, Y.; Li, P.; Liang, T.; Wu, X.; Wang, X.; Cui, Y. A Novel Opportunistic Network Routing Method on Campus Based on the Improved Markov Model. *Appl. Sci.* **2023**, *13*, 5217. [CrossRef]
5. Freire, D.; Borrego, C.; Robles, S. Corpus for Development of Routing Algorithms in Opportunistic Networks. *Appl. Sci.* **2022**, *12*, 9240. [CrossRef]
6. de Toro, M.; Borrego, C.; Robles, S. A Controller-Driven Approach for Opportunistic Networking. *Appl. Sci.* **2022**, *12*, 12479. [CrossRef]
7. Wang, Y.; Yang, L.; Yang, Z.; Liu, Y.; Huang, Z.; Zhang, L. High-Performance Microwave Photonic Transmission Enabled by an Adapter for Fundamental Mode in MMFs. *Appl. Sci.* **2023**, *13*, 1794. [CrossRef]
8. Wang, X.; Sun, H.; Wang, H.; Yang, Z.; Liu, Y.; Huang, Z.; Zhang, L. Heterogeneously Integrated Multicore Fibers for Smart Oilfield Applications. *Appl. Sci.* **2023**, *13*, 1579. [CrossRef]
9. Bazargur, B.; Bataa, O.; Budjav, U. Reliability Study for Communication System: A Case Study of an Underground Mine. *Appl. Sci.* **2023**, *13*, 821. [CrossRef]
10. Duan, F.; Guo, Y.; Gu, Z.; Yin, Y.; Wu, Y.; Chen, T. Optical Beamforming Networks for Millimeter-Wave Wireless Communications. *Appl. Sci.* **2023**, *13*, 8346. [CrossRef]

applied sciences

MDPI

Article

Predictive Modeling of Delay in an LTE Network by Optimizing the Number of Predictors Using Dimensionality Reduction Techniques

Mirko Stojčić [1], Milorad K. Banjanin [2,3,*], Milan Vasiljević [2], Dragana Nedić [1], Aleksandar Stjepanović [1], Dejan Danilović [1] and Goran Puzić [4]

[1] Department of Information and Communication Systems in Traffic, Faculty of Transport and Traffic Engineering Doboj, University of East Sarajevo, Vojvode Mišića 52, 74000 Doboj, Bosnia and Herzegovina; mirko.stojcic@sf.ues.rs.ba (M.S.); dragana.nedic@sf.ues.rs.ba (D.N.); aleksandar.stjepanovic@sf.ues.rs.ba (A.S.); danilovic.dejan@gmail.com (D.D.)

[2] Department of Computer Science and Systems, Faculty of Philosophy Pale, University of East Sarajevo, Alekse Šantića 1, 71420 Pale, Bosnia and Herzegovina; milanvasiljevic84@gmail.com

[3] Faculty of Technical Sciences, University of Novi Sad, Trg Dositeja Obradovića 6, 21102 Novi Sad, Serbia

[4] Faculty of Economics and Engineering Management in Novi Sad, University Business Academy in Novi Sad, Cvećarska 2, 21107 Novi Sad, Serbia; goran.puzic@fimek.edu.rs

* Correspondence: milorad.banjanin@ff.ues.rs.ba or milorad.banjanin@ffuis.edu.ba

Abstract: Delay in data transmission is one of the key performance indicators (KPIs) of a network. The planning and design value of delay in network management is of crucial importance for the optimal allocation of network resources and their performance focuses. To create optimal solutions, predictive models, which are currently most often based on machine learning (ML), are used. This paper aims to investigate the training, testing and selection of the best predictive delay model for a VoIP service in a Long Term Evolution (LTE) network using three ML techniques: Multilayer Perceptron (MLP), Support Vector Machines (SVM) and k-Nearest Neighbors (k-NN). The space of model input variables is optimized by dimensionality reduction techniques: RReliefF algorithm, Backward selection via the recursive feature elimination algorithm and the Pareto 80/20 rule. A three-segment road in the geo-space between the cities of Banja Luka (BL) and Doboj (Db) in the Republic of Srpska (RS), Bosnia and Herzegovina (BiH), covered by the cellular network (LTE) of the M:tel BL operator was chosen for the case study. The results show that the k-NN model has been selected as the best solution in all three optimization approaches. For the RReliefF optimization algorithm, the best model has six inputs and the minimum relative error (*RE*) $RE = 0.109$. For the Backward selection via the recursive feature elimination algorithm, the best model has four inputs and $RE = 0.041$. Finally, for the Pareto 80/20 rule, the best model has 11 inputs and $RE = 0.049$. The comparative analysis of the results concludes that, according to observed criteria for the selection of the final model, the best solution is an approach to optimizing the number of predictors based on the Backward selection via the recursive feature elimination algorithm.

Keywords: delay; dimensionality reduction; LTE; VoIP; Multilayer Perceptron; Support Vector Machines; k-nearest neighbors; Feature Selection; Pareto 80/20 rule

Citation: Stojčić, M.; Banjanin, M.K.; Vasiljević, M.; Nedić, D.; Stjepanović, A.; Danilović, D.; Puzić, G. Predictive Modeling of Delay in an LTE Network by Optimizing the Number of Predictors Using Dimensionality Reduction Techniques. *Appl. Sci.* **2023**, *13*, 8511. https://doi.org/10.3390/app13148511

Academic Editors: Runzhou Zhang, Lin Zhang, Yang Yue, Hao Feng, Zheda Li and Dawei Ying

Received: 28 June 2023
Revised: 19 July 2023
Accepted: 21 July 2023
Published: 23 July 2023

1. Introduction

Sustainable Quality of Service (QoS) for users is one of the main tasks of mobile operators. This orients them to provide comprehensive support for various applications and services with numerous QoS requirements in order to meet the expected levels of user Quality of Experience (QoE) [1,2]. The development of Long Term Evolution (LTE) technology, which today is based on IP network configuration [3], is an example of such an orientation. The target reason is optimal performance, i.e., low delay and high data transfer speed, as well as better optimization of packet transfer. In addition to the mentioned key features of LTE network technology, there is Radio Resource Management (RRM), which can raise network performance almost to the level of the Shannon limit [4]. An important operational technology of LTE is Packet scheduling for assigning a part of the network's resources to each User Equipment (UE) depending on QoS requirements, but also on the impact of delay, channel quality, number of active UEs, throughput, etc. During network congestion, users' QoS requirements increase, and today popular interactive real-time services, such as Voice over IP (VoIP), i.e., Voice over Long Term Evolution (VoLTE), and streaming are the most sensitive and susceptible to degradation in that period. Key network performance indicators during congestion are end-to-end (E2E) delay and jitter, which represents variations in delay [5]. According to the standard 123 107 v12.0.0 (2014) of the European Telecommunications Standards Institute (ETSI) [6], the maximum tolerated delay for VoIP services is defined as 100 ms, and 300 ms for streaming services. End-to-end delay can be defined as the time required for a data packet to be transmitted through a network from a source node to a destination node, and in a VoIP network it consists of the sum of transmission delay, signal propagation delay and packet waiting delay.

Current research in various fields shows that predictive models, which predict events and situations from the present towards the future based on data from the past, have an enormously wide range of applications. Predictive models are most often based on machine learning (ML) techniques, especially in telecommunications. The relevance and application of predictive models using ML techniques are encouraged by a very rapid increase in the amount of multidimensional data: Big Data (BD) publicly available on the Internet. BD increases the complexity of the problem of finding the optimal way to the solution to functional tasks in the network domain. At the same time, the high dimensionality of data, i.e., a large number of variables, often makes it difficult to create a model and jeopardizes the accuracy of prediction results. Among the discovered approaches to solutions for reducing the problem of complexity, data preprocessing by dimensionality reduction techniques is used. Complexity represents a key indicator of the state configuration in the situational dynamics of telecommunication traffic. Data dimensionality reduction implies optimization of the space of input/independent variables and the number of predictors, but with the obligation to preserve relevance and other qualitative attributes of information [7]. Feature Selection is one of the most common and important dimensionality reduction techniques, and, in research papers, it is also known as variable selection, attribute selection or variable subset selection. In this paper, the research focus is on three dimensionality reduction approaches: RReliefF algorithm, Backward selection via the recursive feature elimination algorithm and the Pareto 80/20 rule. The first two approaches belong to Feature selection techniques. The selection of input variables is a process that includes the detection of variables that have a significant impact on the prediction of output, and the removal of redundant variables. As the main benefits achieved by this technique, the following can be highlighted: increasing the speed of data mining algorithms, increasing the accuracy of prediction, reducing the complexity of the model [7,8].

The assumption is that better planning and design of networks and allocation of network resources can be achieved in the future if the value of end-to-end delay is known. Thus, this paper examines the performance of three predictive delay models for a VoIP service in an LTE network based on Multilayer Perceptron (MLP), Support Vector Machines (SVM) and k-Nearest Neighbors (k-NN), whose input set of variables is optimized [9]. As a case study, the geographic area in the Republic of Srpska (RS), Bosnia and Herzegovina (BiH), in the vicinity of a three-segment road between the cities of Banja Luka (BL) and Doboj (Db), which is covered by the cellular network of the M:tel BL operator, is chosen. The main goal is to select an ML model with an optimal number of input variables, which provides the most accurate prediction results.

The most important aims and objectives of this research are the following:

- Reducing the dimensionality of the space of model input variables by optimization with Feature Selection techniques (RReliefF and Backward selection via the recursive feature elimination algorithms) and the Pareto 80/20 rule;

- Training and testing of ML models (MLP, SVM and k-NN) including the selection of the best delay prediction model in the LTE network using accuracy and complexity/interpretability criteria;
- Presentation of the aforementioned approaches to optimizing the number of predictors for LTE KPI predictive modeling, which is, according to the authors' knowledge and the review of former research papers, a particularly innovative solution;
- Implementation of a unique methodology of indirect assessment and calculation of delay values based on the average number of active users in the network;
- Creation of universally applicable predictive modeling of delays in the LTE network based on real research. For the case study, a data space related to one of the most important roads in the geo-road network of RS, BiH, was chosen.

The structure of the paper consists of five sections. Section 1 provides an introduction. Section 2 presents a review of relevant published research papers, and Section 3 contains the materials and methods used in the paper. The main research focus is in Section 4, where the results and discussion are provided, after which the conclusions are drawn in Section 5. The references used are listed in the last section of the paper, after the conclusion.

2. Review of Relevant Published Research

In a previous study [10], the authors created models for end-to-end delay prediction in Cellular Vehicle-to-Everything (C-V2X) communication using different ML techniques. Model training was performed on KPI-related variables, and data was collected from real LTE networks. In this paper, prediction is viewed as a delay classification problem depending on a given threshold. Similar research is conducted in [11] with a focus on delay prediction for V2X applications in Mobile Cloud/Edge Computing systems. The proposed prediction framework in this case consists of a component based on machine learning techniques and a statistical component. Paper [12] presents an algorithm for resource allocation prediction in LTE uplink (UL) connection for machine to machine (M2M) applications. Mathematical models for prediction probability, successful prediction probability, failed prediction probability, resource utilization/underutilization probability and a mean uplink delay model were developed. All these models are validated using a simulation model implemented on the OPNET platform. An original approach based on machine learning for delay prediction in 4G networks is presented in [13]. To create the model, the authors used real data from three different mobile networks. Paper [14] considers a case study related to the Industrial Internet of Things (IIoT), in which the potential of digitization of mines is investigated. For this purpose, a software tool for sending sensor data using the LTE network is presented, and predictive delay models are created in order to evaluate the network performance. Lai and Tang (2013), in their paper [15], developed a Packet Prediction Mechanism (PPM), based on mathematical models, for delay prediction when using real-time services. The main research focus was on a virtual queue concept, which has the function of predicting the behavior of incoming packets in the future based on the packets currently in the queue. Due to the increasing user demand for real-time services, the development of wireless access technologies that provide greater bandwidth is evident every day. Therefore, the same team of authors, in the published research paper [16], proposed and designed an LTE scheduling mechanism and PPM. In doing so, the authors assume that the proposed PPM will increase capacity, reduce resource consumption and thereby increase network efficiency. The assumption is that the monitoring and prediction of QoS indicators are the basic prerequisites for user satisfaction in the use of LTE network services. Thus, delay and average user throughput are considered as key indicators of network performance in [17]. The authors created models to estimate the values of these dependent variables as linear functions of total network traffic and an average Channel Quality Indicator (CQI). In [18], the subject of research is the changes in Round-trip time (RTT) delay and the prediction of the increase in these values in mobile broadband networks. Four classification models based on machine learning were developed, using data from a large number of probes in the network, and the best classification performance was shown by the binary ensemble model.

The essential characteristics of the previously analyzed papers are shown in Table 1. It provides information on reference numbers of the papers, the models and techniques used, the prediction problem being solved (regression/classification), the service/application being observed, some of the dimensionality reduction methods and techniques, if applied, and performance.

Table 1. Overview of important criteria in relevant published research papers.

Ref. No.	Models and Techniques	Regression/ Classification	Service/ Application	Dimensionality Reduction Methods and Techniques	Performance
[10]	Neural Network (NN), Recurrent Neural Network (RNN) with Long Short-Term Memory (LSTM) neurons, Random Forest (RF), SVM	Classification	C-V2X	Maximum Dependency (MD) algorithm	Prediction accuracy (PA): - For NN: PA = 0.8127; - For RNN: PA = 0.8176; - For RF: 0.7967; - For SVM: PA = 0.8107.
[11]	LSTM; k-medoids classification, Epanechnikov Kerne, Moving average functions	Regression and Classification	Delay-sensitive V2X Applications in Mobile Cloud/Edge Computing Systems	-	Best performance in both relative mean error and relative standard deviation. This methodology can reduce the mean error by 45% (which achieves around half of the benchmark)
[12]	Mathematical models	Regression	M2M uplink communication	-	Can reduce the mean uplink delay significantly below the minimum possible for a non-predictive resource allocation algorithm
[13]	Logistic Regression (LR), SVM, Decision Tree (DT)	Classification	Operational 4G Networks Services	Random Forest	The nested cross-validation performance of the SVM, DT, and LR models are $0.664 \pm 0.100\%$, $0.743 \pm 0.004\%$, and $0.609 \pm 0.076\%$, respectively
[14]	Artificial Neural Networks, Decision Tree, Ensemble modeling: Bagging technique with a Decision Tree	Regression	IIoT	Lag features, Window features	The highest accuracy of the prediction is estimated at 90%
[15]	Mathematical models, PPM, virtual queues	Regression	Real time services	-	PPM is able to achieve notable improvement in terms of invalid packet rates and goodputs compared to Maximum Throughput (MT), Proportional Fair (PF), Modified Largest Weighted Delay First (MLWDF), and Exponential-Proportional Fair (EXP-PF) and very low delays
[16]	Mathematical models, PPM, virtual queues	Regression	Real time services	-	Possibility of expired packets can be reduced by the proposed PPM
[17]	Multivariate linear regression technique	Regression	LTE services	-	Plane function very well represents the dependence of average delay on average reported Channel Quality Indicator (CQI) and the total traffic
[18]	Logistic regression, Random forest, Light gradient-boosting machine (LightGBM), Ensemble	Classification	4G and 5G services	-	Model misclassified 20% of the tested samples

Compared to previously analyzed published research papers, the following five contributions stand out as the main improvements and novelties presented in this paper:

- Network delay is investigated by observing a real geospatial and LTE network segment as very important factors affecting KPIs;

- The number of predictors in LTE delay examination is optimized for the first time by simultaneously using three approaches for predictive modeling of delays in the LTE network;
- A complete set of 17 independent/input research variables is used and Dimensionality Reduction is explained in detail;
- The original indirect method of assessment and calculation of the values of the dependent/output variable is applied;
- The optimization of the set of input variables is modeled with Feature Selection techniques and the Pareto 80/20 rule, and the obtained results are compared according to the criteria of prediction accuracy and complexity/interpretability of the model.

3. Materials and Methods

The research process in this paper was completed through several successive steps:

1. Analysis of a real geospatial and network research segment in the case study;
2. Data collection and analysis of independent research variables;
3. Calculation of dependent variable values;
4. Structuring data into input/output vectors;
5. Optimization of a set of independent variables by Feature selection techniques: RReliefF and Backward selection via the recursive feature elimination algorithms;
6. Optimization of a set of independent variables by the Pareto 80/20 rule;
7. Training and testing of predictive delay models over an optimized set of independent variables;
8. Comparative analysis of prediction results and selection of the final model.

3.1. *Geospatial and Network Research Segment—A Case Study*

For the case study in this paper, a three-segment road connected by a geodesic line, in the geo-space of RS, BiH, between the cities of BL and Db, consisting of the following road segments, was chosen:

1. A segment of the 9th January Motorway (M9J), 72 km long, between the Jakupovci toll station, near the city of BL, and the Kladari toll station, near the town of Db;
2. A segment of the M16 Main Road, about 6 km long, on the route Jakupovci (entrance to the city of BL);
3. A segment of the M17 trunk road, about 10 km long, located between the Kladari toll station and the town of Db.

In the observed geo-space, the research focus is on the fourth generation (4G) telecommunications network based on LTE network technology, managed by the M:tel BL provider [1,2]. Figure 1 shows a part of the geographical map (Google Earth) of the RS and BiH with marked areas of road segments, where the area marked in blue is covered by LTE Carrier Aggregation (CA), and the area in green is covered by LTE Frequency Division Duplexing (FDD) technology.

LTE CA is one of the key technologies used to achieve very high data transfer speeds in 4G networks. The principle is based on combining more than one signal carrier (in the same or different bands) in order to increase the bandwidth and channel capacity. In the case study, out of the total geographical area, 14.75% is covered by LTE CA technology, and 85.25% by LTE FDD technology, which enables duplex communication between eNB and UE. It is based on paired spectrums with sufficient spacing between frequency domains to allow simultaneous sending and receiving of data.

3.2. *Analysis of Independent Research Variables and Data Collection*

In this research, the following 17 independent variables or predictors selected from the set of research data provided by the M:tel operator are observed: (1) Cell; (2) Downlink (DL) PRB Usage Rate; (3) Average CQI; (4) DL ReTrans Rate; (5) UL ReTrans Rate; (6) DL IBLER; (7) UL IBLER; (8) Cell Traffic Volume DL; (9) Cell Traffic Volume UL;

(10) Cell Downlink Average Throughput; (11) Cell Uplink Average Throughput; (12) Average DL User Throughput; (13) Average UL User Throughput; (14) UL Average Interference; (15) DL.QPSK.TB.Retrans; (16) DL.16QAM.TB.Retrans; (17) DL.64QAM.TB.Retrans.

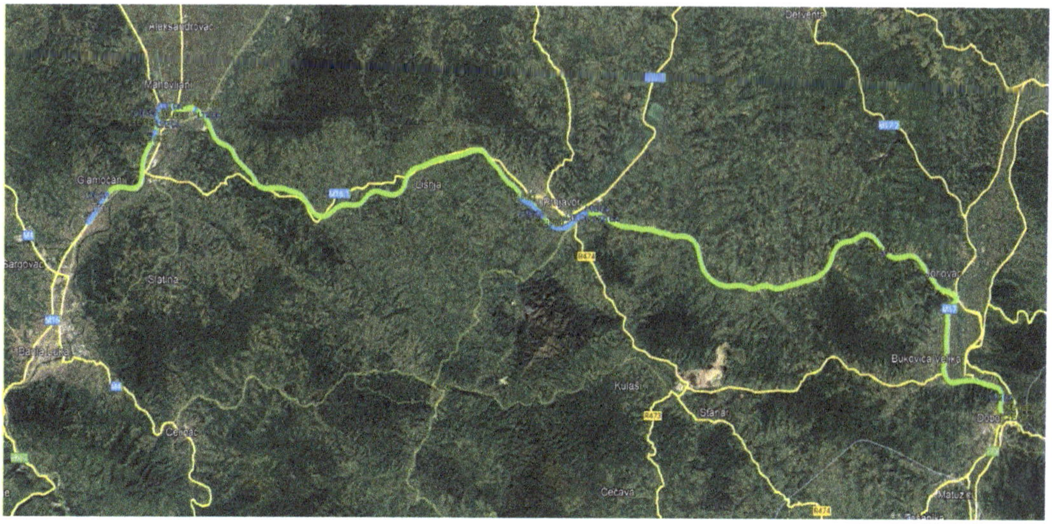

Figure 1. Geographical area of research.

The M:tel BL mobile operator provided the data for research purposes based on the official Request. The Request specified the necessary variables related to KPIs, radio channel properties, utilization of physical resources, number of users, eNodeB parameters, topology and signal parameters in the observed research geo-space [19]. From the obtained database, the values of the variables for the period of data collection between 1 January 2021 and 15 January 2021 and with a one-hour sampling frequency were extracted in an Excel file for the purposes of this research. By inspecting the data, empty cells (missing values), unusual values equal to zero and unusual values even several thousand times less than usual were observed and filtered. The final database was formed and consisted of a total of 31,143 measurements for each of the observed independent variables. According to the supervised learning paradigm, the total data set is divided into two parts: (1) model training data, consisting of 21,756 measurements (instances or vectors) or 70% of the total data set, and (2) model testing data, consisting of 9387 measurements or 30% of the total available data set.

(1) Cell

The access LTE network of the M:tel operator in the area of the observed three-segment road consists of a large number of eNodeBs that provide the connection of the UE with the rest of the 4G network. Their locations are represented by red squares in Figure 2. According to the number of mobile users, it is obvious that the highest density of eNodeB deployment is in the vicinity of BL city [19]. Also, in Figure 2, based on the colors and the map legend, areas with different levels of signal attenuation can be identified. Specifically, it refers to the areas between −126 dB and −90 dB and areas between −90 dB and 0 dB. Each of the eNodeBs covers one or more cells with a signal, and a total of 87 cells can be identified in the area observed.

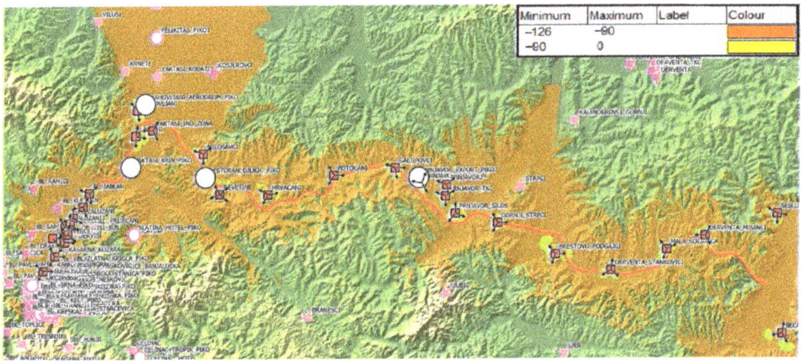

Figure 2. Layout of eNodeB locations with marked signal propagation in the research area.

(2) DL PRB Usage Rate

The smallest unit of radio resources in the LTE network that can be allocated to a user is called a Physical Resource Block (PRB). It consists of 84 resource elements (7 symbols of 0.5 ms duration × 12 subs of 15 kHz each). When available but unused PRBs are not sufficient to serve all active users, it can cause degradation of quality of service (QoS). DL PRB Usage Rate represents the ratio of the average number of used physical blocks in the Physical Downlink Shared Channel (PDSCH) and the total number of DL PRB available, multiplied by 100. PDSCH represents a DL physical shared channel whose priority function is the transmission of user data, but also the transmission of data essential for control, and DL system information [19].

(3) Average CQI

The CQI can have a numerical value between 1 and 15, which the UE sends over the uplink connection to the base station. Based on the received CQI value, the eNodeB selects the appropriate Modulation and Coding Scheme (MCS), thereby defining the data transmission rate in the communication channel. This means that each CQI value is mapped to a specific MCS: Quadrature Phase Shift Keying (QPSK), Quadrature Amplitude Modulation (QAM, 16QAM, 64QAM) [19,20].

(4) DL ReTrans Rate & (5) UL ReTrans Rate

When the communication between the base station and the UE is not established in the first or any subsequent attempt, data resending or retransmission is performed. Data are sent in packets, i.e., in a Transport Block (TB) within one Transmission Time Interval (TTI), with its duration of 1 ms. The DL/UL retransmission rate can be defined as the ratio of retransmitted packets (packets sent with retransmission) to all packets sent via the DL/UL SCH transport [19].

(6) DL IBLER & (7) UL IBLER

Block Error Rate (BLER) shows as a percentage how many blocks with errors were received compared to the total number of blocks sent. The Initial Block Error Rate (IBLER) is an indicator used to evaluate network performance; it shows the relationship between the number of blocks with initial transmission errors and the total number of initially transmitted TBs in the DL and UL direction [19].

(8) Cell Traffic Volume DL & (9) Cell Traffic Volume UL

Cell Traffic Volume DL/UL represents the total aggregated DL/UL traffic in the cell in a period of one hour expressed in Gbit. In LTE networks, the total aggregated traffic represents the sum of traffic in 9 classes, which are identified by the QoS Class Identifier (QCI) [19]. The classes marked with QCI 1—QCI 4 are characterized by a defined and guaranteed throughput of Guaranteed Bit Rate (GBR), and examples of services that belong to them are QCI 1—Conversational Voice; QCI 2—Conversational Video; QCI 3—Real Time Gaming; QCI 4—Non-Conversational Video. Non-GBR classes are marked with QCI 5—QCI 9 and imply a certain risk of packet loss, especially in conditions of network congestion. Examples of services belonging to them are QCI 5—IMS Signaling; QCI 6—Video, TCP-based; QCI 7—Voice, Video, Interactive Gaming; QCI 8 and QCI 9—Video, TCP-based.

(10) Cell Downlink Average Throughput & (11) Cell Uplink Average Throughput

One of the most important indicators of network performance is Throughput, which can be defined as the ratio of the amount of data transferred and the time for which the transfer is made. The variable Cell Downlink/Uplink Average Throughput represents the average value of this indicator for a period of one hour, at the level of one cell in the DL and UL direction. The average throughput value can be determined not only geographically (per spatial unit-cell), but also logically (per service) [19].

(12) Average DL User Throughput & (13) Average UL User Throughput

The average value of Throughput at the user level in the LTE network, in the observed space in the DL and UL direction, is determined by the value of the Average DL/UL User Throughput variable. This value is calculated for a period of one hour [19].

(14) UL Average Interference

The total power of the noise floor and the interference of neighboring cells, received by each PRB, is measured during one TTI in the UL direction. The eNodeB divides the total power of the noise floor and the interference of neighboring cells by the number of PRBs, and the resulting value is used as the sampling result. At the end of the one-hour measurement period, the average of these sampling results expressed in dBm is used as the value of the UL Average Interference variable [21].

(15) DL.QPSK.TB.Retrans, (16) DL.16QAM.TB.Retrans & (17) DL.64QAM.TB.Retrans

The variables DL.QPSK.TB.Retrans, DL.16QAM.TB.Retrans and DL.64QAM.TB.Retrans are related to the variable DL ReTrans Rate and refer to retransmission rates for certain modulation schemes. Their meaning is as follows:

(15) DL.QPSK.TB.Retrans—Number of retransmitted TBs in DL SCH at Quadrature Phase Shift Keying (QPSK) modulation;

(16) DL.16QAM.TB.Retrans—Number of retransmitted TBs in DL SCH at Quadrature Amplitude Modulation (QAM) with 16 carrier states (16QAM);

(17) DL.64QAM.TB.Retrans—Number of retransmitted TBs in DL SCH at QAM with 64 carrier states (64QAM).

3.3. *Calculation of Dependent Variable Values*

End-to-end delay (D_{EtoE}) consists of the sum of the delay at the Medium Access Control (MAC)/Radio Link Control (RLC) layer, which makes up the largest part of D_{EtoE}, then of the delay due to signal propagation at the physical level and the transmission delay between the eNodeB and UE [22]. Therefore, the delay in this case implies "the time duration that starts when a flow is generated by a traffic source, transmitted through the communication system, until it reaches the application layer of the user's equipment—UE" [22].

The values of the variable D_{EtoE} were collected by estimation and calculation based on the results presented in paper [22]. In that paper, Madi et al. (2018) used a simulation method to measure the end-to-end delay for VoIP traffic depending on the number of active UEs in the cell. It involved mobile users' movement speeds of 3 km/h and 120 km/h considered for each of the four observed scheduling algorithms: Exponential Rule (EXP-RULE), EXP-PF, PPM and Delay–based and QoS–Aware Scheduling (DQAS). Based on graphically presented simulation results in [22] (Figure 7 Average D_{E2E} on RT VoIP flows in [22]), for an interval from 10 to 100 active UEs in a cell with a step of 10 and for a speed of 120 km/h, average values of the estimated delays for the four observed scheduling algorithms are calculated in this paper. The values calculated in this way are shown by points in Figure 3, where the regression curve that best describes the functional dependence of the average delay on the number of UEs is given.

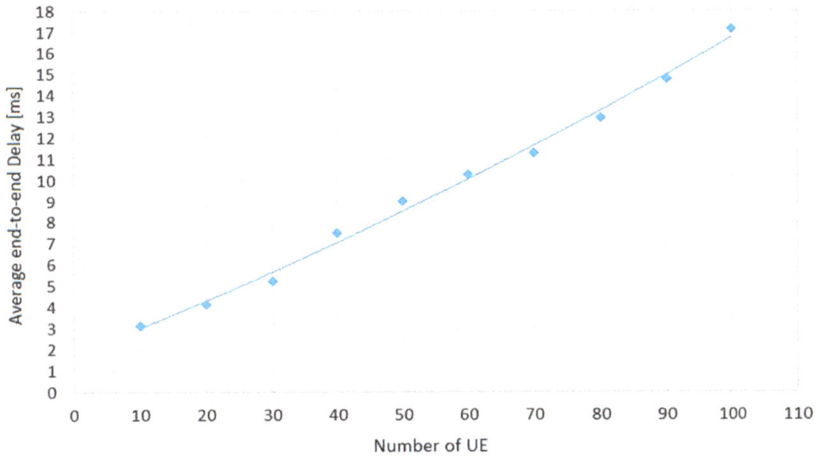

Figure 3. Average values of end-to-end delay of VoIP services calculated for EXP-RULE, EXP-PF, PPM and DQAS scheduling algorithms and for mobile user movement speed of 120 km/h.

The curve shown in Figure 3 has a polynomial form of the second degree and can be represented by a quadratic equation as follows:

$$Delay = 0.0003 \cdot UE^2 + 0.1197 \cdot UE + 1.8071 \quad (1)$$

As an indicator of the quality of this model, a very high coefficient of determination (R^2), $R^2 = 0.9941$, appears. Among other data, the database received from the M:tel operator provided the values of the average number of active UEs in the cells of the observed geographical area. As such, the D_{EtoE} values were calculated indirectly, using the model given by Equation (1). The method of calculating the values of the dependent variable D_{EtoE} is shown graphically in Figure 4.

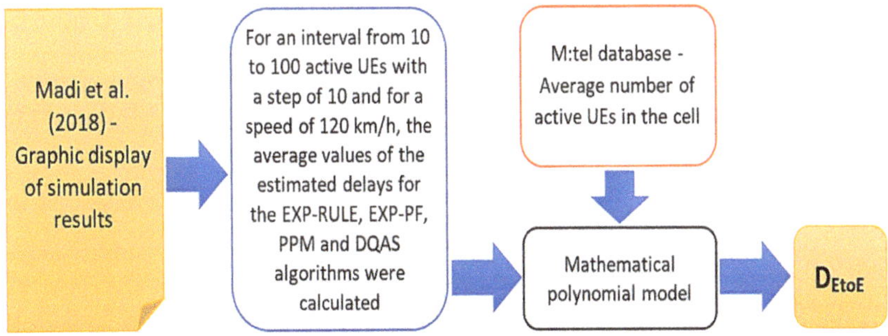

Figure 4. Method of calculating the values of the dependent variable D_{EtoE} [22].

Descriptive statistics for the dependent variable are given in Table 2, and the histogram of the D_{EtoE} variable is shown in Figure 5.

Table 2. Descriptive statistics for the dependent variable D_{EtoE}.

Mean	StDev	Var	Min	Median	Max	Skewness	Kurtosis
4.1503	2.6520	7.0329	1.8081	3.2516	25.8282	2.81	10.17

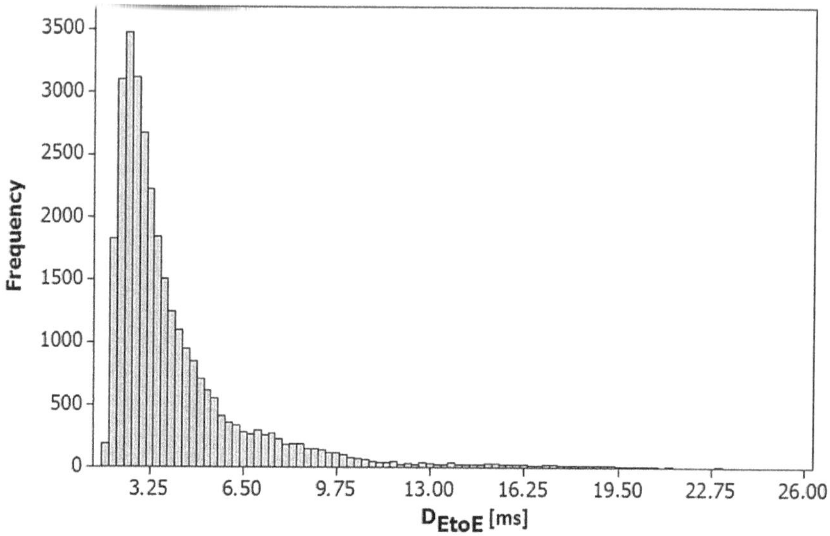

Figure 5. Histogram of the dependent variable D_{EtoE}.

From a more detailed analysis of the histogram shown in Figure 5 (in Minitab 14 software), it is concluded that D_{EtoE} values in the range between 2.375 ms and 2.625 ms have the highest frequency (3477 repetitions). The arithmetic mean of the delay values is 4.1503 ms, and of the median is 3.2516 ms.

3.4. *Structuring Data into Input/Output Vectors*

Considering that data processing in this research is performed with ML techniques, it is necessary to structure the values of independent variables and the values of the dependent variable into input/output vectors [9]. This kind of data structure enables the training of the ML model according to the supervised learning paradigm, where the independent variables have the role of inputs to the model, and the dependent variable has a function of an output from the model. One input-output vector, in this case, represents a one-dimensional array, where the first 17 numbers represent the values of the independent/input variable (input vector), and the last number refers to the value of the dependent/output variable D_{EtoE}. In the IBM SPSS Statistics Data file, a total of 31,143 input-output vectors are structured as described; of them, in this paper, 70% are used for training and 30% for model testing.

3.5. *Optimization of a Set of Independent Variables by Feature Selection Techniques*

A large number of inputs or predictors can make the ML model very complex. It can also complicate its interpretability, requires increased memory space in the system and increases the chances of overfitting to training data. However, the problem of poor accuracy of prediction and classification is often solved precisely by including additional parameters or variables. It means that achieving a compromise (optimum) between simplicity and accuracy is one of the most important goals when creating an ML model [23,24].

In many cases, more inputs to the model does not mean better model performance. Feature selection represents one of the techniques for reducing the dimensionality of a data set (Dimensionality Reduction) by filtering certain predictors that are redundant or not relevant in the ML model. By excluding such independent variables, the prediction accuracy or classification performance of the model can be significantly improved [25]. For this purpose, three basic variants of the Feature selection technique are available:

- Filter technique—This is based on measuring the importance of variables considering features such as variance and relevance to the output variable. Predictors are selected according to the desired level of importance or relevance, after which an ML model is created using the selected set of inputs [26].
- Wrapper technique—Model training is performed using a selected subset or the entire set of independent variables, and then individual predictors are added or removed based on a certain criterion that measures the change in model performance. Model training and testing are repeated until predefined stopping criteria are met [26].
- Embedded technique—Assessing the importance of the predictor is, in this case, an integral part of a model training process.

3.5.1. RReliefF Algorithm

The RReliefF algorithm belongs to the Filter technique for optimizing a set of variables. Relief (Kira and Rendell, 1992 [27,28]) and its extension ReliefF (Kononenko, 1994 [29]) are "context-aware" algorithms that assess the quality of model variables for solving classification problems where there is strong interdependence among predictors [30]. Unlike the previous two, the Regression ReliefF (RReliefF) algorithm is not limited to category dependent variables only. It is used for regression tasks in which it "penalizes" predictors that give different prediction values for adjacent observations with the same values of the dependent variable. In this case, the observation represents one row in the input data matrix, i.e., one input vector. On the other hand, this algorithm "rewards" predictors that give different prediction values for neighboring observations with different output values [31]. RReliefF uses intermediate weights to calculate the final predictor weight coefficients; if the two nearest neighbors are considered, the following notation is used:

- W_j is the weighting coefficient of the predictor F_j;
- W_{dy} is the weighting coefficient for different values of the dependent variable y;
- W_{dj} is the weighting coefficient for different predictor values F_j;
- $W_{dy \wedge dj}$ is the weighting coefficient for different values of y and different values of the predictor F_j [31].

The weighting coefficients, W_{dy}, W_{dj}, $W_{dy \wedge dj}$ and W_j, are equal to zero at the beginning of the algorithm. The algorithm iteratively selects a random observation x_r and a k-nearest observation for x_r. For each nearest neighbor x_q, intermediate weights are updated as follows [31]:

$$W_{dy}^{i} = W_{dy}^{i-1} + \Delta_y(x_r, x_q) \cdot d_{rq} \tag{2}$$

$$W_{dj}^{i} = W_{dj}^{i-1} + \Delta_j(x_r, x_q) \cdot d_{rq} \tag{3}$$

$$W_{dy \wedge dj}^{i} = W_{dy \wedge dj}^{i-1} + \Delta_y(x_r, x_q) \cdot \Delta_j(x_r, x_q) \cdot d_{rq} \tag{4}$$

In the mathematical expressions (2), (3) and (4), i and $i-1$ denote the ordinal numbers of a total of m specified iterations. The expression $\Delta_y(x_r, x_q)$ represents the difference between the values of the dependent variable for observations x_r and x_q, and can be calculated as follows [31]:

$$\Delta_y(x_r, x_q) = \frac{|y_r - y_q|}{\max(y) - \min(y)} \tag{5}$$

where y_r and y_q are the values of the dependent variable for observations x_r and x_q, respectively. The difference of the values of the predictor F_j for the observations x_r and x_q is defined by the expression $\Delta_j(x_r, x_q)$ [31]. When x_r represents the value of the j-th predictor for the observation x_r, and x_{qj} is the value of the j-th predictor for the observation x_q, then

$$\Delta_j(x_r, x_q) = \frac{|x_{rj} - x_{qj}|}{\max(F_j) - \min(F_j)} \tag{6}$$

After updating all intermediate weights, RReliefF calculates the weighting coefficients of the predictor W_j according to Equation [31]:

$$W_j = \frac{W_{dy \wedge dj}}{W_{dy}} - \frac{W_{dj} - W_{dy \wedge dj}}{m - W_{dy}} \tag{7}$$

In order to select the optimal set of predictors in the model in addition to the values of weighting coefficients, it is necessary to define the Relevance Threshold (*RT*) as the limit of the significance of independent variables [32]. According to the criterion set in this way, all predictors with $W_j \geq RT$ participate in the creation of the model. Generally, that threshold has a value in the interval between 0 and 1, and more precisely, its value is calculated according to the following expression based on Chebyshev's inequality [32]:

$$0 < RT < \frac{1}{\sqrt{\alpha \cdot t}} \tag{8}$$

where α is the probability of accepting an insignificant feature as significant (type I errors or first type error) and t is the number of training observations for updating W_j, out of a total of n observations. Within the stated limits, the selection of *RT* is arbitrary, where there is a probability that not all variables with W_j above the defined threshold will necessarily be significant because some unimportant variables are expected to have a positive weighting coefficient by chance [32].

3.5.2. Backward Selection via the Recursive Feature Elimination Algorithm

The application of the Wrapper technique for the selection of an optimal set of input variables in this research is based on the Backward selection via the recursive feature elimination algorithm, which was presented in [33]. In Figure 6, this algorithm is graphically represented by a flowchart. In the initial step, all 17 independent variables are used as inputs to the ML model, after which multiple predictive models are trained and tested. At the same time, it is necessary to determine the importance or influence of each predictor on the prediction results. In the next step, the input variable of least importance is eliminated, and the training and testing procedure is repeated over the subset obtained in this way, as well as the performance analysis of the solutions created. As long as the current subset of input variables consists of more than two inputs, it is necessary to eliminate individually each input variable with the next lowest importance from the ranked list, and so on. The elimination procedure is shown in a loop in Figure 6. When the input subset is reduced to two predictors, the performance of the created models is compared for each subset. Finally, for the optimal solution, the subset of inputs used to create the most accurate predictive models is selected.

The performance of predictive models is measured with the Relative Error (*RE*) prediction criterion. *RE* can be calculated as follows:

$$RE = \frac{\sum_{i=1}^{n} (D_{EtoEi} - D_{PREDi})^2}{\sum_{i=1}^{n} (D_{EtoEi} - D_{AVGi})^2} \tag{9}$$

where:

- D_{EtoEi} is a calculated end-to-end delay value for the i-th input/output vector,
- D_{PREDi} is a prediction value of D_{EtoEi}, and
- D_{AVGi} is the arithmetic mean of the variable D_{EtoEi}.

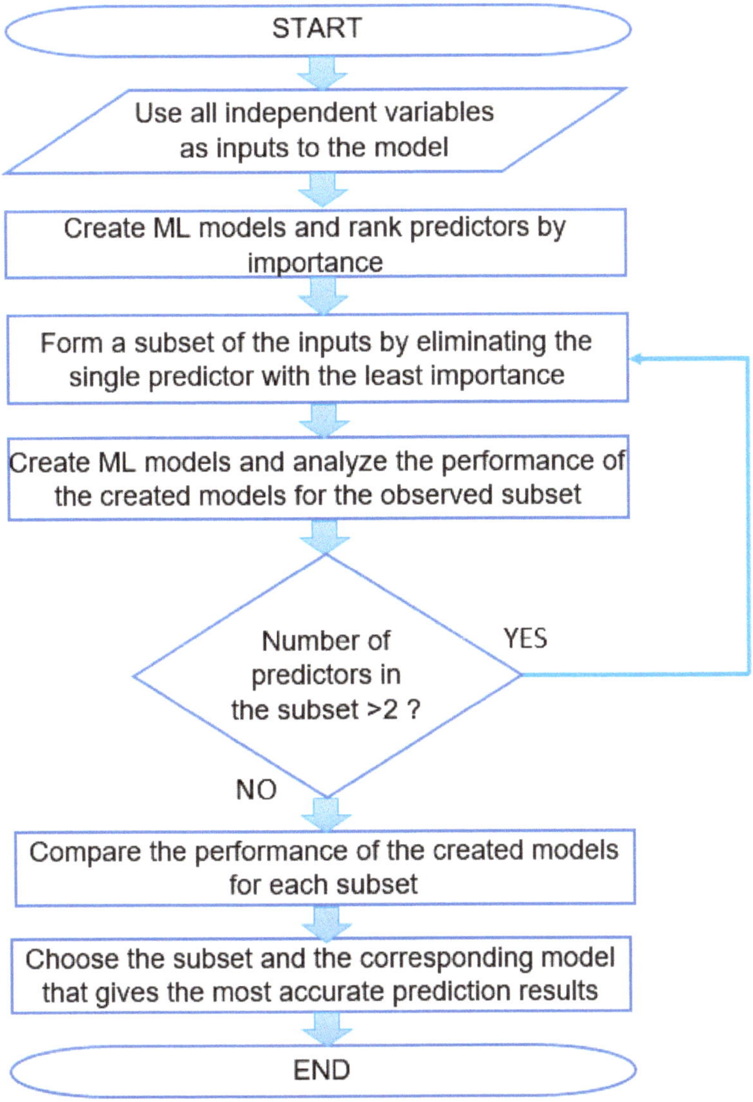

Figure 6. Backward selection via the recursive feature elimination algorithm to optimize the number of predictors.

3.6. Optimization of a Set of Independent Variables by the Pareto 80/20 Rule

Another applied approach to optimization within the dimensionality reduction technique is the Pareto principle. Since optimality means the best combination of relevant factors, the Pareto principle is based on the strategic assumption that 80% of problems or effects in solutions come from 20% of causes. This is why it is often referred to as the "80/20 rule". The Pareto principle has proven its applicability in various fields, even though it has its roots in economics [34,35]. In this paper, based on the created Pareto diagram, the optimal number of input variables is chosen so that their cumulative PI value is equal to or greater than 0.8 or 80%. For all alternative actions in predictive decision-making, available relevant information is used, and possible solutions for selecting one of the alternatives can be presented in matrix form. Machine learning is viewed as a multi-objective task. However, most often only one goal is observed, cost function optimization, or multiple objectives are aggregated into a scalar cost function. Using the Pareto principle to solve multi-objective tasks has proven to be one of the most effective approaches. In the Pareto-based approach to multi-objective optimization, the objective function is not a scalar value, but a vector. Therefore, several Pareto-optimal solutions are created instead of one, which can significantly improve the predictive performance of a model [36].

3.7. Creating Predictive Models Using the ML Method of Automatic Modeling

In the IBM SPSS Modeler software environment, optimized sets of predictors are brought to the input of the Auto Numeric node. Auto Numeric represents a method for automatic modeling where training and testing of multiple models is performed in just one step on the basis of different ML techniques [9]. As a result, the software analyzes the performance, ranks and offers the user the best solutions and sorts the input variables according to the influence (importance) on the prediction results. Based on the aims and objectives of this research, three machine learning techniques are in focus: MLP, k-NN and SVM [9,37]. These models are some of the most popular predictive models in research. In addition, they are universal, as they can be used for classification and regression tasks and are of different levels of interpretability/complexity. These are the main reasons why they were observed in this research.

Hyperparameter optimization was not used in this paper, but these values were automatically set to default:

- The MLP model automatically determines the required number of hidden layers (one or two) and the number of neurons in each of them; the maximum training time of 15 min is used as a stopping criterion.
- SVM stopping criterion has a value of 10^{-3}; Regularization parameter (C) = 10; Regression precision (epsilon) = 0.1; Kernel type is Radial Basis Function (RBF); RBF gamma = 0.1; Gamma value = 1; Bias = 0; Degree = 3.
- For the k-NN model, k is automatically determined between three and five; Distance Computation is based on the Euclidean metric.

Optimization of model hyperparameters, assessment of its performance and avoidance of overfitting are most often achieved with the help of cross-validation techniques. The most common techniques include K-Fold Cross-Validation, Stratified K-Fold Cross-Validation, Leave-One-Out Cross-Validation (LOOCV), Leave-P-Out Cross-Validation (LPOCV), Time Series Cross-Validation and Repeated K-Fold Cross-Validation. The main effect achieved by their application is the stability of the predictive model, which is reflected in reliability, good generalization on new data and preservation of performance over time and in different circumstances. They are most often used in medical research and medical statistics. However, these techniques have limitations, especially in cases where the data evolves, resulting in differences between the training set and the validation set. Based on the assumption that in this research the set of available data is simple and large enough, but also due to time and resource requirements, a simple train-test split was used in the paper. The model's predictive performance and stability were evaluated on a test dataset that remained unseen to the ML models during their training.

3.8. Comparative Analysis of Prediction Results and Selection of the Final Model

The comparative analysis of the prediction results and the selection of the final model represents the last step in the research process. Based on the prediction performance expressed through the relative error criterion, one of the most accurate models is selected for each of the three observed approaches to optimizing the set of input variables. The main goal of this procedure is to test the statistical significance of the differences in prediction results for three ML models, i.e., for three approaches to predictor set optimization. Given that the same data set is used for testing in all three cases, the prediction results are compared using statistical methods, specifically by the ANOVA test with Repeated Measures and the Friedman test.

In addition to relative error, as one of the key indicators of prediction performance, special attention is paid to its complexity and interpretability when selecting an ML model [5]. According to the conducted studies, models with more complex ML algorithms are more demanding for interpretation. The dimensionality of the space of input variables, and the complexity of the functions that the models need to learn, in addition to the algorithms, can affect the complexity of the model [5]. Figure 7 shows the methodological steps, from the optimization of a set of independent variables in each of the three investigated predictive models to the comparative analysis and selection of the final ML model.

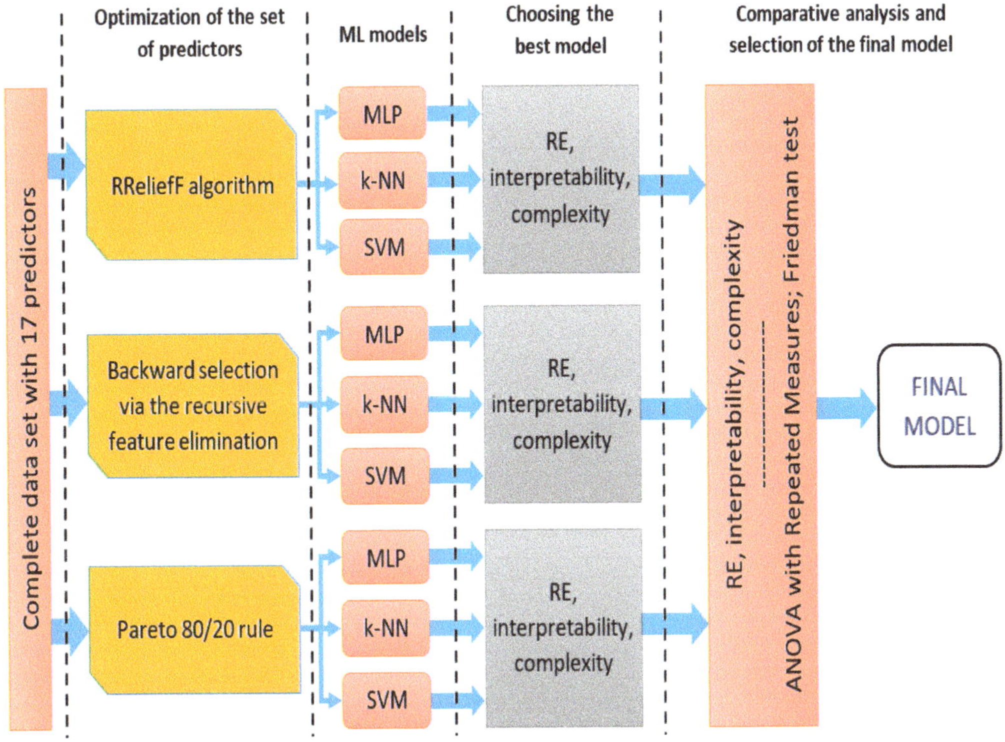

Figure 7. Steps of the methodological procedure from the optimization of the set of predictors to the selection of the final model.

According to the results of numerous studies, priority is given to simpler, more interpretable solutions, although complex predictive models usually provide better performance [38–40]. In paper [39], several definitions of the concept of interpretability are listed, among which the following stands out: "interpretability in ML is a degree to which a human can understand the cause of a decision from an ML model". For this reason, in recent years, a relatively new field, Interpretable Machine Learning (IML), has appeared. Within it, methods are investigated to transform ML models, the so-called black boxes, into white box models [5,39,41]. Figure 8 shows common models ranked according to accuracy and interpretability in relatively recently published research papers [42–46]. In the figure, the accuracy from the lowest to the highest value is given in a down-up orientation, while the interpretability with a growing trend is oriented in the Top-down direction.

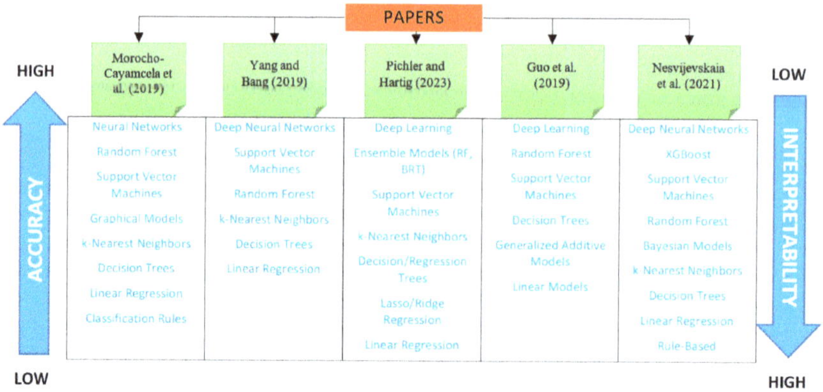

Figure 8. ML models ranked by accuracy and interpretability in various published research papers [42–46].

4. Results and Discussion

4.1. Predictive ML Models Created over a Set of Predictors Optimized by the RReliefF Algorithm

In order to compare performance, the optimization results of the set of independent variables by the RReliefF algorithm for different values of k-Nearest Neighbors are shown in Table 3. The values of this parameter are chosen empirically. For $k = 10$, $k = 15$ and $k = 20$, the algorithm ranked the independent variables by weighting coefficients in the same ranking. The most influential predictor for all three cases is DL.16QAM.TB.Retrans, while Average_DL_User_Throughput is in last place with a negative weighting coefficient for each k.

Table 3. Optimization results of the set of independent variables by the RReliefF algorithm.

Rank	Independent Variable or Predictor	Predictor Weighting Coefficients for Individual Values of k		
		$k = 10$	$k = 15$	$k = 20$
1	DL.16QAM.TB.Retrans	0.0061	0.0065	0.007
2	DL.QPSK.TB.Retrans	0.006	0.0064	0.0067
3	Cell_Traffic_Volume_UL	0.0041	0.0044	0.0045
4	DL_PRB_Usage_Rate	0.0037	0.004	0.0043
5	Cell_Traffic_Volume_DL	0.0033	0.0035	0.0038
6	UL_Average_Interference	0.0028	0.0031	0.0033
7	DL.64QAM.TB.Retrans	0.0027	0.0028	0.0029
8	Cell	0.0024	0.0025	0.0027
9	UL_IBLER	0.001	0.001	0.0012
10	UL_ReTrans_Rate	0.0009	0.001	0.0011
11	Cell_Uplink_Average_Throughput	0.0006	0.0006	0.0007
12	Average_UL_User_Throughput	0.0001	0.0001	0.0001
13	Average_CQI	−0.0008	−0.0008	−0.0009
14	DL_ReTrans_Rate	−0.0013	−0.0013	−0.0014
15	DL_IBLER	−0.0015	−0.0016	−0.0017
16	Cell_Downlink_Average_Throughput	−0.0019	−0.002	−0.0021
17	Average_DL_User_Throughput	−0.0027	−0.0029	−0.003

According to expression (8), with the conventional value $\alpha = 0.05$ and the default value m = 31,143, the RT value is selected in the interval $0 < RT < 0.025$. However, in practice, instead of a certain value of RT, and in accordance with the limitations, a few of the most important predictors that affect the prediction of the dependent variable are often chosen. Considering that the number of variables with weighting coefficients greater than 0 is equal to 12 in the observed case, the first six ranked predictors, according to Table 3, are selected as the final number of inputs. The RT threshold value that can be set hypothetically, and which can correspond to this selection of the optimal set of variables, is $RT = 0.0028$. This threshold applies to all three values of k.

Table 4 shows the ranked RE values and correlations for the three tested models that were created over the data set optimized by the RReliefF algorithm. Pearson's correlation coefficients r are calculated as follows:

$$r = \frac{\sum(D_{EtoEi} - D_{EtoEAVG})(D_{PREDi} - D_{PREDAVG})}{\sqrt{\sum(D_{EtoEi} - D_{EtoEAVG})^2 \sum(D_{PREDi} - D_{PREDAVG})^2}} \quad (10)$$

where $D_{PREDAVG}$ is the arithmetic mean of the variable D_{PREDi}.

Table 4. Results of testing the models created over the data set optimized by the RReliefF algorithm.

Model	*RE*	Correlation
1. k-NN	0.109	0.944
2. MLP	0.159	0.917
3. SVM	0.205	0.893

According to Table 4, the best predictive performance is shown by the model based on k-NN, which has $RE = 0.109$ and the correlation coefficient equal to 0.944. That is why this model is selected as the best solution in the approach to predictor set optimization with the RReliefF algorithm. The SVM model has the highest relative error, which is $RE = 0.205$, but also the lowest correlation value, which is equal to 0.893.

4.2. ML Predictive Models Created over a Set of Predictors Optimized by the Backward Selection via the Recursive Feature Elimination Algorithm

In accordance with the first step of the algorithm shown in Figure 6, all 17 independent variables are used as inputs to the ML models. Automatic training and testing of predictive models based on MLP, SVM and k-NN techniques are performed using the Auto Numeric method. As one of the results of this step, Figure 9 shows the input variables ranked by PI value [47]. The first two variables with the highest PI value are related to packet retransmission, which directly and negatively affects the end-to-end delay.

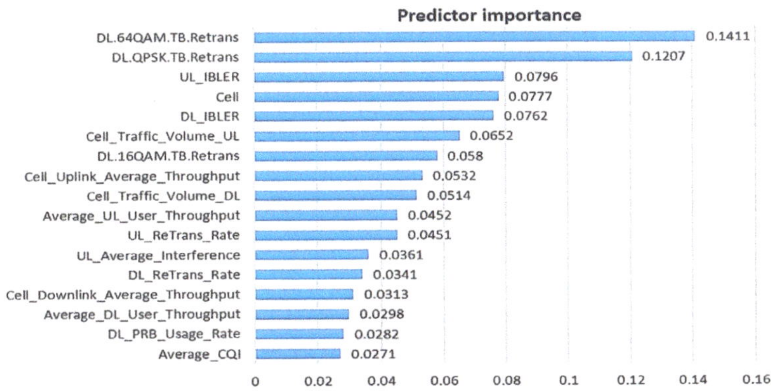

Figure 9. Independent variables ranked by PI.

By multiple executions of the loop of the algorithm given in Figure 6, the RE values of the model testing are obtained, as shown in Figure 10. From the figure, it can be concluded that the best predictive performance is shown by the model based on k-NN, which has the smallest relative error ($RE = 0.04$) for the five most influential input variables sorted according to Figure 10. Nevertheless, due to lower complexity, the k-NN model with four inputs is selected as the best solution; its relative error is slightly higher and amounts to

RE = 0.041. In addition, it is evident that the prediction performance decreases drastically with a further reduction in the number of inputs to three and two variables.

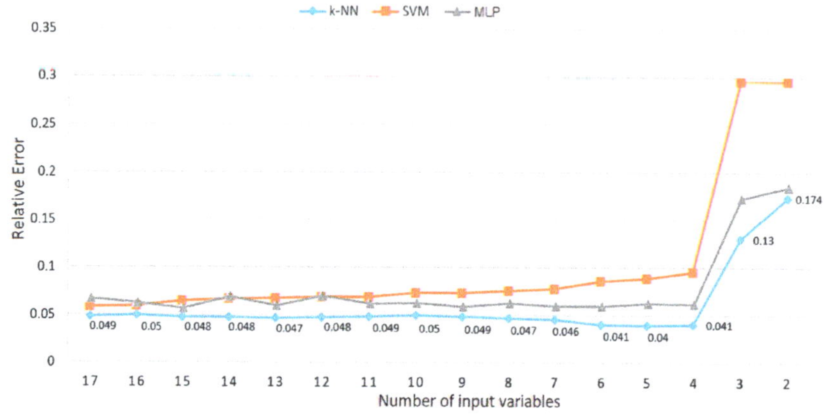

Figure 10. Relative error of ML models testing.

The performance of the tested predictive models, in addition to the relative error, can also be expressed by correlation, which is shown in Figure 11. It is concluded that the values of the Pearson correlation coefficients of the prediction results with the real data from the test set are "inverse" in relation to the *RE* values shown in Figure 10. Accordingly, the model with five inputs has the highest correlation coefficient (0.98), but due to the reasons mentioned above, the k-NN model with four inputs, whose correlation coefficient is equal to 0.979, was selected as the best solution. One of the main reasons for the better performance of the k-NN model when compared to the SVM and MLP models lies in the fact that the k-NN model is oriented towards simpler data sets, such as the one observed.

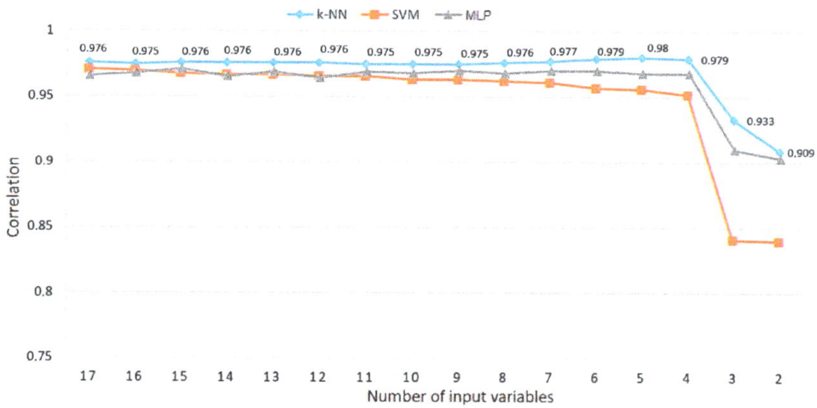

Figure 11. Correlation of ML models prediction results with data test set.

4.3. *Predictive ML Models Created over a Set of Predictors Optimized by the Pareto 80/20 Rule*

Figure 12 shows a Pareto diagram where the observed input variables are ordered according to the value of PI, from the highest to the lowest, by the ranking shown in Figure 9 [48]. The value of the cumulative curve for any input variable is equal to the sum of the PI values of individual predictors up to the observed variable, moving from the left to the right side of the diagram. According to the Pareto 80/20 rule, the goal is to find the first point on the curve with a cumulative value equal to or greater than 80%. This optimal point is marked in Figure 12, and the cumulative percentage in it is 81.34% for 11 input variables. According to the results shown in Figure 10, the k-NN model is selected as the best solution in this optimization approach, whose relative error at that point is $RE = 0.049$, while the correlation is equal to 0.975 (Figure 11).

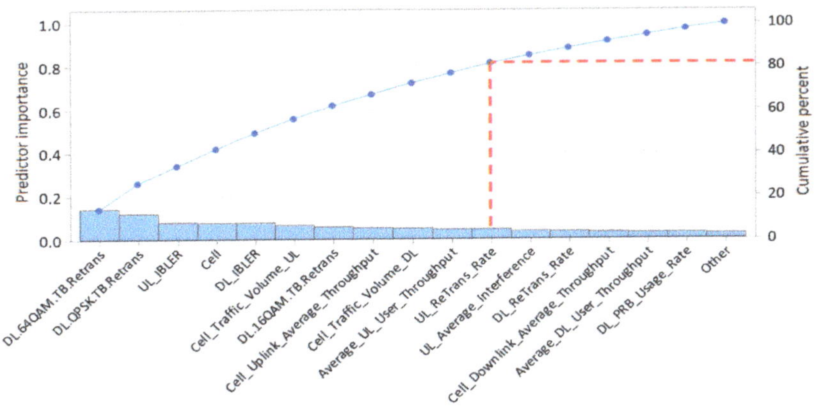

Figure 12. Pareto diagram for predictor set optimization.

4.4. *Comparative Analysis of Results Using Statistical Methods and Selection of the Final Model*

Comparative analysis compares the delay prediction results of three ML models, each of which was selected as the best solution in one of the three observed approaches to optimizing the input set of variables. The main goal is to determine the statistical significance of the differences between the prediction results, which is the reason for testing the null hypothesis:

Hypothesis 0 (H0). $\mu_1 = \mu_2 = \mu_3$, *where* μ_1, μ_2 *and* μ_3 *are the arithmetic means of delay prediction values for k-NN models selected as the best solutions in the approach based on the RRelieff algorithm, Backward selection via the recursive feature elimination algorithm, and the Pareto 80/20 rule, respectively. In other words, this hypothesis represents the assumption that there are no significant statistical differences in the arithmetic means of the delay prediction results for the three observed models.*

In contrast, the alternative hypothesis can be stated as follows:

Hypothesis 1 (H1). *There are significant statistical differences in the prediction results between at least two models, i.e., two optimization approaches.*

The parametric statistical test that tests the null hypothesis is ANOVA with Repeated Measures [49]. However, it is first necessary to test one of the basic conditions for the application of this test, which is the normality of the distribution of the dependent variable in groups. The results of the Kolmogorov–Smirnov normality test for the observed models are given in Table 5.

Table 5. Tests of Normality with summarized optimization and prediction results.

An Approach to Optimization of a Set of Input Variables	ML Model Selected	Number of Inputs	*RE*	Kolmogorov-Smirnov		
				Statistic	df	Sig.
RReliefF algorithm	k-NN	6	0.109	0.188	31,143	0.000
Backward selection via the recursive feature elimination algorithm	k-NN	4	0.041	0.191	31,143	0.000
Pareto 80/20 rule	k-NN	11	0.049	0.189	31,143	0.000

The obtained significance value of the Sig. test for all three cases has the same value (Sig. = 0.000). It means that the assumption about the normality of the distribution of the dependent variable in groups can be rejected. This conclusion can be confirmed graphically on the basis of the Q-Q plots shown in Figure 13. On the diagrams, it is obvious that there are significant deviations of the points from the line representing the normal distribution.

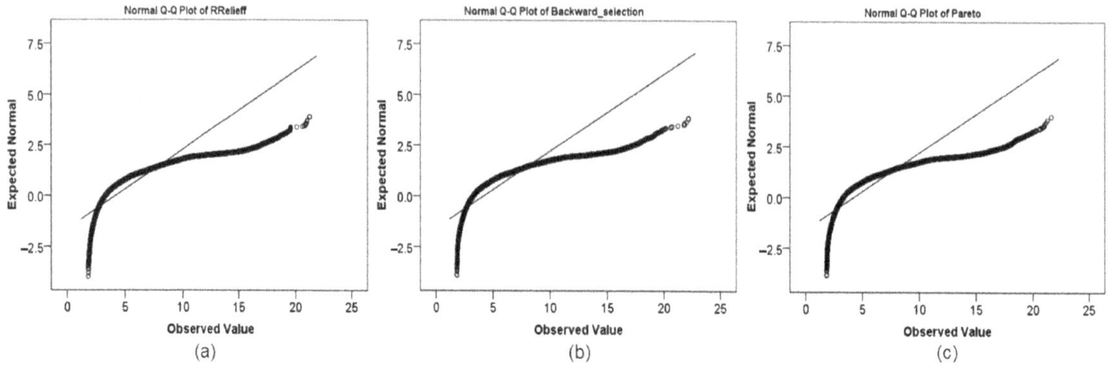

Figure 13. Normality tests of delay prediction results: (**a**) RReliefF algorithm; (**b**) Backward selection via the recursive feature elimination algorithm; (**c**) Pareto 80/20 rule.

Given the non-fulfillment of the conditions from the aspect of normality of distribution, it is necessary to test the hypotheses with the Friedman test, which is a non-parametric alternative to ANOVA with Repeated Measures. Table 6 shows the results of the Friedman test performed in IBM SPSS Statistics [50]. In addition to the sample size (N), a statistical test (Chi-Square), degree of freedom (df) and significance level (Asymp. Sig.) are given in the table. Based on the value of Asymp. Sig., which is less than the α = 0.05 level, it is concluded that there are statistically significant differences in the prediction results for the three models, i.e., for three approaches to optimizing the set of input variables.

Table 6. Results of the Friedman test.

N	31,143
Chi-Square	268.019
df	2
Asymp. Sig.	0.000

The results given in Table 6 do not show the information for which pair of combined optimization techniques there is a significant statistical difference. The answer to this question is obtained with a Post Hoc statistical test. Table 7 shows the results of the Wilcoxon signed-rank post hoc test with the value of Z and Asymp. Sig. for each of the three combinations of approaches.

Table 7. Wilcoxon signed-rank post hoc test results.

	Pairs for Comparison		
	RReliefF—Pareto 80/20 Rule	**Backward Selection via the Recursive Feature Elimination—RReliefF**	**Backward Selection via the Recursive Feature Elimination—Pareto 80/20 Rule**
Z	−3.077	−7.848	−18.727
Asymp. Sig. (2-tailed)	0.002	0.000	0.000

In order to interpret the results obtained, it is necessary to calculate the adjusted Bonferroni level of significance as the ratio of level α = 0.05 and the number of pairs being compared, which as a result provides a value of 0.017. Given that Asymp. Sig. < 0.017

applies to all combinations, it is concluded that there are statistically significant differences among the delay prediction results for all three pairs of approaches to optimizing the input set of variables.

Based on the presented results, the k-NN model whose number of inputs is optimized to four by the algorithm Backward selection via the recursive feature elimination is chosen as the final model. Figure 14 shows the prediction results of the dependent variable D_{EtoE} using this model based on the data from the test set. The diagram also shows the regression line equation that explains the linear relationship between the actual and the predicted D_{EtoE} values.

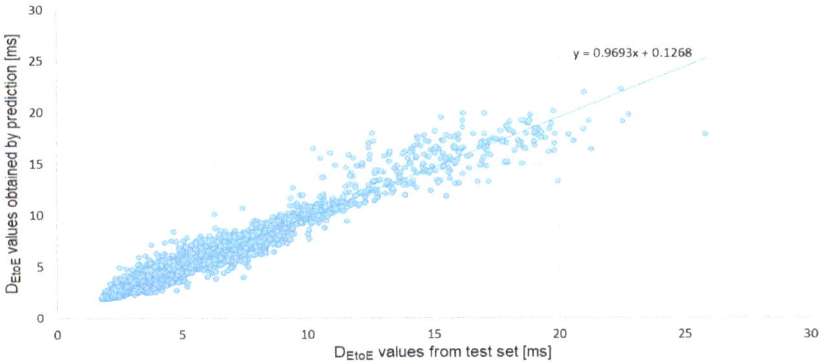

Figure 14. Scatter plot of actual D_{EtoE} values and D_{EtoE} values obtained by prediction using the final k-NN model.

State-of-the-art methods and techniques in network delay prediction are based mainly on machine learning and deep learning. In particular, Graph Neural Networks (GNN) stand out, which are adapted for processing data structured in the form of graphs. Other popular state-of-the-art techniques include Autoregressive Integrated Moving Average (ARIMA), LSTM, Gated Recurrent Unit (GRU), RNN, Convolutional Neural Network (CNN) models. Some of the most important advantages that distinguish the final selected k-NN model in this research with the mentioned state-of-the-art technique are the simplicity of interpretation and application, it maintains good performance with small data sets, it is adaptable to changes in the data set and it does not require time stationarity data.

5. Conclusions

For a long period of time, not only the amount of data, called BD, but also the number of users of network services and the range of user requests for higher QoS has been increasing drastically. Telecommunications operators face increasingly complex technical and technological problems in a domain of network traffic management, adequate planning and modern design of all dimensions of network resource quality, including their allocation and performance—KPI. This is especially important for services such as VoIP and traffic streaming. Predictive modeling of required solutions currently is most often based on the techniques of the ML method. Numerous studies of different approaches to certain solutions for indicated problems are analyzed in this paper and presented in Section 2. Using the above and other experiences and theoretical findings of more comprehensive studies, the paper presents original approaches to predictive modeling of end-to-end delay of data packets through a real 4G LTE network. The network is in the geo-space covered by the M:tel BL mobile operator with a focus on the area of a three-segment road in the road network of RS, BiH. In the LTE architecture, a total of 87 cells are located in the observed area, which provide users with a continuous and permanent network connection.

In the paper, the aims and objectives of the research have been fulfilled. It includes reducing the dimensionality of the space of input variables in the optimization model with Feature Selection techniques (RReliefF and Backward selection via the recursive feature elimination algorithms) and the Pareto 80/20 rule. It is followed by training and testing of ML models with MLP, SVM and k-NN techniques including the selection of the best delay prediction model in the LTE network according to criteria of accuracy and complexity/interpretability. Then, the implementation of a unique methodology of indirect assessment and calculation of dependent variable values based on the average number of active users in the network has been performed. At the same time, a universally applicable predictive model of delay in the LTE network, based on the research in the real space of Big Data (BD) with input-output vectors, has been created. In the opinion of the team of authors, the presented approaches to the optimization of the number of predictors by end-to-end delay ML modeling techniques in LTE networks by reducing the dimensions of BD and connecting independent variables in pairs with the calculation of KPI are a particularly important innovative contribution to the research of telecommunications traffic provided in this paper. It also involves the methodology of presenting and interpreting textual, algorithmic, graphic, photo-documentation, mathematical and computer-generated solutions. An optimal explanatory strategy has also been used in creating a system of clarification of the presented methodology and results referring to similarities in the structure of what is being investigated in this paper with already known facts that, among other things, were published in the cited papers and other authors' solutions. Also, familiar systems of relations that are used as models which can be useful to understand the new experience in the systematic scientific research of telecommunications traffic are taken into account, and the similarities created in analogies and in hypotheses have led to the proven quality of the results presented.

The research results show that the k-NN model has been selected as the best solution in all three approaches to the optimization of the input set of variables. For the RReliefF optimization algorithm, the best model has 6 inputs and $RE = 0.109$; for Backward selection via the recursive feature elimination algorithm, the best model has 4 inputs and $RE = 0.041$; and for the Pareto 80/20 rule, the best model has 11 inputs and $RE = 0.049$. The comparative analysis of the results concludes that according to both observed criteria for the selection of the final model, the best solution is an approach to optimizing the number of predictors based on the Backward selection via the recursive feature elimination algorithm. In other words, the k-NN model created within this approach has the lowest *RE* value and the lowest number of input variables of all tested ones.

Cross-validation techniques were not applied in this paper, which in the strictest sense can be considered as a possible limitation of this work. Nevertheless, the results showed that the model has performance stability with deafult hyperparameters. The test data set was not made visible, and information was not leaked to the models during training and was only used to evaluate their generalization abilities.

Author Contributions: Conceptualization, M.K.B., M.S. and M.V.; methodology, M.K.B. and M.S.; software, M.S. and M.V.; validation, M.K.B., D.N. and A.S.; formal analysis, M.K.B., D.N., M.S. and A.S.; investigation, M.K.B., M.S. and M.V.; resources, M.K.B., M.S., A.S., D.D. and G.P.; data curation, M.K.B., M.S., M.V. and D.D.; writing—original draft preparation, M.K.B. and M.S.; writing—review and editing, M.K.B., M.S. and M.V.; visualization, M.K.B., M.S., D.D. and G.P.; supervision, M.K.B. and M.S.; project administration, D.D., D.N., G.P. and M.V.; funding acquisition, D.D., G.P., A.S., D.N. and M.V. All authors have read and agreed to the published version of the manuscript.

Funding: This research received no external funding.

Institutional Review Board Statement: Not applicable.

Informed Consent Statement: Not applicable.

Data Availability Statement: The dataset is available upon request.

Acknowledgments: The authors gratefully acknowledge the mobile operator M:tel Banja Luka for their support by providing research data.

Conflicts of Interest: The authors declare no conflict of interest.

References

1. Banjanin, M.K.; Maričić, G.; Stojčić, M. Multifactor Influences on the Quality of Experience Service Users of Telecommunication Providers in the Republic of Srpska, Bosnia and Herzegovina. *Int. J. Qual. Res.* **2022**, *17*, 369–386. [CrossRef]
2. Banjanin, M.K.; Stojčić, M.; Danilović, D.; Ćurguz, Z.; Vasiljević, M.; Puzić, G. Classification and Prediction of Sustainable Quality of Experience of Telecommunication Service Users Using Machine Learning Models. *Sustainability* **2022**, *14*, 17053. [CrossRef]
3. Mesbahi, N.; Dahmouni, H. Delay and jitter analysis in LTE networks. In Proceedings of the 2016 International Conference on Wireless Networks and Mobile Communications (WINCOM), Fez, Morocco, 26–29 October 2016; IEEE: Amsterdam, The Netherlands, 2016; pp. 122–126. [CrossRef]
4. Yaqoob, J.I.A.Y.; Pang, W.L.; Wong, S.K.; Chan, K.Y. Enhanced exponential rule scheduling algorithm for real-time traffic in LTE network. *Int. J. Electr. Comput. Eng. (IJECE)* **2020**, *10*, 1993–2002. [CrossRef]
5. Stojčić, M.; Banjanin, M.K.; Vasiljević, M.; Stjepanović, A.; Ćurguz, Z. PCA modeling of extraction and selection of variables influencing LTE network delay in urban mobility conditions. In Proceedings of the International Conference on Advances in Traffic and Communication Technologies ATCT 2023, Sarajevo, Bosnia and Herzegovina, 11–12 May 2023.
6. *ETSI TS 123 107 v12.0.0*; Digital Cellular Telecommunications System (Phase 2+); Universal Mobile Telecommunications System (UMTS); LTE; Quality of Service (QoS) Concept and Architecture. European Telecommunications Standards Institute: Sophia Antipolis, France, 2014. Available online: https://www.etsi.org/deliver/etsi_ts/123100_123199/123107/12.00.00_60/ts_123107v120000p.pdf (accessed on 26 June 2023).
7. Kumar, V.; Minz, S. Feature selection: A literature review. *SmartCR* **2014**, *4*, 211–229. [CrossRef]
8. Cai, J.; Luo, J.; Wang, S.; Yang, S. Feature selection in machine learning: A new perspective. *Neurocomputing* **2018**, *300*, 70–79. [CrossRef]
9. Đukić, A.; Bjelošević, R.; Stojčić, M.; Banjanin, M.K. Network Model of Multiagent Communication of Traffic Inspection for Supervision and Control of Passenger Transportation in Road and City Traffic. In Proceedings of the Croatian Society for Information, Communication and Electronic Technology–MIPRO 2023 46th (Hybrid) Convention, Opatija, Croatia, 22–26 May 2023; pp. 1352–1357.
10. Torres-Figueroa, L.; Schepker, H.F.; Jiru, J. QoS evaluation and prediction for C-V2X communication in commercially-deployed LTE and mobile edge networks. In Proceedings of the 2020 IEEE 91st Vehicular Technology Conference (VTC2020-Spring), Antwerp, Belgium, 25–28 May 2020; IEEE: Amsterdam, The Netherlands, 2020; pp. 1–7. [CrossRef]
11. Zhang, W.; Feng, M.; Krunz, M.; Volos, H. Latency prediction for delay-sensitive v2x applications in mobile cloud/edge computing systems. In Proceedings of the GLOBECOM 2020–2020 IEEE Global Communications Conference, Taipei, Taiwan, 7–11 December 2020; IEEE: Amsterdam, The Netherlands, 2020; pp. 1–6. [CrossRef]
12. Brown, J.; Khan, J.Y. A predictive resource allocation algorithm in the LTE uplink for event based M2M applications. *IEEE Trans. Mob. Comput.* **2015**, *14*, 2433–2446. [CrossRef]
13. Khatouni, A.S.; Soro, F.; Giordano, D. A machine learning application for latency prediction in operational 4g networks. In Proceedings of the 2019 IFIP/IEEE Symposium on Integrated Network and Service Management (IM), Arlington, VA, USA, 8–12 April 2019; IEEE: Amsterdam, The Netherlands, 2019; pp. 71–74.
14. Zhohov, R.; Minovski, D.; Johansson, P.; Andersson, K. Real-time performance evaluation of LTE for IIoT. In Proceedings of the 2018 IEEE 43rd Conference on Local Computer Networks (LCN), Chicago, IL, USA, 1–4 October 2018; IEEE: Amsterdam, The Netherlands, 2018; pp. 623–631. [CrossRef]
15. Lai, W.K.; Tang, C.L. QoS-aware downlink packet scheduling for LTE networks. *Comput. Netw.* **2013**, *57*, 1689–1698. [CrossRef]
16. Lai, W.K.; Hsu, C.W.; Kuo, T.H.; Lin, M.T. A LTE downlink scheduling mechanism with the prediction of packet delay. In Proceedings of the 2015 Seventh International Conference on Ubiquitous and Future Networks, Sapporo, Japan, 7–10 July 2015; IEEE: Amsterdam, The Netherlands, 2015; pp. 257–262. [CrossRef]
17. Nasri, M.; Hamdi, M. LTE QoS parameters prediction using multivariate linear regression algorithm. In Proceedings of the 2019 22nd Conference on Innovation in Clouds, Internet and Networks and Workshops (ICIN), Paris, France, 19–21 February 2019; IEEE: Amsterdam, The Netherlands, 2019; pp. 145–150. [CrossRef]
18. Ahmed, A.H.; Hicks, S.; Riegler, M.A.; Elmokashfi, A. Predicting High Delays in Mobile Broadband Networks. *IEEE Access* **2021**, *9*, 168999–169013. [CrossRef]
19. Banjanin, M.K.; Stojčić, M.; Drajić, D.; Ćurguz, Z.; Milanović, Z.; Stjepanović, A. Adaptive Modeling of Prediction of Telecommunications Network Throughput Performances in the Domain of Motorway Coverage. *Appl. Sci.* **2021**, *11*, 3559. [CrossRef]
20. Loshakov, V.A.; Al-Janabi, H.D.; Al-Zayadi, H.K. Adaptive control signal parameters in LTE technology with MIMO. *Telecommun. Probl.* **2012**, *2*, 78–90. Available online: http://openarchive.nure.ua/handle/document/430 (accessed on 27 March 2023).
21. Ren, J.; Zhang, X.; Xin, Y. Using Deep Convolutional Neural Network to Recognize LTE Uplink Interference. In Proceedings of the 2019 IEEE Wireless Communications and Networking Conference (WCNC), Marrakesh, Morocco, 15–18 April 2019; IEEE: Amsterdam, The Netherlands, 2019; pp. 1–6. [CrossRef]
22. Madi, N.K.; Hanapi, Z.M.; Othman, M.; Subramaniam, S.K. Delay-based and QoS-aware packet scheduling for RT and NRT multimedia services in LTE downlink systems. *EURASIP J. Wirel. Commun. Netw.* **2018**, *180*, 180. [CrossRef]
23. Kuhn, M.; Johnson, K. *Feature Engineering and Selection: A Practical Approach for Predictive Models*, 1st ed.; Chapman and Hall/CRC: Boca Raton, FL, USA, 2019; ISBN 978-1-13-807922-9.
24. Li, J.; Cheng, K.; Wang, S.; Morstatter, F.; Trevino, R.P.; Tang, J.; Liu, H. Feature selection: A data perspective. *ACM Comput. Surv. (CSUR)* **2017**, *50*, 1–45. [CrossRef]

25. Wah, Y.B.; Ibrahim, N.; Hamid, H.A.; Abdul-Rahman, S.; Fong, S. Feature Selection Methods: Case of Filter and Wrapper Approaches for Maximising Classification Accuracy. *Pertanika J. Sci. Technol.* **2018**, *26*, 329–340.
26. MathWorks. Introduction to Feature Selection. Available online: https://www.mathworks.com/help/stats/feature-selection.html (accessed on 27 March 2023).
27. Kira, K.; Rendell, L.A. A practical approach to feature selection. In Proceedings of the Machine learning proceedings, Aberdeen, UK, 1–3 July 1992; pp. 249–256. [CrossRef]
28. Kira, K.; Rendell, L.A. The feature selection problem: Traditional methods and a new algorithm. In Proceedings of the Tenth National Conference on Artificial Intelligence—AAAI'92, San Jose, CA, USA, 12–16 July 1992; pp. 129–134. Available online: https://cdn.aaai.org/AAAI/1992/AAAI92-020.pdf (accessed on 27 March 2023).
29. Kononenko, I. Estimating Attributes: Analysis and extensions of RELIEF. In *Machine Learning: ECML-94*; Bergadano, F., De Raedt, L., Eds.; Springer: Berlin/Heidelberg, Germany, 1994; Volume 784. [CrossRef]
30. Robnik-Šikonja, M.; Kononenko, I. An adaptation of Relief for attribute estimation in regression. In Proceedings of the Machine Learning: Proceedings of the Fourteenth International Conference (ICML'97), Nashville, TN, USA, 8–12 July 1997; pp. 296–304.
31. MathWorks. Relief. Available online: https://www.mathworks.com/help/stats/relieff.html (accessed on 24 April 2023).
32. Urbanowicz, R.J.; Meeker, M.; La Cava, W.; Olson, R.S.; Moore, J.H. Relief-based feature selection: Introduction and review. *J. Biomed. Inform.* **2018**, *85*, 189–203. [CrossRef] [PubMed]
33. Guyon, I.; Weston, J.; Barnhill, S.; Vapnik, V. Gene selection for cancer classification using support vector machines. *Mach. Learn.* **2002**, *46*, 389–422. [CrossRef]
34. Okorie, O.; Salonitis, K.; Charnley, F.; Turner, C. A systems dynamics enabled real-time efficiency for fuel cell data-driven remanufacturing. *J. Manuf. Mater. Process.* **2018**, *2*, 77. [CrossRef]
35. Hugh, J. *Engineering Design, Planning, and Management*, 2nd ed.; Academic Press: Cambridge, MA, USA, 2021; ISBN 978-0-12-821055-0.
36. Jin, Y.; Sendhoff, B. Pareto-based multiobjective machine learning: An overview and case studies. *IEEE Trans. Syst. Man Cybern. Part C (Appl. Rev.)* **2008**, *38*, 397–415. [CrossRef]
37. Lee, S.H.; Mazumder, J.; Park, J.; Kim, S. Ranked feature-based laser material processing monitoring and defect diagnosis using k-NN and SVM. *J. Manuf. Process.* **2020**, *55*, 307–316. [CrossRef]
38. Ahmad, M.A.; Eckert, C.; Teredesai, A. Interpretable machine learning in healthcare. In Proceedings of the 2018 ACM International Conference on Bioinformatics, Computational Biology, and Health Informatics, Washington, DC, USA, 29 August–1 September 2018; pp. 559–560.
39. Abdullah, T.A.; Zahid, M.S.M.; Ali, W. A review of interpretable ML in healthcare: Taxonomy, applications, challenges, and future directions. *Symmetry* **2021**, *13*, 2439. [CrossRef]
40. Dherin, B.; Munn, M.; Rosca, M.; Barrett, D. Why neural networks find simple solutions: The many regularizers of geometric complexity. In Proceedings of the Thirty-Sixth Conference on Neural Information Processing Systems-NeurIPS, New Orleans, LA, USA, 28 November–9 December 2022; Volume 35, pp. 2333–2349.
41. Stiglic, G.; Kocbek, P.; Fijacko, N.; Zitnik, M.; Verbert, K.; Cilar, L. Interpretability of machine learning-based prediction models in healthcare. *Wiley Interdiscip. Rev. Data Min. Knowl. Discov.* **2020**, *10*, e1379. [CrossRef]
42. Morocho-Cayamcela, M.E.; Lee, H.; Lim, W. Machine learning for 5G/B5G mobile and wireless communications: Potential, limitations, and future directions. *IEEE Access* **2019**, *7*, 137184–137206. [CrossRef]
43. Yang, Y.J.; Bang, C.S. Application of artificial intelligence in gastroenterology. *World J. Gastroenterol.* **2019**, *25*, 1666. [CrossRef]
44. Pichler, M.; Hartig, F. Machine learning and deep learning—A review for ecologists. *Methods Ecol. Evol.* **2023**, *14*, 994–1016. [CrossRef]
45. Guo, M.; Zhang, Q.; Liao, X.; Chen, Y. An interpretable machine learning framework for modelling human decision behavior. *arXiv* **2019**, arXiv:1906.01233.
46. Nesvijevskaia, A.; Ouillade, S.; Guilmin, P.; Zucker, J.D. The accuracy versus interpretability trade-off in fraud detection model. *Data Policy* **2021**, *3*, e12. [CrossRef]
47. Chowdhury, M.Z.I.; Turin, T.C. Variable selection strategies and its importance in clinical prediction modelling. *Fam. Med. Community Health* **2020**, *8*, e000262. [CrossRef] [PubMed]
48. Wang, J.; Jiang, C.; Zhang, H.; Ren, Y.; Chen, K.C.; Hanzo, L. Thirty years of machine learning: The road to Pareto-optimal wireless networks. *IEEE Commun. Surv. Tutor.* **2020**, *22*, 1472–1514. [CrossRef]
49. Yu, Z.; Guindani, M.; Grieco, S.F.; Chen, L.; Holmes, T.C.; Xu, X. Beyond t test and ANOVA: Applications of mixed-effects models for more rigorous statistical analysis in neuroscience research. *Neuron* **2022**, *110*, 21–35. [CrossRef]
50. Balali, A.; Valipour, A. Identification and selection of building façade's smart materials according to sustainable development goals. *Sustain. Mater. Technol.* **2020**, *26*, e00213. [CrossRef]

applied sciences

Article

Intelligent TCP Congestion Control Policy Optimization

Hanbing Shi and Juan Wang *

College of Mechatronics Engineering, Xi'an University of Architecture and Technology, Xi'an 710055, China; shbing0601@163.com
* Correspondence: juanwang618@126.com

Abstract: Network congestion control is an important means to improve network throughput and reduce data transmission delay. To further optimize the network data transmission capability, this research suggests a proximal policy optimization-based intelligent TCP congestion management method, creates a proxy that can communicate with the real-time network environment, and abstracts the TCP congestion control mechanism into a partially observable Markov decision process. Changes in the real-time state of the network are fed back to the agent, and the agent makes action commands to control the size of the congestion window, which will produce a new network state, and the agent will immediately receive a feedback reward value. To guarantee that the actions taken are optimum, the agent's goal is to obtain the highest feedback reward value. The state space of network characteristics should be designed so that agents can observe enough information to make appropriate decisions. The reward function is designed through a weighted algorithm that enables the agent to balance and optimize throughput and latency. The model parameters of the agent are updated by the proximal policy optimization algorithm, and the truncation function keeps the parameters within a certain range, reducing the possibility of oscillation during gradient descent and ensuring that the training process can converge quickly. Compared to the traditional CUBIC control method, the results show that the TCP-PPO$_2$ policy reduces latency by 11.7–87.5%.

Keywords: network congestion; congestion control; internet; proximal policy optimization

Citation: Shi, H.; Wang, J. Intelligent TCP Congestion Control Policy Optimization. *Appl. Sci.* **2023**, *13*, 6644. https://doi.org/10.3390/app13116644

Academic Editors: Christos Bouras, Runzhou Zhang, Lin Zhang, Yang Yue, Hao Feng, Zheda Li and Dawei Ying

Received: 13 April 2023
Revised: 24 May 2023
Accepted: 28 May 2023
Published: 30 May 2023

1. Introduction

With the accelerated growth of mobile broadband network technology and the increase in the number of users in recent years, the huge amount of network information transmission will cause network congestion, and network congestion may lead to a slow transmission speed, high delay, high loss rate, etc., and seriously lead to network failure. To realize the reliable transmission of network data, it is necessary to build an efficient and reliable network transmission protocol, and congestion control is the key technology to achieve efficient and reliable transmission.

NewReno [1] and CUBIC [2] use packet loss to detect congestion and reduce the congestion window length after congestion is detected. Westwood [3] is an adaptation of NewReno, and the transmission capacity-based congestion control mechanism uses the prediction of the link transmit capacity as the basis for congestion control.

Traditional congestion control mechanisms use defined control rules to adjust congestion windows, making it difficult to adapt to the complexity and real-time changes in modern networks. Therefore, the researchers propose a congestion control algorithm based on reinforcement learning. Van et al. used reinforcement learning algorithms to adaptively change parameter configurations [4], thereby improving the quality of the video stream. Cui et al. proposed a custom congestion control algorithm Hd-TCP [5], which applies deep reinforcement learning to deal with the poor network experience caused by frequent network switching on a high-speed rail from the perspective of the transport layer. Lin et al. improved the applicability of virtual network functions using a model-assisted deep reinforcement learning framework [6]. Xie et al. proposed a congestion window length for 5G mobile edge computing based on deep learning [7]. TCP-Drinc is a model-free intelligent congestion control algorithm based on deep reinforcement learning [8], which obtains eigenvalues from past network states and experiences and adjusts the congestion window length based on the set of these eigenvalues. The Rax algorithm uses online reinforcement learning [9] to maintain the optimal congestion window length based on the given reward function and network conditions. QTCP is based on Q-learning for congestion control [10,11], which improves throughput to a certain extent. MPTCP [12] uses Q-learning and Deep Q-Networks (DQN) for multipath congestion control, which is able to learn to take the best action based on the runtime state. However, the Q-learning algorithm is slow to learn and difficult to converge. The reinforcement learning algorithm based on the policy gradient can solve the shortcomings of the Q-learning algorithm such as slow learning speed and difficult convergence.

To further improve the communication capability of the congestion control strategy in the unknown network environment, this paper analyzes the characteristics of the four stages of congestion control and proposes a congestion control strategy based on the proximal policy optimization algorithm, which is one of the best policy gradient algorithms [13]; this strategy can save a lot of model training time, make full use of the training data, and finally realize the reliable transmission of data. Compared with the traditional CUBIC congestion control strategy, the proposed algorithm is feasible and effective in improving network transmission performance.

2. Related Work

2.1. Fundamentals of Congestion Control

Network congestion is a phenomenon that often occurs in the operation of computer networks, and from its manifestation, network congestion is the phenomenon that the cache in the router drops packets because of overflow. When the packet arrives at the router, the packet is forwarded according to the configured forwarding rules and output to the corresponding link. Due to limited network link resources (including cache size, fixed bandwidth, processing power, etc.), queues form in the link and network congestion occurs when packets arrive too quickly. Increasing the buffer area can absorb excess packets and prevent packet loss, but if you blindly increase the size of the buffer without improving the link bandwidth and processor capacity, this will cause the waiting time in the queue to greatly increase, and the upper protocol can only retransmit them, so simply expanding the cache space can not solve the network congestion problem, but will cause a waste of network resources. In addition, network nodes process thousands of data streams per second, sharing bandwidth between data streams, and the maximum rate of data transmission is limited by the bottleneck link. Congestion occurs when a network node needs to process more data than it can handle. Therefore, the job of congestion control is to prevent data senders from sending large amounts of data into the network, causing the transmission link to be overloaded. The principle of congestion control is shown in Figure 1.

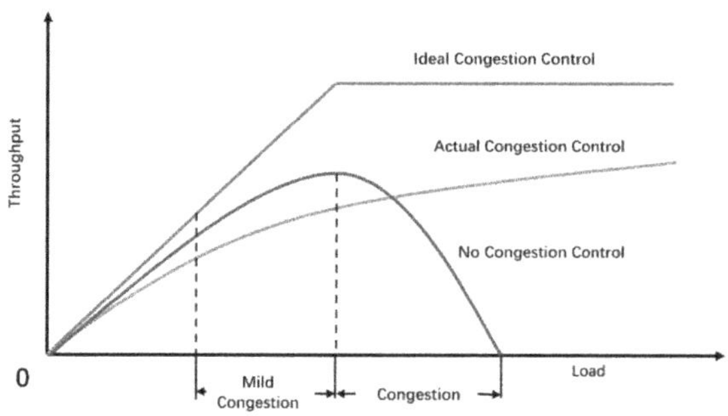

Figure 1. The role of congestion control.

2.2. Deep Reinforcement Learning-Based Congestion Control

The congestion control strategy framework based on deep reinforcement learning is shown in Figure 2. Deep reinforcement learning requires the construction of environments and agents. Taking the network environment as the environment, by collecting the real-time state of the network environment, the strategy function used by the agent is constructed, the agent responds after learning, and the strategy function is fitted by an artificial neural network. The agent makes the optimal control strategy according to the output of the policy function, controls the congestion window length, and changes the TCP sending policy. After the agent makes an action, deep reinforcement learning will judge the action according to the state, so as to output the reward value and depending on the reward value, modify the parameters of the artificial neural network so that the agent can maximize the reward.

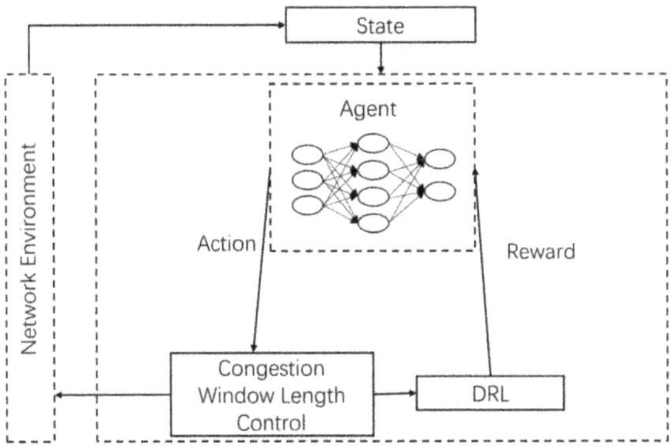

Figure 2. DRL-based congestion control algorithm framework.

The QTCP algorithm is an algorithm based on the Q-learning framework. Its state space is continuous, and its state includes average RTT, average interval time between sending and receiving packets, and discrete action spaces including increasing by 10 bytes, decreasing by 1 byte, and remaining unchanged. Next, the system compares the size of the utility function of the current time period and the previous time period to determine whether it is a positive reward or a negative reward, where the utility function $U = a \cdot \log(throughput) - b \cdot \log(RTT)$. QTCP algorithms perform better than traditional algorithms in a variety of network scenarios, but the effect is still linked to network scenarios and cannot completely avoid the shortcomings of traditional congestion control algorithms. The congestion control algorithm Indigo, based on imitation learning, sets the network scene information to expert knowledge, and the decision network is a single-layer LSTM network, to achieve congestion control in the current training scenario. However, the performance of the algorithm can only play a superior performance in the trained scenario, so the practical application is limited. In contrast, DRL-based congestion control algorithms only require a simple neural network and combine historical information based on multiple time slice states before the current moment to obtain performance beyond traditional algorithms, so we will take a closer look at DRL-based congestion control algorithms.

3. Methods

3.1. Deep Reinforcement Learning

3.1.1. Background

Deep reinforcement learning is a combination of deep learning and reinforcement learning. Deep learning uses representation learning to refine data [14] and does not have to choose features, compressed dimensions, conversion formats, or other data processing techniques, offering better feature representation capabilities than conventional machine learning techniques and providing a distributed representation of data by mixing low-level features to create more abstract high-level features. Reinforcement learning originated from the optimal control theory in cybernetics [15], which is mainly used to solve the problem of timing decision making, by ongoing environmental interaction and trial-and-error, and finally obtains the optimal strategy for a specific task and maximizes the cumulative expected return of the task. The mainstream methods of traditional reinforcement learning mainly include the Monte Carlo class method and the time difference classification method [16]. The former is an unbiased estimate with a larger variance, while the latter uses a finite step bootstrapping method with a smaller variance but introduces bias. Deep reinforcement learning combines the structure of deep learning with the idea of reinforcement learning for solving decision-making problems. With the help of the powerful representation ability of deep neural networks, any component of reinforcement learning can be fitted, including state value functions, action value functions, strategies, models, etc., and the weights in deep neural networks can be used as fitting parameters. DRL is mainly used to solve high-dimensional state action space tasks, integrating deep learning's powerful understanding ability in feature representation problems and reinforcement learning's decision-making ability to achieve end-to-end learning. The emergence of deep reinforcement learning has made reinforcement learning technology truly practical and it can solve complex problems in real-world scenarios. The most representative deep Q-network (DQN) is an extension of the Q-learning algorithm [17], which uses neural networks to approximate the action–value function, and the optimization goal is the minimization loss function:

$$L_i(\theta_i) = E_{\pi_{\theta_i}}\left[(y_i - Q(s,a;\theta_i))^2\right] \quad (1)$$

where i represents the ith iteration, $y_i = r + \gamma \max_{a'} Q(s', a'; \theta_{i-1})$.

Van et al. proposed the use of the deep double Q-network (DDQN) [18]. DDQN selects the action of the target Q value based on the current Q-network and uses the target Q-network to calculate the corresponding Q value of the action.

$$y_i = r + \gamma Q(s', \text{argmax}_a Q(s', a; \theta_i); \theta_i^-) \quad (2)$$

3.1.2. Proximal Policy Optimization Algorithms

The description of reinforcement learning is usually based on the Markov decision process [19], which is a mathematical formalization of sequential decision making, in which immediate rewards and subsequent states of the system are affected by behavior, causing changes in future rewards, whose tasks correspond to the multivariate array E = <S, A, P, R>, where:

S—is a state space.
A—is an action space.
P—is state transition probability.
R—is the reward function.

Cumulative discount returns are often used to define state returns at t moment:

$$R_t = \sum_{i=t}^{T} \gamma^{i-1} r(s_i, a_i) \quad (3)$$

γ is the discount factor, which indicates that the farther away the return, the less impact it has on the assessment of the current state, $r(s_i, a_i)$ represents the value of the return obtained by selecting the action a_i in the state s_i; the initial state is s_i, and under a certain policy π, the state distribution obeys ρ_π, then the task of reinforcement learning is to learn a policy π so that the desired initial state return is maximized.

3.1.3. Policy Gradient

The Policy Gradient (PG) method works by calculating the estimator of the strategy gradient and inserting it into the stochastic gradient ascent algorithm. The most commonly used gradient estimators have this form:

$$\hat{g} = \hat{E}_t[\nabla_\theta \log \pi_\theta(a_t|s_t)\hat{A}_t] \quad (4)$$

where π_θ is the stochastic strategy and $\hat{A}_t$ is the estimator of the dominant function at time step t. Here, the expectation $\hat{E}_t[\ldots]$ represents the empirical mean of the finite batch sample, in algorithms alternating between sampling and optimization, using the implementation of automatic discrimination software to work by constructing an objective function with a gradient as a gradient estimator of the strategy gradient. By deriving the target, the estimator $\hat{g}$ is obtained.

$$L^{PG}(\theta) = \hat{E}_t[\log \pi_\theta(a_t|s_t)\hat{A}_t] \quad (5)$$

While it is advantageous to use a uniform trajectory to perform multiple optimization steps for L^{PG}, it is not reasonable and, empirically, often leads to a large number of policy updates being disrupted.

3.1.4. Trust Region Methods

In TRPO, the objective function is maximized but is limited by the size of the policy update. The details are as follows:

$$\underset{\theta}{maximize}\, \hat{E}_t[\frac{\pi_\theta(a_t|s_t)}{\pi_{\theta_{old}}(a_t|s_t)}\hat{A}t] \quad (6)$$

$$Subject\ to\ \hat{E}_t[KL[\pi_{\theta_{old}}(\cdot|s_t), \pi_\theta(\cdot|s_t)]] \leq \delta \quad (7)$$

θ_{old} is the vector of policy parameters before the update. After linear approximation of the target and quadratic approximation of the constraint, the conjugate gradient algorithm can be used to effectively approximate the problem.

This theory proves that TRPO actually recommends using penalties rather than constraints, i.e., solving unconstrained optimization problems.

$$\underset{\theta}{maximize}\hat{E}_t\left[\frac{\pi_\theta(a_t|s_t)}{\pi_{\theta_{old}}(a_t|s_t)}\hat{A}t - \beta KL[\pi_{\theta_{old}}(\cdot|s_t), \pi_\theta(\cdot|s_t)]\right] \quad (8)$$

The coefficients β follow the fact that a defined agent targets from the lower bound of the strategy π, and TRPO uses hard constraints rather than penalties because it is difficult to choose a single β value that performs well in different problems or single problems, where features change during the learning process. Therefore, it is not enough to simply choose a fixed penalty coefficient β using SGD to optimize the penalty target (8).

3.1.5. Proximal Policy Optimization Algorithm

The Proximal Policy Optimization (PPO) algorithm, which is one of the most effective model-free policy gradient methods, achieves state-of-the-art performance in many reinforcement learning continuous control benchmarks [13]. It is derived from the TRPO algorithm [20], but is easier to implement and has better sample complexity. Therefore, PPO is used to train a defined congestion-controlled reinforcement learning strategy. PPO is an actor-critical algorithm, so it uses a multi-step return of TD(λ) as a function of training values and a generalized advantage estimator (GAE) to compute the policy gradient [21].

3.1.6. PPO_2 Principle

The reinforcement learning algorithm needs to design the strategy function $\pi(a_t|s_t)$ so that it can generate the probability of performing some action a_t under state s_t. Artificial neural networks can theoretically fit arbitrary functions, so the current reinforcement learning algorithm fits the strategy function $\pi(a_t|s_t)$ through artificial neural networks, and the neural network parameters are denoted as θ. The goal of reinforcement learning is to make each action achieve the maximum reward value, and the core is how to judge the quality of the selected action. To do this, the following advantage function is defined:

$$\hat{L}_{\pi_\varphi,t} = \hat{R}_t - V_\phi(s_t) \quad (9)$$

where $V_\phi(s_t)$ is a function of the value of the state and reflects all the cumulative reward values that are expected to be achieved after the end of the round under state s_t. The dominance function reflects the advantage of selecting an action a_t relative to the average action a_t at moment t. If the value v_t corresponding to all states s and action a is a two-dimensional table, the large value range of states s_t will cause the storage space of the two-dimensional table to be large and difficult to store. Therefore, the artificial neural network is also chosen to approximate the value function $V_\phi(s_t)$. Finally, the optimization objective function for reinforcement learning is defined as follows:

$$L^{MSE} = E_{\pi_{\theta_k}}(\hat{R}_t - V_\phi(s_t))^2 \quad (10)$$

The goal of the (10) function is to update the strategy function parameter θ so that each action can obtain a larger reward value. However, the problem with the objective function L^{MSE} is that if the parameter θ is updated too much, it will cause repeated oscillations when the gradient rises without fast convergence to the best advantage. For this reason, the PPO_2 algorithm redefines the following objective function formula:

$$L^{clip}(\theta) = \hat{E}_t[\min(r_t(\theta)\hat{A}_t, clip(r_t(\theta), 1-\varepsilon, 1+\varepsilon)\hat{A}_t)] \quad (11)$$

The clip function is a truncation function, defined as:

$$clip(r, 1-\varepsilon, 1+\varepsilon) = \begin{cases} r & 1-\varepsilon < r < 1+\varepsilon \\ 1-\varepsilon, & r \leq 1-\varepsilon \\ 1+\varepsilon, & r \geq 1+\varepsilon \end{cases} \quad (12)$$

$r_t(\theta)$ is a probability ratio function, defined as:

$$r_t(\theta) = \frac{\pi_\theta(a_t|s_t)}{\pi_{\theta_{old}}(a_t|s_t)} \quad (13)$$

$r_t(\theta)$ reflects the magnitude of parameter updates, the larger $r_t(\theta)$ is, the larger the amplitude of the update parameter, and vice versa. The goal of Equation (10) is to obtain a biased estimate of the value function $V_\phi(s_t)$, so the commonly used least squares method is used to define the objective function, and the squared operation ensures that the objective function is non-negative. When the value of the advantage function in Equation (11) is positive, it means that the reward value obtained by the current action is higher than the average, and the objective function optimization goal is to let the agent choose such actions as much as possible. When the dominance function is negative, it means that the reward value obtained by the current action is lower than the average, and the agent should avoid selecting this action. The $L^{clip}(\theta)$ function avoids excessive update fluctuations by intercepting $r_t(\theta)$ to limit it to $[1 - \varepsilon, 1 + \varepsilon]$. The $L^{clip}(\theta)$ function is schematically shown in Figure 3. When the dominant function $L > 0$ (Figure 3a), if $r_t(\theta)$ is greater than $1 + \varepsilon$, it is truncated so that it is not too large. Similarly, when $L < 0$ (Figure 3b), if $r_t(\theta)$ is less than $1 - \varepsilon$, it is also truncated so that it is not too small. The $L^{clip}(\theta)$ function (Figure 3c) guarantees that $r_t(\theta)$ does not fluctuate sharply.

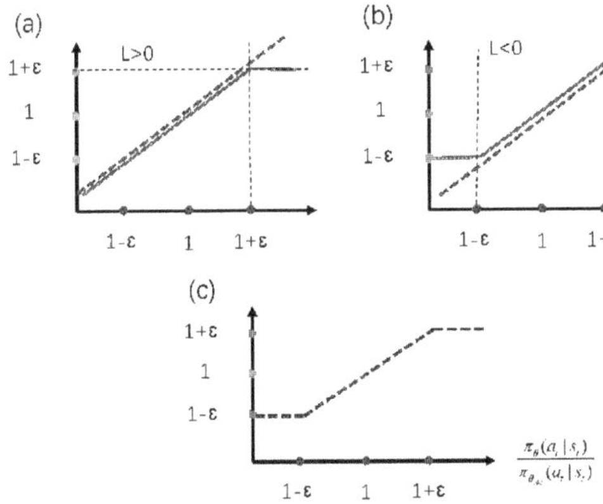

Figure 3. Schematic diagram of the truncation function; (**a**) is the range of values when the value of the dominant function > 0; (**b**) is the range of values when the value of the dominant function < 0; and (**c**) is the final range of values allowed by the truncation function.

3.2. State Space Design

A reasonable state s_t is crucial for efficient reinforcement learning implementation, and only by observing enough information can the reinforcement learning algorithm make the correct action choice. Excessive state information can lead to slower learning and increased computational demands. Therefore, this paper refers to the state parameters and design state s_t required by mainstream TCP algorithms such as CUBIC for decision making. s_t contains the following parameters.

(1) The current relative time t_r. Described as the amount of time that has passed when TCP first established the connection up to the present. In algorithms such as CUBIC, the window length is designed as a third-degree function of time t_r. Consequently, t_r plays a crucial role in determining the congestion window.

(2) The size of the current congestion window. The adjustment of the new window value in the congestion control algorithm should be based on the present congestion window length, which can be increased at a faster rate if the current congestion window length is small, and stopped or increased more slowly if the window is large.

(3) The number of bytes is not acknowledged. Defined as the number of bytes transmitted but not yet acknowledged by the receiver. The unacknowledged bytes can be metaphorically compared to water stored in a pipe, where the network link is similar to the pipe. This parameter is also an important parameter that the congestion control algorithm needs to refer to. If the amount of water in the pipe is sufficient, it should stop or reduce the injection of water into the pipe; if the amount of water in the pipe is small, the amount of water injection into the pipe should be increased, and the volume of water in the pipe may be used to calculate the water injection rate (congestion window duration).

(4) The quantity of ACK packets obtained. When the normal amount of ACK packets is received, the network is functioning well without congestion and the congestion window length can gradually be increased. If the network is congested, with a reduced number of ACK packets received, the congestion window length should either be kept constant or reduced.

(5) RTT. Latency refers to the total time it takes for a packet to be sent to the receiving acknowledgment packet, which can be figuratively understood as the time it takes for the data to make a round trip from the sender to the receiver. Network congestion and latency are strongly associated, and when network congestion is bad, latency increases a lot. As a result, the delay can be an indicator of network congestion, and

the congestion control algorithm can modify the congestion window in response to the delay.

(6) Throughput rate. Described as the number of data bytes the receiver acknowledges each second. A high throughput rate indicates that enough packets have been transmitted in the present connection; alternatively, it shows that there is more available network capacity and that more packets may be sent to the link. This parameter directly reflects the network circumstances.

(7) The number of packet losses. The higher the number of packet losses, the more serious the current network congestion is, and the congestion window size needs to be reduced; a small number of packet losses suggests that the current network is not congested, and the congestion window length should be increased.

3.3. Action Space Design

a_t is the control action made at the moment t for the congestion window. This document defines the action to increase the congestion window length c by n segment length.

$$c = c_{old} + ns' \tag{14}$$

The idea of Equation (14) is to provide a generalization formula that determines the rate of growth of the congestion window length based on the observed state parameter information. Different policies should be selected in different network scenarios. In a high-bandwidth environment, $n > 1$ should be adjusted to increase the congestion window length at an exponential rate; in a low-bandwidth environment, $n = 1$ should be adjusted to make the congestion window grow at a linear speed. When network congestion occurs, $n \leq 0$ should be adjusted to maintain or reduce the length of the congestion window and reduce the pressure of network congestion.

3.4. Reward Function Design

The reward from the environment at time t is referred to as reward r_t, and the design reward letter is as follows:

$$r_t = \alpha\left(\frac{O}{O_{\max}}\right) - (1-\alpha)\frac{l_{\min}}{l} \tag{15}$$

where O is the currently observed throughput rate and $O_{\max}$ is the maximum throughput rate observed in history, and the ratio of the two reflects the throughput rate effect that can be increased by the action a_t; l represents the average delay during the observation period and $l_{\min}$ represents the smallest delay observed in history, and the ratio of the two reflects the delay effect of action a_t improvement; and α, a hyperparameter that measures the weight ratio of throughput rate and delays to the reward, is a weight factor. α defines whether the congestion control algorithm's optimization objective is more concerned with throughput rate or delay. In this example, $\alpha = 0.5$ is selected to balance throughput and latency. In addition, the minimum throughput rate and the maximum delay of the history are saved. When it is observed that the current throughput rate is less than or equal to the minimum throughput rate or greater than or equal to the maximum delay, the reward is set to -10 to avoid reaching these two extreme states.

3.5. Algorithm Description

The input of the Algorithm 1 is the current state of the network s_t, and the output is the congestion window length c_{new}. The pseudocode is as follows.

Algorithm 1. PPO2

1. Input: s_t = {congestion window length, the number of ACK packets, latency, throughput rate, packet loss rate}.
2. Initialize the policy parameters $\theta_0 = \theta_{old} = \theta_{new}$.
3. Run strategy $\pi_{\theta k}$ for a total of T time steps, collect $\{s_t, a_t\}$.
4. $\theta_{old} \leftarrow \theta_{new}$
5. $r_t = \alpha\left(\frac{O}{O_{\max}}\right) - (1-\alpha)\frac{l_{\min}}{l}$.
6. $\hat{R} = \sum_{t=0}^{T} \gamma^t r_t$.
7. $L^{clip}(\theta) = \hat{E}_t[\min(r_t(\theta)\hat{A}_t, clip(r_t(\theta), 1-\varepsilon, 1+\varepsilon)\hat{A}_t)]$.
8. Update the parameter θ by the gradient ascent method so that $L^{clip}(\theta)$ is the maximum.
9. Output: The length of the new congestion window after adjustment is $c = c_{old} + ns'$.

4. Experiment

4.1. Experimental Environment

4.1.1. Hardware and Software Environment

The test was performed on a powerful server, and the specific configuration is as follows: ① CPU, AMD Ryzen 7 5800H with Radeon Graphics @3.19 GHz; ② Memory, 16 GB DDR4; ③ GPU, NVIDIA GeForce RTX 3060 Laptop 6 GB; and ④ Operating System, Windows 11.

The data space topology is simulated by the Mininet simulator, and the TCP-PPO_2 and TCP-CUBIC algorithms are implemented, which are compared with the traditional TCP congestion control algorithms.

4.1.2. PPO Algorithm Parameter Settings

The main parameters of PPO_2 are set as follows: the neural network's hidden layers total 2, the number of neurons in the two layers is 32 and 16, respectively, the discount factor is 0.99, the learning rate is 0.00025, the ε is 0.2, and the number of training steps when running each update is 128.

4.2. Bandwidth Sensitivity Comparison

The amount of data that may be transmitted per unit of time is referred to as network bandwidth (generally 1 s). The greater the bandwidth, the greater its traffic capacity. Figure 4 shows the bandwidth sensitivity comparison, and the link bandwidth in our Mininet is set to between 1 Mbps and 1000 Mbps, with 0 ms latency, 1000 packet queues, and 0% random loss. TCP-PPO_2 works well in this bandwidth range; when the link bandwidth is less than 18 Mbps, both traditional and deep reinforcement learning-based congestion control protocols produce large delays, mainly due to the narrow bandwidth and too long queue length causing queuing problems, but when the bandwidth is higher than 50 Mbps, TCP-PPO_2 can achieve extremely low latency.

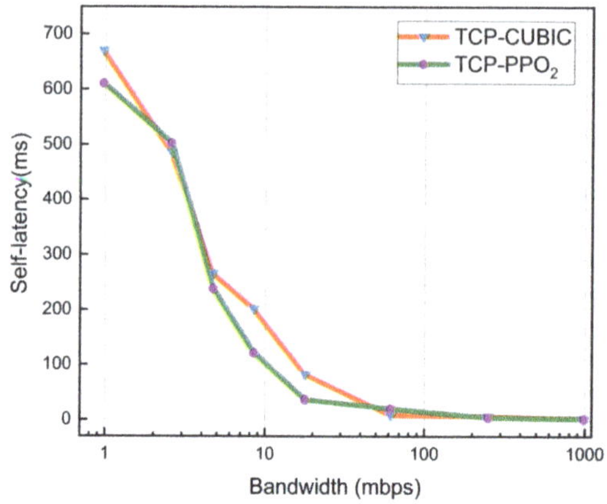

Figure 4. Under the condition that only the bandwidth size is changed, comparison of the latency of the traditional congestion control strategy and the congestion control strategy based on deep reinforcement learning.

4.3. Latency Sensitivity

RTT represents the time it takes for a packet to be sent from sent to receive to acknowledged, reflecting the current network latency. Figure 5 shows a comparison of latency sensitivity, with the link bandwidth set to 700 Mbps, a queue of 1000 packets, and 0% random loss in our Mininet. When the link delay is less than 50 ms, the system latency of the traditional congestion control protocol is much higher than that of the intelligent congestion control strategy, indicating that TCP-PPO_2 can adapt to today's low-latency networks, thereby ensuring more efficient data transmission. When the link delay is higher than 90 ms, the delay of both is greatly reduced, but the delay of the intelligent system is still lower than the system delay of the TCP-CUBIC protocol.

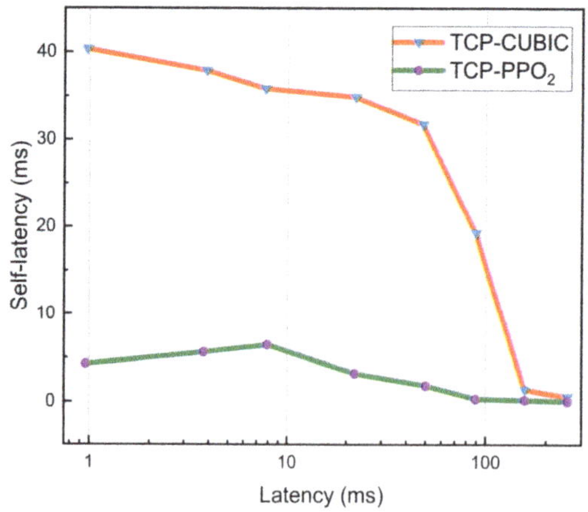

Figure 5. Under the condition that only the network latency is changed, comparison of the latency of the traditional congestion control strategy and the congestion control strategy based on deep reinforcement learning.

4.4. Queue Sensitivity

The queue represents the size of the packets sent at one time; Figure 6 shows the queue sensitivity comparison, and we changed the queue size between 1 and 10,000 packets. Other configurations of the links in Mininet are a bandwidth of 700 Mbps, a latency of

40 ms, and a random loss of 0%. When the queue size is greater than 10, the delay of the traditional congestion control strategy is significantly higher than that of TCP-PPO$_2$.

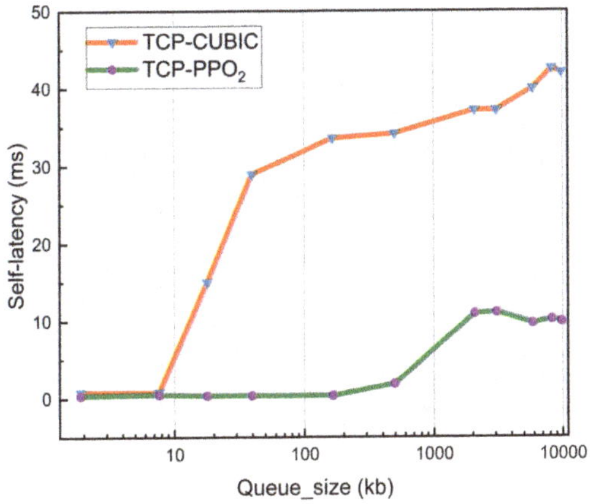

Figure 6. Under the condition that only the queue size is changed, comparison of the latency of the traditional congestion control strategy and the congestion control strategy based on deep reinforcement learning.

4.5. Packet Loss Sensitivity

The packet loss rate is an important indicator to measure the reliability of network transmission protocols, and Figure 7 shows the packet loss sensitivity comparison, setting a random loss rate of up to 8%. Other configurations of links in Mininet are a bandwidth of 700 Mbps, a latency of 40 ms, and queues of 1000 packets. When the random loss rate increases from zero, the latency of both congestion control strategies decreases rapidly, while the delay at the beginning of TCP-PPO2 is significantly lower than that of traditional congestion control strategies.

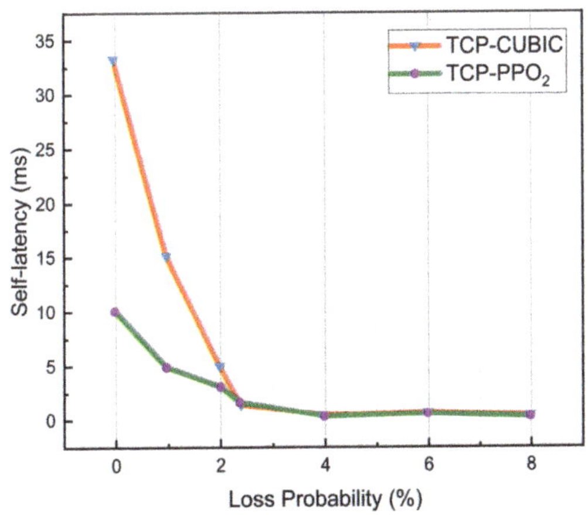

Figure 7. Under the condition that only the packet loss rate is changed, comparison of the latency of the traditional congestion control strategy and the congestion control strategy based on deep reinforcement learning.

5. Conclusions

Aiming at the problems of poor adaptability of mainstream TCP congestion control algorithms and inability to effectively use virtual data space network borrowing, a TCP congestion control strategy based on deep reinforcement learning is proposed, which effectively improves the data transmission efficiency. The main conclusions of this paper are as follows:

(1) Optimize the traditional TCP congestion control strategy by using the near end policy optimization algorithm, map the system's send rate to the behavior of deep reinforcement learning, set the reward function by balancing throughput, latency, and packet loss, and use a simple deep neural network to approximate the final strategy. Through the comparison of a large number of experimental data, the parameters such as the number of neural network layers, the number of neurons, and the length of the history were determined, and the optimization of TCP-PPO2 was successfully realized.

(2) Through Mininet simulation experiments, it is determined that the TCP congestion control algorithm based on the proximity policy optimization adapts to network changes faster than the traditional TCP congestion control algorithm, changes the real-time congestion window size, improves transmission efficiency, and reduces the data transmission delay by 11.7–87.5%.

Author Contributions: Conceptualization, H.S.; methodology, H.S.; software, H.S.; validation, H.S. and J.W.; formal analysis, H.S. and J.W.; investigation, H.S. and J.W.; resources, H.S. and J.W.; data curation, H.S.; writing—original draft preparation, H.S.; writing—review and editing, J.W.; visualization, H.S.; supervision, J.W.; project administration, J.W.; funding acquisition, J.W. All authors have read and agreed to the published version of the manuscript.

Funding: This research was funded by Xi'an Key Laboratory of Clean Energy: 2019219914SYS014C G036; Key R&D Program of Shaanxi Province: 2023-ZDLGY-24.

Institutional Review Board Statement: Not applicable.

Data Availability Statement: The data are not publicly available due to privacy.

Conflicts of Interest: The authors declare no conflict of interest.

References

1. Henderson, T.; Floyd, S.; Gurtov, A. The NewReno Modification to TCP's Fast Recovery Algorithm. 2012. Available online: https://www.rfc-editor.org/rfc/rfc6582.html (accessed on 12 April 2023).
2. Ha, S.; Rhee, I.; Xu, L. CUBIC: A new TCP-friendly high-speed TCP variant. *ACM SIGOPS Oper. Syst. Rev.* **2008**, *42*, 64–74. [CrossRef]
3. Mascolo, S.; Casetti, C.; Gerla, M.; Sanadidi, M.Y.; Wang, R. TCP Westwood: Bandwidth estimation for enhanced transport over wireless links. In Proceedings of the 7th Annual International Conference on Mobile Computing and Networking, New York, NY, USA, 16 July 2001.
4. Van Der Hooft, J.; Petrangeli, S.; Claeys, M.; Famaey, J.; Turck, F. A learning-based algorithm for improved bandwidth-awareness of adaptive streaming clients. In Proceedings of the 2015 IFIP/IEEE International Symposium on Integrated Network Management (IM), Ottawa, ON, Canada, 11–15 May 2015.
5. Cui, L.; Yuan, Z.; Ming, Z.; Yang, S. Improving the congestion control performance for mobile networks in high-speed railway via deep reinforcement learning. *IEEE Trans. Veh. Technol.* **2020**, *69*, 5864–5875. [CrossRef]
6. Gu, L.; Zeng, D.; Li, W.; Guo, S.; Zomaya, A. Intelligent VNF orchestration and flow scheduling via model-assisted deep reinforcement learning. *IEEE J. Sel. Areas Commun.* **2019**, *38*, 279–291. [CrossRef]
7. Xie, R.; Jia, X.; Wu, K. Adaptive online decision method for initial congestion window in 5G mobile edge computing using deep reinforcement learning. *IEEE J. Sel. Areas Commun.* **2019**, *38*, 389–403. [CrossRef]
8. Xiao, K.; Mao, S.; Tugnait, J.K. TCP-Drinc: Smart congestion control based on deep reinforcement learning. *IEEE Access* **2019**, *7*, 11892–11904. [CrossRef]
9. Bachl, M.; Zseby, T.; Fabini, J. Rax: Deep reinforcement learning for congestion control. In Proceedings of the ICC 2019—2019 IEEE International Conference on Communications (ICC), Shanghai, China, 20–24 May 2019.
10. Li, W.; Zhou, F.; Chowdhury, K.R.; Meleis, W. QTCP: Adaptive congestion control with reinforcement learning. *IEEE Trans. Netw. Sci. Eng.* **2018**, *6*, 445–458. [CrossRef]
11. Watkins, C.J.; Daya, P. Technical Note: Q-Learning. *Mach. Learn.* **1992**, *8*, 279–292. [CrossRef]
12. Ha, T.; Masood, A.; Na, W.; Cho, S. Intelligent Multi-Path TCP Congestion Control for Video Streaming in Internet of Deep Space Things Communication. 2023. Available online: https://www.sciencedirect.com/science/article/pii/S2405959523000231 (accessed on 12 April 2023).

13. Schulman, J.; Wolski, F.; Dhariwal, P.; Radford, A.; Klimov, O. Proximal policy optimization algorithms. *arXiv* **2017**, arXiv:1707.06347.
14. Goodfellow, I.; Bengio, Y.; Courville, A. *Deep Learning*; MIT Press: Cambridge, MA, USA, 2016.
15. Sutton, R.S.; Barto, A.G. *Reinforcement Learning: An Introduction*; MIT Press: Cambridge, MA, USA, 2018.
16. Rummery, G.A.; Niranjan, M. *On-Line Q-Learning Using Connectionist Systems*; University of Cambridge: Cambridge, UK, 1994.
17. Mnih, V.; Kavukcuoglu, K.; Silver, D.; Andrei, A.; Rusu, J.; Marc, B.; Alex, G.; Martin, R.; Andreas, F.; Georg, O.; et al. Human-level control through deep reinforcement learning. *Nature* **2015**, *518*, 529–533. [CrossRef] [PubMed]
18. Van Hasselt, H.; Guez, A.; Silver, D. Deep reinforcement learning with double q-learning. In Proceedings of the AAAI Conference on Artificial Intelligence, Palo Alto, CA, USA, 22 February–1 March 2016.
19. Puterman, M.L. Markov decision processes. *Handb. Oper. Res. Manag. Sci.* **1990**, *2*, 331–434.
20. Schulman, J.; Levine, S.; Abbeel, P.; Jordan, M.; Moritz, P. Trust Region Policy Optimization. In Proceedings of the International Conference On Machine Learning. 2015. Available online: https://proceedings.mlr.press/v37/schulman15.html (accessed on 12 April 2023).
21. Schulman, J.; Moritz, P.; Levine, S.; Jordan, M.; Abbeel, P. High-dimensional continuous control using generalized advantage estimation. *arXiv* **2015**, arXiv:150602438.

applied sciences

MDPI

Article

A Novel Opportunistic Network Routing Method on Campus Based on the Improved Markov Model

Yumei Cao [1,2,3,4], Peng Li [1,2,3,4,5,*], Tianmian Liang [1,2,3,4], Xiaojun Wu [1,2,3,4,5], Xiaoming Wang [1,2,3,4,5] and Yuanru Cui [1,2,3,4]

1 Key Laboratory of Intelligent Computing and Service Technology for Folk Song, Ministry of Culture and Tourism, Xi'an 710119, China; cym41712121@snnu.edu.cn
2 School of Computer Science, Shaanxi Normal University, Xi'an 710119, China
3 Engineering Laboratory of Teaching Information Technology of Shaanxi Province, Xi'an 710119, China
4 Key Laboratory of Modern Teaching Technology, Ministry of Education, Xi'an 710062, China
5 Xi'an Key Laboratory of Cultural Tourism Resources Development and Utilization, Xi'an 710062, China
* Correspondence: lipeng@snnu.edu.cn

Abstract: Opportunities networks' message transmission is significantly impacted by routing prediction, which has been a focus of opportunity network research. The network of student nodes with smart devices is a particular type of opportunity network in the campus setting, and the predictability of campus node movement trajectories is also influenced by the regularity of students' social mobility. In this research, a novel Markov route prediction method is proposed under the campus background. When two nodes meet, they share the movement track data of other nodes stored in each other's cache in order to predict the probability of two nodes meeting in the future. The impact of the node within the group is indicated by the node centrality. The utility value of the message is defined to describe the spread degree of the message and the energy consumption of the current node, then the cache is managed according to the utility value. By creating a concurrent hash mapping table of delivered messages, the remaining nodes are notified to delete the delivered messages and release the cache space in time after the messages are delivered to their destinations. The method suggested in this research can successfully lower the packet loss rate, minimize transmission latency and network overhead, and further increase the success rate of message delivery, according to experimental analysis and algorithm comparison.

Keywords: routing prediction; opportunistic network; utility value; cache management; centrality

Citation: Cao, Y.; Li, P.; Liang, T.; Wu, X.; Wang, X.; Cui, Y. A Novel Opportunistic Network Routing Method on Campus Based on the Improved Markov Model. *Appl. Sci.* **2023**, *13*, 5217. https://doi.org/10.3390/app13085217

Academic Editors: Yang Yue, Runzhou Zhang, Hao Feng, Zheda Li, Lin Zhang and Dawei Ying

Received: 8 March 2023
Revised: 16 April 2023
Accepted: 17 April 2023
Published: 21 April 2023

1. Introduction

1.1. Background

An opportunistic network is a type of self-organizing network [1], which employs node movement to create opportunities for encounter and communication rather than requiring an end-to-end complete connection. The increased usage of portable electronics, including smartphones and tablets, opens up numerous possibilities for the growth of opportunistic networks. Opportunistic networks use the store-carry-forward routing mode and send messages hop by hop between nodes [2]. In order to increase network communication efficiency and achieve timely message transmission, it is now essential to accurately predict node destinations, choose the best next-hop nodes, and reduce the number of message copies in the network, owing to the limited memory and power of portable devices. The transmission path can be planned based on the nodes' geographical locations to address the issues of limited energy and storage overhead in the network [3]. The issue of excessive network resource consumption can be resolved by managing data congestion [4,5] and restricting the transmission range of nodes [6].

1.2. Motivation

A wireless self-organizing network made up of learner nodes using smart devices on campus, the campus opportunistic network allows each learner node to communicate with other nodes. Although many academics have made progress in this area, few have concentrated on the campus context and have failed to suggest efficient routing algorithms or manage the cache in a way that is suitable for the specific movement of nodes in the campus. The predictability of node movement trajectories is also determined by how regularly learner nodes travel around campus.

The blind nature of message delivery is one of the major issues when performing message transmission. If the node delivers the message directly to its neighbors, this results in a significant message transmission delay, as messages must pass through numerous relay nodes in order to reach their destinations and they cannot deliver to the recipient before the message survival time ends sometimes. On the other hand, network resources are wasted because some nodes carrying copies of messages do not eventually connect with the destination node, resulting in a large number of duplicate messages in the network that are truly unnecessary. Additionally, since copies of individual messages are forwarded numerous times, they can take up a lot of memory, preventing the reception of recently arrived messages. Since nodes typically have limited memory, it is important to manage the cache because a multi-copy-based routing strategy leaves a lot of copies in the network. This can even have a significant impact on the success rate of message delivery, making it a vital study in this paper. A certain amount of message copies exists in the network to ensure that messages get to their destination quickly, the network needs a certain number of message copies, so when the node cache is too small to accommodate new messages, it should choose which messages to delete initially to make room for new arrivals.

1.3. Contribution

We suggest improved Markov path prediction and cache management algorithms on the basis of some of the aforementioned difficult issues. When designing the algorithm, we make an effort to deliver messages as quickly as feasible. Considering that device memory is typically small, then we suggest a proven cache management technique. The following are the major contributions of this paper.

- In this study, we distinguish between intra-group forwarding and extra-group forwarding when it comes to messaging. When a message needs to be sent between groups, we use a novel Markov model to determine the probability that the sender and the recipient will be in the same place. We then send the message to the nodes with a higher probability of doing so. The message only needs to be delivered within the group when the recipient and the source node are both members of the same group. This not only gets the message to its target quickly but also saves a significant amount of cache space by sending the message to those nodes that have high centrality within the group.
- The utility value of a message is defined in terms of both the message's degree of diffusion and the present node's energy usage. According to our theory, if a message has a high degree of diffusion, there are likely already some copies of it in the network. As a result, priority should be given to receiving messages with a lower degree of diffusion. Moreover, if a message requires a lot of energy from the current node that node might not be the best choice to serve as its relay.
- The node also keeps track of both the message list and the delivered message list, prompting the node to remove any messages that have already been delivered.
- Our suggested strategy enhances network performance in terms of packet delivery rate, average delivery delay, average cost and overhead when compared to current methods.

The remaining portions of the paper are structured as follows: related work is discussed in Section 2. Section 3 improves Markov's method and mentions the way messages are forwarded in the network. Section 4 defines message utility values and proposes cache management methods. Section 5 evaluates the experimental results. Section 6 concludes the paper.

2. Related Works

In opportunity networks, predicting node paths, managing caches reasonably, and reducing energy consumption have become research hotspots. Researchers at home and abroad have made many research results in related directions and proposed some solutions to the problems of path prediction and cache management in opportunity networks.

Singh et al. [7] proposed a social-based opportunistic routing algorithm, which only spreads the content when the social relationship between the next relay node and the destination node is closer than all previously encountered nodes and uses the social relationship to determine the most suitable node to forward the message, which significantly reduces the overhead in the message routing process. The routing method given in [8] used a secure routing protocol based on blockchain to design an integrated protocol, which can effectively protect data security. This routing method can effectively prevent eavesdropping, camouflage, wormholes, black holes, and fabrication attacks. In [9], Sharma et al. fully automated the routing process of opportunistic networks by using an iterative strategy algorithm, modeling the network environment as Markov decision processes, and using strategy iteration to solve the optimal strategy obtained by Markov decision processes to optimize the routing process and maximize the possibility of message delivery. Kumar et al. [10] put forward an innovative routing strategy based on node activity, which considers the previous operation of the node on the message, calculates the confidence level of the node through the past behavior and activity of the node to determine whether it is a good candidate to forward a specific message and reduces the message dropping rate. Gou et al. [11] presented a social network evolution analysis method based on triple, including a prediction algorithm and a quantization algorithm. The algorithm reduces the blindness of message forwarding and unnecessary waste of resources by predicting the connection probability between nodes in the network. Chunyue et al. [12] put forward an algorithm combined with the sleep mechanism, which mainly solves the problem of judging the conditions of sleep state and wake-up time, forces the nodes in a low-energy state to sleep and avoids the nodes from consuming energy quickly. Derakhshanfard et al. [13] proposed a method based on a bitmap, which uses a routing tree based on the bitmap to find the path, and when the tree receives a request to send a message to a designated node, it directly sends a packet, which effectively improves the message delivery rate. Chithaluru et al. [14] studied an energy-efficient opportunistic routing protocol based on adaptive ranking. The residual energy and geographical location of nodes are used to calculate the level, and an efficient forwarding mode based on node level is determined, which improves the effective use of energy in the process of data transmission. Hernández-Orallo et al. [15] proposed a method to minimize the consumption of network resources, using an epidemic diffusion model to evaluate the impact of message expiration time on message transmission, calculating the optimal expiration time and dynamically setting the expiration time, which significantly reduced the buffer usage and energy consumption. Raverta et al. [16] proposed the routing under an uncertain contact plan, extended the single copy routing in the Markov decision-making process to multiple copies, and used multiple copies to model the network state, which effectively improved the message delivery rate. Das et al. [17] used special monitoring nodes to check the behavior of other nodes and routed messages to nodes with sufficient residual energy levels, so that most of the forwarded messages are proportional to the energy level of the receiver, effectively solving the problem that nodes in the network are paralyzed due to energy consumption. Kang et al. [18] proposed an improved hybrid routing protocol combining mobile ad hoc networks and latency-tolerant networks. When the routing path to the destination node is not successfully established by using the ad hoc network protocol, the virtual source node is selected according to the predictability of the delivery of the destination node by the Prophet protocol, and then the delivery rate is effectively increased at the cost of overhead again. Pirzadi et al. [19] explored a reducted-delivery delay routing (RDR) strategy in disaster relief operations, using a simulated annealing algorithm to optimize the message delivery process in the network and achieve an optimal routing method for efficient message distribution. Mao et al. [20] proposed a fair credit-based routing incentive mechanism (FCIM) that uses incentives to the selfishness problem of nodes, uses some trust mechanisms to avoid nodes from cheating the network and ensures fairness among nodes.

A few of the algorithms discussed in this paper include the following:

- Epidemic [21], which is a flooding-based routing method where a node passes a message copy to every node it encounters. By creating numerous message duplicates, it increases the probability that the message will be delivered when it comes across the destination node. However, a lot of copies use up network resources, such as cache space and node energy.
- Prophet [22] is a method that is frequently used to send messages based on predictions. Two nodes exchange vectors of transmission probabilities for recognized destinations when they come into contact. Messages can be sent to nodes that meet regularly by updating the transmission probability between nodes based on how long it has been since their last encounter. Nevertheless, it ignores the location information of the nodes and the number of encounters between them.
- RDR, which chooses the next-hop node based on the node's estimated latency, estimated speed variation, the direction of motion, available space in the buffer, and previously sent messages. It provides a constrained amount of replicas, reducing the network resource footprint. With this approach, the amount of network resources used can be drastically decreased, and the size of the cache area has less of an impact. However, messages may not be delivered for a long time, and it requires a longer message survival time.
- FCIM, where each relay node is rewarded with some points when the source node sends a message to its target, increases the message delivery rate by motivating selfish nodes to actively participate in message forwarding. Nodes are permitted to engage in some acceptable selfish behaviors under this strategy, such as rejecting messages when the cache is full. However, no more properties are considered, such as the energy consumption of nodes to forward messages.

Three types of node misconduct are discussed by Rehman et al. [23], and they look into how these types of misconduct may affect nine VDTN routing algorithms. The third category of misconduct of nodes is specifically presented. The node reduces the message TTL by storing the message for a long time in its own memory after it has been received. An incentive and punishment strategy is suggested by Rehman et al. [24–26] to incentivize the selfish cluster nodes to forward messages. An active node can raise its reputation by passing messages and engaging in conversations. Nodes that exhibit selfish behavior repeatedly are punished. The reward and punishment mechanism can effectively enhance the degree of cooperation among nodes and improve the probability of packet delivery.

In this paper, some novel studies are made in relation to some of the methods previously mentioned. The innovation points of this paper are specified in Table 1.

The literature listed above has looked at social relationship analysis, route prediction, and energy conservation, but it has not developed a workable route prediction method based on the regularity of student node mobility from the context of campus opportunity networks. These schemes are not applicable to the case of regular group movement of nodes in campus opportunity networks. This paper proposes a novel method for routing campus opportunity networks based on improved Markov which predicts node paths by collecting historical movement trajectories of nodes and models message utility values reasonably based on message diffusion and consumption of node energy to improve cache utilization by deleting delivered messages in time. The efficiency of the suggested method is confirmed in the experimental section of this study by comparison with the RDR [19] and FCIM [20] algorithms, as well as the classical methods Epidemic and Prophet.

Table 1. Novelties of this paper.

Limitations of Existing Works	Novelties of This Paper
The previous section describes how nodes can reduce network resource usage by providing a restricted number of copies, but messages with a short survival time may not be delivered.	In this paper, we transmit messages based on the probability that the nodes will meet at the next location which can guarantee the successful transmission of messages in a short time.
The prediction-based routing presented above takes into account the encounter interval of the nodes.	We consider the probability that nodes will meet one another at various places and the number of contacts between nodes.
FCIM considers the caching of networks.	Description of the message's energy consumption and the network's degree of message spread was added to the node.
They encourage selfish nodes to engage in collaboration.	Skip selfish nodes to avoid being impacted by them.

3. Materials and Methods

3.1. Markov-Based Next Destination Prediction

The storage-forward model of messages in opportunity networks relies on human contact, and the potential for message delivery arises as people move around. There are more options for message delivery when nodes are on their way to the same location as the destination node of a message. This is so that these nodes can send the message to the intended node faster. Some of the symbols used in the text and what they represent are listed in Table 2.

Table 2. Lists of the notations used in this paper to represent variables.

Notation	Description
N_a, N_b	Node N_a and node N_b
$fc(N_a, N_b)$	Probability N_a and N_b meet
m	Message m
DC	Centrality degree of node
D_m	Destination node of message m

According to the way people move in the campus opportunity network, we make the following provisions:

The number of times a node chooses a location as its next destination in time period, 0 to t is denoted as $X(t)$, which is a stochastic process with $X(0) = 0$. Assume:

- Within mutually exclusive time intervals, the number of times that nodes choose the place as a destination point is independent of one another;
- The probability distribution of the number of times $X(s+t) - X(s)$ that a node chooses this location in period $(s, s+t]$ is independent of s, where $s \geq 0$;
- $o(\Delta t)$ is the likelihood that a node will choose the same place more than once in a sufficiently little period of time.

According to the above rules, the number of times $X(t)$ for a node to select the location before time t is a stochastic process, and the number of times to select the location before $t_v(t_v > t)$ in the future only depends on the number of times to select the location at time t. The overall number the location is selected in period $[0, t]$ and $(t, t_v]$ is equal to the number of times the location is selected in $[0, t_v]$. From the assumption (1) that the number of times the location is selected in period $[0, t]$ and period $(t, t_v]$ are independent of each other, it is known that $X(t)$ has no posteriority and belongs to the Markov process. The parameter set for the number of choices $X(t)$ in the above Markov process is $T = [0, \infty)$ and the state space is $E = \{0, 1, 2, \ldots\}$. Therefore, $X(t)$ is a Markov process with continuous time and discrete states.

In addition, $X(t)$ satisfies the following conditions:

- For a sufficiently small Δt;

$$P_1(t, t+\Delta t) = P\{X(t, t+\Delta t) = 1\} = \lambda \Delta t + o(\Delta t)$$

where the constant λ is called the intensity of process $X(t)$, and $o(\Delta t)$ is the high-order infinitesimal about Δt when $\Delta t \to 0$.

- Furthermore;

$$\sum P_j(t, t+\Delta t) = \sum P\{X(t, t+\Delta t) = j\} = o(\Delta t)$$

That is, for a sufficiently small Δt, the probability of meeting twice or more in $(t, t+\Delta t]$ period can be ignored compared with the probability of meeting once.

- $X(0) = 0$.

For this process, it can be seen that $\sum_{j=0}^{\infty} P_j(t, t+\Delta t) = 1$, and combined with (2) and (3) we have:

$$P_0(t, t+\Delta t) = 1 - P_1(t, t+\Delta t) - \sum P_j(t, t+\Delta t) = 1 - \lambda\Delta t + o(\Delta t)$$

Conclusion 1: $p_{ij}^{n_k}(t, t+s)$ denotes the transfer probability function of the above Markov process, i.e., the probability can also be written as $p_{\mathrm{ij}}^{n_k}(s)$ that the number of times the nth node chooses location k as its destination from 0 to t period is i, and the number of times it chooses location k after s time is j, where $s > 0$.

From conclusion 1, it follows that:

$$\sum_j p_{ij}^{n_k}(s) = 1,\ i = 1, 2, \ldots \tag{1}$$

And stipulates that:

$$p_{ij}(0) = \delta_{ij} = \begin{cases} 1, i = j \\ 0, i \neq j \end{cases} \tag{2}$$

Conclusion 2: $q_{ij}^{n_k}$ denotes the rate function of the above Markov process and describes the rate of change of the transfer probability function $p_{\mathrm{ij}}^{n_k}(s)$ at the zero moments.

From conclusion 2 it follows that:

$$\lim_{t \to 0+} \frac{p_{ij}(t) - \delta_{ij}}{t} = q_{ij}, i, j = 0, 1, 2 \ldots N \tag{3}$$

According to the definition of the derivative, we can get:

$$\frac{dp_{0j}(t)}{dt} = \lambda p_{0j-1}(t) - \lambda p_{0j}(t), j = 1, 2, \ldots \tag{4}$$

$$\begin{aligned} P\{X(s+t) &= j | X(s) = i\} \\ &= \tfrac{P\{X(s)=i, X(s+t)=j\}}{P\{X(s)=i\}} \\ &= \tfrac{P\{X(s)=i, X(s+t)-X(s)=j-i\}}{P\{X(s)=i\}} \\ &= \tfrac{P\{X(s)=i\}P\{X(s+t)-X(s)=j-i\}}{P\{X(s)=i\}} \\ &= P\{X(s+t) - X(s) = j - i\} \end{aligned} \tag{5}$$

By assumption (1), the transfer probability function is independent of s. Therefore, *X(t)* is a time-Ziemarkov process.

Calculate q_{ij} based on Conclusion 1 and Conclusion 2:

$$\begin{aligned} p_{ij}(\Delta t) &= P\{X(t+\Delta t) = j | X(t) = i\} \\ &= P\{X(t+\Delta t) = j, X(t) = i | X(t) = i\} \\ &= P\{\text{the number of times the location is selected as} \\ &\quad \text{the next destination in } (t, t+\Delta t] \text{ period is } j - i | X(t) = i\} \\ &= P\{\text{the number of times the location is selected as} \\ &\quad \text{the next destination in } (t, t+\Delta t] \text{ period is } j - i\} \\ &= \begin{cases} \lambda\Delta t + o(\Delta t), j = i+1 \\ 1 - \lambda\Delta t + o(\Delta t), j = i \\ o(\Delta t), j > i+1 \\ 0, j < i \end{cases} \end{aligned} \tag{6}$$

From conclusion 2, it follows that

$$q_{ij} = \lim_{\Delta t \to 0+} \frac{p_{ij}(\Delta t) - \delta_{ij}}{\Delta t} = \begin{cases} \lambda, \ j = i+1 \\ -\lambda, j = i \\ 0, \ j < i \, or \, j > i+1 \end{cases} \tag{7}$$

Substituting into conclusion 1 and taking $i = 0$, we get

$$\begin{cases} \frac{dp_{0j}(t)}{dt} = \lambda p_{0j-1}(t) - \lambda p_{0j}(t), j = 1, 2, \ldots \\ \frac{dp_{0j}(t)}{dt} = -\lambda p_{00}(t) \end{cases} \tag{8}$$

The solution of this system of equations satisfying the initial condition $p_{0j}(0) = \delta_{0j}$ is:

$$p_{0j}(t) = \frac{(\lambda t)^j}{j!} e^{-\lambda t} \tag{9}$$

The final solution of this equation satisfying the initial condition $p_{ij}(0) = \delta_{ij}$ can be found as:

$$p_{ij}(t) = \frac{(\lambda t)^{j-i}}{(j-i)!} e^{-\lambda t}, j = i, i+1, i+2, \ldots \tag{10}$$

Find λ based on $pij(t)$:

$$\begin{aligned} E(X(t)) &= \sum_{n=0}^{\infty} (j-i) \frac{(\lambda t)^{j-i}}{(j-i)!} e^{-\lambda t} \\ &= (\lambda t) e^{-\lambda t} \sum_{n=0}^{\infty} \frac{(\lambda t)^{j-i-1}}{(j-i-1)!} \\ &= (\lambda t) e^{-\lambda t} e^{-\lambda t} \\ &= \lambda t \end{aligned} \tag{11}$$

Then λ is the number of times per unit time interval that a node selects location k as its destination.

A time-continuous Markov chain can be used to describe the entire transfer procedure when considering two nodes in a network. We can symbolize the state (u_0, v_0) in the Markov chain when node N_a is at position u_0 and node N_b is at position v_0. After a period of time, nodes N_a and N_b are shifted to arbitrary positions. If N_a moves to u_k, N_b moves to v_l, the current state is (u_k, v_l) and the transfer rate of this process is $q_{ij}(N_a)$ and $q_{ij}(N_b)$. In specific, the Markov chain enters the absorbing state $A(u_k, v_k)$ if N_a and N_b are moved to the same location. The state transfer process of the node is shown in Figure 1.

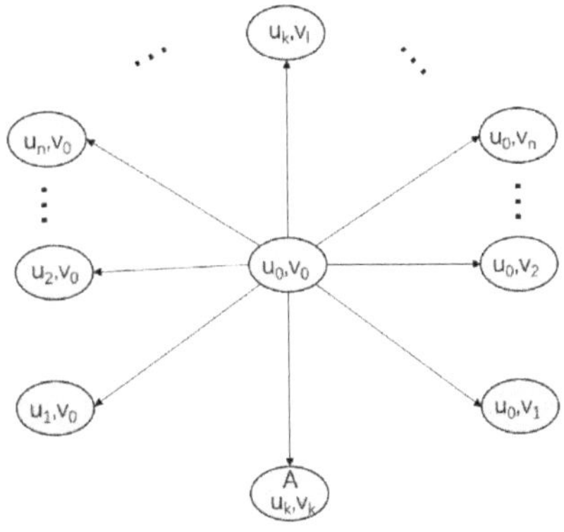

Figure 1. State Transfer Diagram.

From Equation (10), the probability that a node chooses to move to any location in time t is $p_{ij}^{nk}(t)$, where the probability of going to location k once is expressed as f_1^k.

$$f_1^k = \{p_{ij}^{nk} = \frac{(\lambda t)^{j-i}}{(j-i)!}e^{-\lambda t}|j = i+1\} \tag{12}$$

The probability of traveling m times to a particular place k is written as Equation (13).

$$f_m^k = \{p_{ij}^{nk} = \frac{(\lambda t)^{j-i}}{(j-i)!}e^{-\lambda t}|j = i+m\} \tag{13}$$

Therefore, the probability of moving to location k is $f^k = f_1^k + f_2^k + \ldots + f_m^k = 1 - f_0^k$.

It can be considered that two nodes will meet if node N_b arrives before node N_a leaves location k. To remove the exponential restriction of the traditional Markov process, matrix P can be used to record the node's choice of the next location when it is at a different location, and the elements p_{uk} in matrix P denote the probability that the next location is k when the node is at location u. The linked table w records the dwell time at each location, and $w_k(t)$ denotes the probability that the node's dwell time at location k is greater than or equal to t.

In conclusion, the probability that a node will be active within the kth location in the upcoming period t is indicated by p^k.

$$p^k = (p_{uk} + f^k) \cdot w_k(t)/2 \tag{14}$$

Therefore, the probability that nodes N_a and N_b meet is:

$$fc(N_a, N_b) = p_{N_a}^1 \cdot p_{N_b}^1 + p_{N_a}^2 \cdot p_{N_b}^2 + \ldots + p_{N_a}^m \cdot p_{N_b}^m = \sum_{i=1}^{m} p_{N_a}^i \cdot p_{N_b}^i \tag{15}$$

3.2. Node Centering Degree

Nodes move continuously and make contact with other nodes in the network as a result. The destination node is thought to be more likely to be encountered by those who have more contact with other nodes, whereas those who have more contact with the destination node are thought to be closer to it because they meet more frequently. When the current node and the destination node of the message are in the same group, the message only needs to be forwarded within the group because the nodes in the same group are in contact more frequently. This not only gets the message to the destination quickly but also saves a lot of space. In summary, we express the centrality degree DC of a node as the following Equation (16), which indicates the size of the node's ability to deliver messages in this community.

$$DC = \partial(\sum_{i=0}^{n} CN^{i}_{current}/CN_{total}) + \beta(\sum_{j=0}^{n} CT^{j}_{currentcurrent}/\sum_{k=0}^{groupsize}\sum_{j=0}^{n} CT^{j}_{k}) \quad (16)$$

where, $\sum_{i=0}^{n} CN^{i}_{current}$ is the total number of nodes in this group that the current node has contacted, if the current node has contacted with the ith node $CN^{i}_{current} = 1$, and CN_{total} is the total number of nodes in the network. $\sum_{j=0}^{n} CT^{j}_{current}$ is the total number of contacts between the current node and other nodes in in this group, $CT^{j}_{current}$ is the number of contacts between the current node and the jth node, $\sum_{k=0}^{groupsize}\sum_{j=0}^{n} CT^{j}_{k}$ is the total number of contacts between the nodes and other nodes in this group and $groupsize$ is the number of nodes contained in this group. ∂ and β are the control coefficients, which are $1/2$, respectively.

3.3. Historical Information Exchange

Each node must be aware of the historical movement trajectory of other nodes in order to predict its own movement path. The historical movement trajectory of a node is defined as a quintet $(nodeID, P, W, T_{cur}, Location_{cur})$, where P is the transfer probability matrix of the node, W is a chain of dwell times at each location, T_{cur} is the time to update the quintet and $Location_{cur}$ is the current location of the node. The historical movement information spreads epidemically throughout the network. When two nodes meet, the nodes first store each other's movement trajectory locally and update the information of the other node in the local cache if it already exists. When exchanging the history traces of other nodes stored by a node, the quintet with the latest update time replaces the old quintet based on the comparison of the T_{cur} stored by both sides. As a result of the remarkable regularity of student movement on campus, the P matrix will be sparse, demonstrating the strong predictability of student node trajectories.

3.4. Forwarding Strategy

In-group forwarding and out-group forwarding are the two ways that nodes forward messages. In-group forwarding is used when the source and destination nodes of the message are in the same group, and out-group forwarding is used otherwise. Two stages make up the message delivery process: first, the message is delivered to the group node, and then it is transmitted from the group node to the message destination.

1. Out-group forwarding

When the sender and the recipient of the communication are not in the same group, forwarding is determined by the probability that the sender and the recipient are going to meet at the same place. If node N_a carries message m and encounters node N_b, and $fc(N_b, D_m)$ is not less than $fc(N_a, D_m)$, where D_m is the destination node of message m. It shows that if the encounter probability between N_b and D_m is greater than N_a, then N_a will forward the message m to N_b. Otherwise, it will not.

2. In-group forwarding

When a message's source and destination nodes are both members of the same group, the message is only transmitted within that group and is forwarded in accordance with the node's centrality. If $DC(N_b)$ is not less than $DC(N_a)$, then N_a forwards the message m to N_b, otherwise it is not forwarded.

4. Routing Algorithms Based on Node Path Prediction and Cache Management

4.1. Utility Value of the Message

Nodes carry a large number of message copies in the cache as they transmit messages. When there is not enough room in the cache for new messages to be received, older messages with lower utility values can be discarded to make room. Messages with high diffusion and high energy consumption are classified as having low utility value based on the global dissemination of messages and the energy consumption of messages to the current node.

The utility value of the message can be calculated by the following equation.

$$U^m = 1/(\sum_{i=1}^{n} node_i^m / node_{all} + \sum tran^m / \sum_{j=1}^{buffsize} tran^j) \tag{17}$$

where $\sum_{i=1}^{n} node_i^m$ is the number of nodes that the message m passes through during transmission, the larger the value of $\sum_{i=1}^{n} node_i^m$ means the higher the diffusion of the message, $node_{all}$ is the total number of nodes in the network, $tran^m$ is the number of times the current node forwards the message m, the larger the value means the greater the energy consumption of the current node, $\sum_{j=1}^{buffsize} tran^j$ is the number of times the current node forwards all messages in the cache and $buffsize$ is the size of the current node cache, according to the above formula, the utility value of any message at any point can be calculated.

To increase the node transmission success rate and reduce energy consumption, when a new message arrives, it is sorted according to the cache space utility value, and the messages with lower utility values are removed from the cache in priority to release the cache, and when the message m is moved out of the cache due to its lower utility value, no more messages m forwarded by other nodes are received.

4.2. Scheduled Cache Management Mechanism

When more time passes throughout the network's message dispersion process, there will be a lot more copies available. Nodes should swiftly tell other nodes to delete the copies of the message they receive in order to decrease cache occupation and erase unnecessary message copies.

Each node maintains a two-column collection of delivered messages with each element stored in the collection as a key–value pair. Concurrent hash mapping table storage is utilized to enable the simultaneous updating of delivered messages in many connections since a node may generate connections to multiple nodes at once. The delivered collection is constructed as {message ID1: delivery time; message ID2: delivery time; ...}. When two nodes connect and send a message, if the receiving node is the message's destination node and the information contained in the message does not already exist in the current delivered collection, the message's ID and delivery time are added to the collection, and the message that was previously stored in the cache is deleted. Figure 2 displays the set of nodes' delivered messages that are stored in a structure consisting of an array, a chain table, and a red-black tree. Each element of the array is a joint, and each writes operation locks the joint for this operation, and the data is stored by hashing the elements into the chain table or red-black tree of which node. A red-black tree develops from a message chain table when there are more elements in the chain table, while a chain table develops from a red-black tree when there are fewer elements.

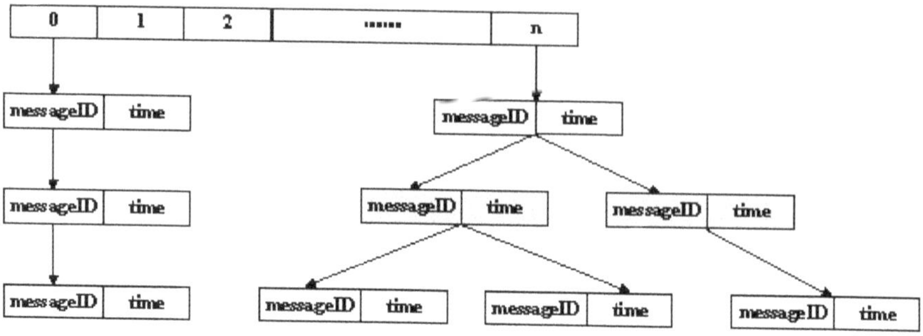

Figure 2. Delivered Message Collection.

When two nodes come together, they check their own caches to see if any messages from the other node's delivered collection are present, and if they do, they delete this message. Then, they update the delivered messages that are not stored in the current node according to the other node's collection, and alert other nodes in the network to delete any messages that have already reached their destinations in order to free up memory and maximize cache space.

After the experimental test, the survival time of each message is set to one and a half cycles with a better effect. This can ensure the prompt removal of useless message copies and maintenance of a certain number of useful copies to ensure the timely delivery of messages. One cycle is the average delivery time of messages, and the average delivery time is determined using Equation (18) below. Create a timer for each node's collection, check for expired delivered messages every minute and delete them in time to release the cache. Just before the simulation is through, remove all messages from the double-column collection of delivered messages and recycle the memory in time.

$$T = \sum (AT_m - GT_m)/N_m \tag{18}$$

where AT_m is the delivery time of the message, GT_m is the message generation time and N_m is the number of messages.

4.3. *Markov Path Prediction and Cache Management*

The probability that any node N_a will arrive at any place v at any time t can be calculated using the approach described above. A destination set $\mathrm{V} = \{\mathrm{v}_1, \mathrm{v}_2, \mathrm{v}_3, \ldots, \mathrm{v}_n\}$ is created based on the places that students frequently visit.

To accurately model the movement characteristics of learner nodes, several representative locations on campus are selected, including dormitory, classroom, canteen, playground, supermarket and library. The node arrives at a region and stays there for a random period and then chooses the next region, assuming that transferring between regions is not time consuming. Two main parameters affect the nodes' mobile behavior, mobile trajectory offset probability p_{offset} and dwell time w_{time}, indicating that the nodes move to the pre-set location with a probability of p_{offset} and to any location with a probability of $1 - p_{offset}$. To reflect the predictability of students' movement behavior, p_{offset} is set to 0.8, where $w_{time} \in [1h, 2h]$.

This study suggests a Modified Markov Path Prediction and Cache Management Routing (MPCM) based on the classification of destinations mentioned above. The learner nodes are grouped to reflect the collaboration between the nodes and to minimize resource waste by forecasting the learner nodes' trajectories and forwarding the messages. The nodes in this experiment are split into five groups, and each group chooses one of five possible destinations. Once at the destination, the nodes travel arbitrarily within the range of the destination. The pseudo code for the node forwarding process is shown in Algorithm 1.

Algorithm 1. MPCM strategy

INPUT: node N_a, node N_b
OUTPUT: Messages
START:
WHILE (node N_a carries message m & node N_b don't carries message m)
IF (the size of free cache space of node $N_b \geq$ the size of m)
IF (node N_b is D_m
call N_b receive m;
END IF;
IF (node N_a has already transmitted the message m to node N_b)
m cannot be transmitted
END IF;
IF (N_a and D_m are in the same group)
IF (node N_b and message m are in the same group & $DC(N_b) \geq DC(N_a)$)
call N_b receive m;
END IF;
ELSE
IF ($fc(N_b, D_m) \geq fc(N_a, D_m)$)
call N_b receive m;
IF (N_b and D_m are in the same group)
call N_b receive m;
ELSE
m cannot be transmitted;
END IF;
END IF;
END IF;
IF (the free cache space of node $N_b <$ the size of m)
Calculate the utility value U^m of the message in the node N_b cache;
Delete the message with the lowest utility value until the message m can be put down;
END IF;
END

If the free cache space of a node is insufficient to keep the message received this time when two nodes meet in the opportunity network, the utility value U^m of the message in the cache of node N_b is calculated and the message with the lowest utility value is deleted until the message m can be put. If the free cache space of node N_b is large enough to accommodate this received message: If node N_a carries message m and node N_b does not, the message is forwarded to node N_b if it is the intended recipient. Otherwise, node N_a checks to see if it has already transmitted this message to node N_b, if it has, the message cannot be transmitted.

This message will be forwarded only within the group if nodes N_a and D_m are members of the same group. The message is forwarded if nodes N_b and D_m are in the same group and centrality of N_b is greater than N_a, otherwise, it is not forwarded. Conversely, if N_a, N_b and D_m are not in the same group, and the probability of meeting between N_b and D_m is not less than N_a, the message is forwarded, otherwise it is not forwarded. Additionally, the message is forwarded if nodes N_a and D_m are not in the same group but nodes N_b and D_m are.

5. Results

5.1. Experimental Scheme Design

ONE (the Opportunity Network Environment) is used as the experimental environment to test the algorithm presented in this paper, and the real data set haggle6-infocom6 [27] is used for performance verification. A total of 3 days of data transmission between 78 mobile Bluetooth devices and 20 fixed devices were collected in the Infocom06 dataset. The experimental parameters are established in Table 3 below, and the Prophet,

Epidemic, RDR [19] and FCIM [20] algorithms are evaluated alongside the algorithm suggested in this research.

Table 3. Simulation parameters.

Parameter	Value
dataset	haggle6-infocom6
simulation time/h	72
simulation area/m^2	4500 × 3400
number of nodes	98
message generation interval/s	100
message size/kb	50 k~5000 k
message TTL/h	5

The performance of each of the five methods is evaluated under identical circumstances but with varying cache sizes, message generation intervals and message survival times. In this paper, we take into account the following four fundamental metrics: message delivery success rate, average latency, routing overhead and the number of packets dropped. A higher message delivery success rate, a lower average latency, a lower routing overhead and a lower number of packet drops signify a routing algorithm's superior performance. We consider the effect of three parameters: cache spaces, message generation intervals and time to live of messages.

Data normalization is used in the processing of the results of the experiment. The first processing is Equation (19) because a greater message delivers success rate is preferable, and second processing is Equation (20) as smaller average delay, overhead, and packet drops are preferable.

$$x' = \frac{x - \min(x)}{\max(x) - \min(x)} \tag{19}$$

$$x'' = 1 - \frac{x - \min(x)}{\max(x) - \min(x)} \tag{20}$$

The algorithm's overall score is added by the normalized values under various indexes, and among them, $\min(x)$ and $\max(x)$ indicate the minimum and maximum values in a collection of data.

5.2. Experimental Results Analysis

5.2.1. Different Cache Spaces

This section will examine the effects of buffer size from 10 M to 100 M on delivery success rate, average latency, routing overhead and a number of packet losses because cache space size has a significant impact on network performance.

The following Figure 3 demonstrates that in contrast to other protocols, cache space size has a small impact on the delivery success rate of MPCM. Rather than sending messages blindly, MPCM makes every effort to deliver messages to nodes that may come together by predicting those nodes' locations based on their movement patterns. Hence, MPCM performs better in networks with constrained cache space. The one routing protocol most impacted by cache size is Epidemic, which employs a flooding technique to message delivery with massive message copies throughout the network. Under various cache spaces, the MPCM message latency will vary significantly, but this effect can be disregarded. Significantly fewer messages are discarded in Epidemic as cache capacity increases, the messages will be stored there for a long period and latency will rise noticeably. MPCM intra-group message forwarding and forwarding based on transmission probability may effectively regulate the number of message copies in the network, and the influence of cache size on MPCM network overhead is likewise minimal. In conclusion, the routing overhead falls dramatically, the number of messages successfully delivered by each routing algorithm rises as the cache size grows, and MPCM always maintains a low overhead level.

The amount of packet drops in MPCM is already close to zero when the cache size exceeds 60 M, and the number of drops continuously declines as more messages may be put in the Epidemic cache.

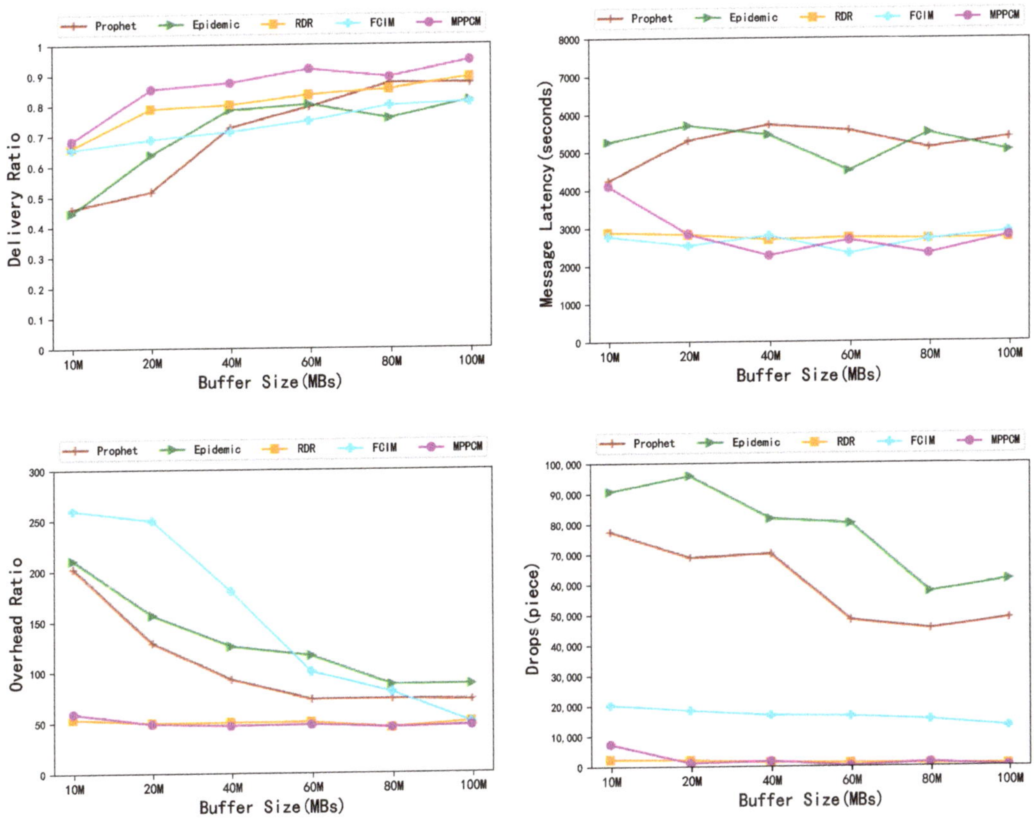

Figure 3. Comparison of success rate, latency, overhead and packet loss with different buffer size.

5.2.2. Different Message Generation Intervals

The impact on each metric is observed by varying the message generation interval at a cache size of 50 M. Each routing technique performs noticeably better as the generation interval gets longer.

As shown in Figure 4, when there are too many messages, MPCM moves out messages with lower utility values in accordance with the message utility values, leaving enough space for messages with high utility values, and on this basis, a certain success rate will be guaranteed. MPCM maintains a better state at intervals greater than the 40s, and the delivery success rate is no longer affected. Several other algorithms perform poorly because they are unable to handle excessively packed communications. With an increase in the message generation interval, MPCM performs fairish in terms of latency, and its overall latency is lower than that of several other three algorithms. Since the messages stay in the cache for a long time as the generation interval rises, the network overhead also rises gradually. When the generation interval exceeds the 20s, MPCM and RDR are largely unaffected, but the overhead of the other three algorithms continues to rise. The Epidemic algorithm has the highest number of dropped packets, and a large number of message copies in the network are discarded due to cache limitations and long survival times.

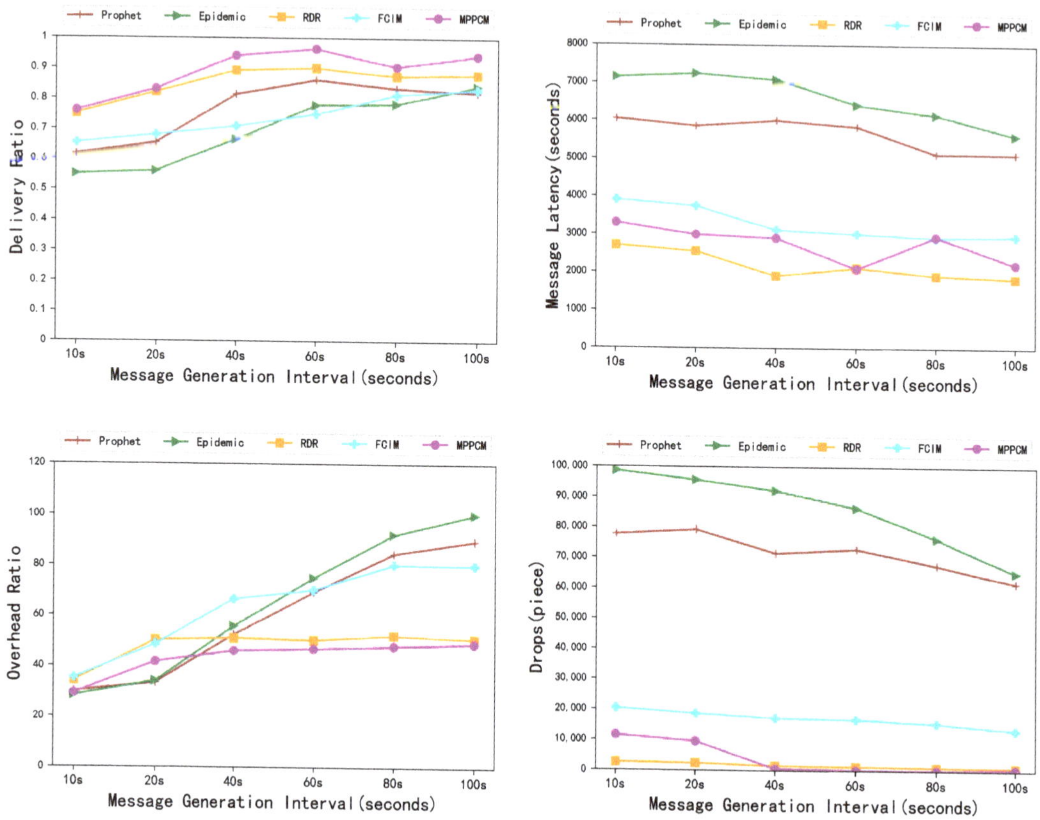

Figure 4. Comparison of success rate, latency, overhead and packet loss with different message generation intervals.

5.2.3. Different Time to Live of Messages (TTL)

By altering the time to live (TTL) of various messages at a cache size of 50 M and a message generation interval of 100s, the impact on the metrics is shown in Figure 5. The delivery success rate of each method has a trend of increasing and then decreasing as the message TTL increases. The message initially has adequate time to reach the destination node as the TTL increases. However, when the TTL rises further, a huge number of message copies will exist in the network, lowering the success rate. MPCM shows better performance when the TTL is greater than 3 h. The success rate of message delivery gradually declines as the TTL increases because the Epidemic algorithm causes a significant increase in copies and no more room to receive new messages. Each algorithm's latency will rise as the TTL rises, the MPCM will not rise any further until it reaches a more stable value, and the overall latency is very low. The network overhead rises as a result of the inability to clear message copies in time as the TTL gets longer. The TTL has the biggest impact on Epidemic's overhead, while MPCM's overhead is consistently kept at a minimal level. When the TTL is greater than 3 h, the number of packet losses in MPCM is close to 0, and most of the messages can be delivered to the destination node before the expiration date.

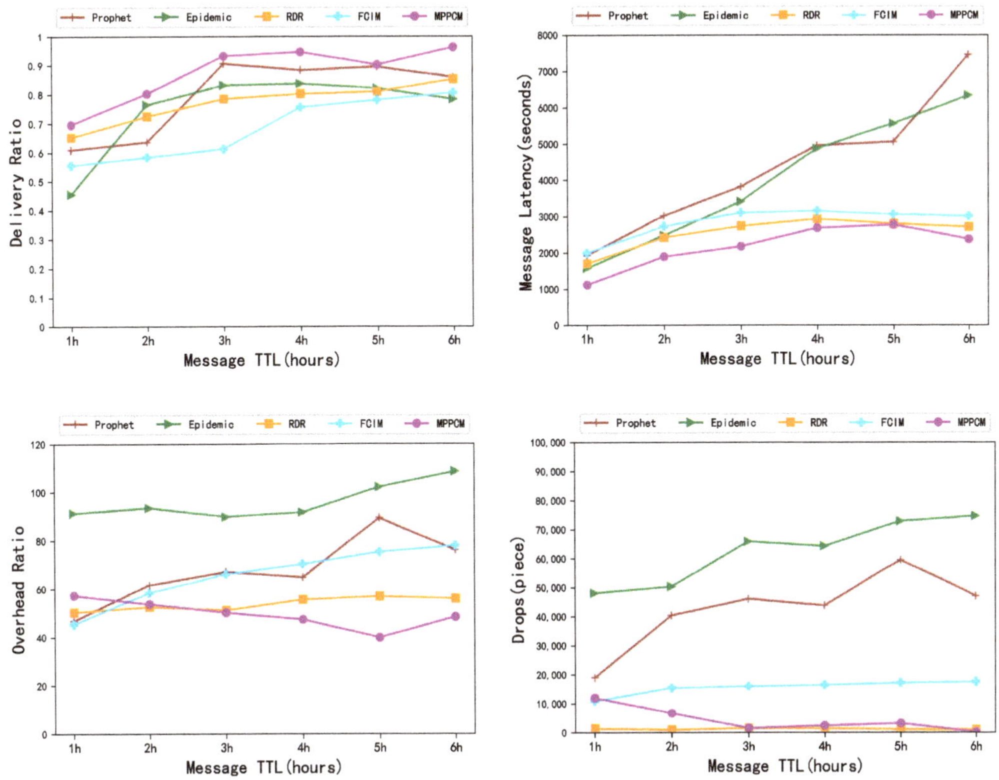

Figure 5. Comparison of success rate, latency, overhead, and packet loss with different message TTL.

The MPCM algorithm presented in this paper performs reasonably well when compared to numerous other algorithms in various cache spaces, message generation intervals and varied message TTLs. Under all circumstances, the delivery success rate of MPCM exhibits the best performance. It also performs exceptionally well in terms of latency, overhead and packet drop which is mostly unaffected by the amount of cache capacity. When messages are excessively dense, the success rate decreases, but it still performs better than many other algorithms in general. Overall, MPCM can adapt to the campus context with a limited node cache, dense message generation and short message TTL. Experiments show that the proposed method in the paper achieves better results in terms of delivery rate, average delivery latency, overhead and several packets drop. The average values of each measure are processed using data normalization, and the experimental results under the influence of three parameters, different cache spaces, different message generation intervals and different message survival times, are tabulated in Tables 4–6 below.

Table 4. Normalized scores under different caches.

Algorithm /Score	Success Rate	Overhead	Latency	Packet Drops	Total Score
Epidemic	0	0.2687	0	0	0.2687
Prophet	0.0034	0.4460	0.0092	0.2374	0.696
RDR	0.6362	0.9897	0.9639	1	3.5898
FCIM	0.1840	0	1	0.7987	1.9827
MPCM	1	1	0.9388	0.9935	3.9323

Table 5. Normalized scores for different message generation intervals.

Algorithm /Score	Success Rate	Overhead	Latency	Packet Drops	Total Score
Epidemic	0.3631	0	0	0	0.3631
Prophet	0	0.2107	0.2123	0.1656	0.5886
RDR	0.8048	0.776	1	1	3.5808
FCIM	0.2226	0.0354	0.7517	0.8169	1.8266
MPCM	1	1	0.8713	0.9753	3.8466

Table 6. Normalized scores with different TTL.

Algorithm /Score	Success Rate	Overhead	Latency	Packet Drops	Total Score
Epidemic	0.6064	0	0.1506	0	0.757
Prophet	0.34914	0.6095	0	0.3269	1.28554
RDR	0.4633	0.9084	0.8289	1	3.2006
FCIM	0	0.6547	0.6942	0.765	2.1139
MPCM	1	1	1	0.9492	3.9492

The suggested algorithm in this paper has the highest score under the influence of the three parameters, yielding the best outcome.

Our proposed approach works well in the campus context because student nodes are more regular in their movement and the message-forwarding process is not blind, which effectively limits the message copies in the network and maximizes cache space utilization. Moreover, messages are forwarded within groups based on centrality, which means that messages can be transmitted to their destinations through those relays that are more influential. However, because of their potential similarity in sparse networks, it may be impossible to identify which nodes have high centrality.

6. Conclusions

This research suggests a modified Markov path prediction algorithm for nodes in campus opportunity networks with specific movement patterns. Students are typically thought to travel in small groups and repeat themselves. The proposed routing strategy is more effective according to these two forwarding methods we present in the paper for nodes: in-group forwarding and out-group forwarding. We first allow the message to reach its group as quickly as possible, then it is further forwarded based on the influence of the node within the group, as nodes with higher influence have more access to the destination node of the message. Moreover, we discovered that storage capacity of nodes is constrained and typically small. As a result, we propose a cache management strategy in this paper. When the node's own cache space is insufficient, the message utility value is calculated based on the message diffusion and the energy consumption to the current node, and the messages with high utility value will be reserved first, achieving a reasonable cache allocation.

The communication between nodes in the same group will be closer as the suggested method is for nodes on campus, and we hope that the following work will result in greater cooperation between nodes in the same group.

Author Contributions: Conceptualization, Y.C. (Yumei Cao) and P.L.; methodology, Y.C. (Yumei Cao); software, Y.C. (Yumei Cao); validation, Y.C. (Yumei Cao); formal analysis, P.L.; investigation, T.L.; resources X.W. (Xiaojun Wu); data curation, X.W. (Xiaoming Wang); writing—original draft preparation, Y.C. (Yumei Cao); writing—review and editing, Y.C. (Yumei Cao); visualization, Y.C. (Yumei Cao); supervision, Y.C. (Yuanru Cui); project administration, P.L.; funding acquisition, P.L. All authors have read and agreed to the published version of the manuscript.

Funding: This work is partly supported by the National Key R and D Program of China under grant No. 2020YFC1523305, Key Laboratory Funds of the Ministry of Culture and Tourism under grant No 2022-13, the National Natural Science Foundation of China under Grant No. 61877037, 61872228, 61977044, the Shaanxi Key Science and Technology Innovation Team Project under Grant No. 2022TD-26.

Institutional Review Board Statement: Not applicable.

Informed Consent Statement: Not applicable.

Data Availability Statement: Not applicable.

Acknowledgments: We appreciate the anonymous reviewers' and editorial team members' suggestions and comments. Thanks for the support of the fund projects mentioned above.

Conflicts of Interest: The authors declare no conflict of interest.

References

1. Sachdeva, R.; Dev, A. Review of opportunistic network: Assessing past, present, and future. *Int. J. Commun. Syst.* **2021**, *34*, e4860. [CrossRef]
2. LI, P.; Wang, X.M.; Zhang, L.C.; LU, J.L.; Zhu, T.J.; Zhang, D. A Novel Method of Video Data Fragmentary and Progressive Transmission in Opportunistic Network. *Acta Electonica Sin.* **2018**, *46*, 2165.
3. Bagirathan, K.; Palanisamy, A. Opportunistic routing protocol based EPO–BES in MANET for optimal path selection. *Wirel. Pers. Commun.* **2022**, *123*, 473–494. [CrossRef]
4. Gautam, T.; Dev, A. Improving Packet Queues Using Selective Epidemic Routing Protocol in Opportunistic Networks (SERPO) BT—Advances in Computing and Data Sciences. In *Advances in Computing and Data Sciences, 4th International Conference, ICACDS 2020, Valletta, Malta, 24–25 April 2020*; Revised Selected Papers 4; Springer: Singapore, 2020; pp. 382–394.
5. Bansal, A.; Gupta, A.; Sharma, D.K.; Gambhir, V. Iicar-inheritance inspired context aware routing protocol for opportunistic networks. *J. Ambient. Intell. Humaniz. Comput.* **2019**, *10*, 2235–2253. [CrossRef]
6. Sharma, D.K.; Kukreja, D.; Chugh, S.; Kumaram, S. Supernode routing: A grid-based message passing scheme for sparse opportunistic networks. *J. Ambient. Intell. Humaniz. Comput.* **2019**, *10*, 1307–1324. [CrossRef]
7. Singh, J.; Obaidat, M.S.; Dhurandher, S.K. Location based Routing in Opportunistic Networks using Cascade Learning. In Proceedings of the 2021 International Conference on Computer, Information and Telecommunication Systems (CITS), Istanbul, Turkey, 29–31 July 2021; pp. 1–5.
8. Dhurandher, S.K.; Singh, J.; Nicopolitidis, P.; Kumar, R.; Gupta, G. A blockchain-based secure routing protocol for opportunistic networks. *J. Ambient. Intell. Humaniz. Comput.* **2022**, *13*, 2191–2203. [CrossRef]
9. Sharma, D.K.; Rodrigues, J.J.P.C.; Vashishth, V.; Khanna, A.; Chhabra, A. RLProph: A dynamic programming based reinforcement learning approach for optimal routing in opportunistic IoT networks. *Wirel. Netw.* **2020**, *26*, 4319–4338. [CrossRef]
10. Kumar, P.; Chauhan, N.; Chand, N. Node activity based routing in opportunistic networks. In Proceedings of the International Conference on Futuristic Trends in Network and Communication Technologies, Taganrog, Russia, 14–16 October 2019; pp. 265–277.
11. Gou, F.; Wu, J. Triad link prediction method based on the evolutionary analysis with IoT in opportunistic social networks. *Comput. Commun.* **2022**, *181*, 143–155. [CrossRef]
12. Chunyue, Z.; Hui, T.; Yaocong, D. An Energy-Saving Routing Algorithm for Opportunity Networks Based on Sleeping Mode. In Proceedings of the 2019 20th International Conference on Parallel and Distributed Computing, Applications and Technologies (PDCAT), Gold Coast, Australia, 5–7 December 2019; pp. 13–18.
13. Derakhshanfard, N.; Soltani, R. Opportunistic routing in wireless networks using bitmap-based weighted tree. *Comput. Netw.* **2021**, *188*, 107892. [CrossRef]
14. Chithaluru, P.; Tiwari, R.; Kumar, K. AREOR–Adaptive ranking based energy efficient opportunistic routing scheme in Wireless Sensor Network. *Comput. Netw.* **2019**, *162*, 106863. [CrossRef]
15. Hernández-Orallo, E.; Borrego, C.; Manzoni, P.; Marquez-Barja, J.M.; Cano, J.C.; Calafate, C.T. Optimising data diffusion while reducing local resources consumption in Opportunistic Mobile Crowdsensing. *Pervasive Mob. Comput.* **2020**, *67*, 101201. [CrossRef]
16. Raverta, F.D.; Fraire, J.A.; Madoery, P.G.; Demasi, R.A.; Finochietto, J.M.; D'argenio, P.R. Routing in Delay-Tolerant Networks under uncertain contact plans. *Ad. Hoc. Netw.* **2021**, *123*, 102663. [CrossRef]
17. Das, P.; Nishantkar, P.; De, T. SECA on MIA-DTN: Tackling the Energy Issue in Monitor Incorporated Adaptive Delay Tolerant Network Using a Simplistic Energy Conscious Approach. *J. Netw. Syst. Manag.* **2019**, *27*, 121–148. [CrossRef]
18. Kang, M.W.; Chung, Y.W. An improved hybrid routing protocol combining MANET and DTN. *Electronics* **2020**, *9*, 439. [CrossRef]
19. Pirzadi, S.; Pourmina, M.A.; Safavi-Hemami, S.M. A novel routing method in hybrid DTN–MANET networks in the critical situations. *Computing* **2022**, *104*, 2137–2156. [CrossRef]
20. Mao, Y.; Zhou, C.; Qi, J.; Zhu, X. A fair credit-based incentive mechanism for routing in DTN-based sensor network with nodes' selfishness. *EURASIP J. Wirel. Commun. Netw.* **2020**, *2020*, 1–18. [CrossRef]

21. Vahdat, A.; Becker, D. Epidemic Routing for Partially-Connected Ad Hoc Networks. In *Handbook of Systemic Autoimmune Diseases*; Elsevier: Amsterdam, The Netherlands, 2000.
22. Lindgren, A.; Doria, A.; Schelén, O. Probabilistic Routing in Intermittently Connected Networks. *ACM Sigmobile Mob. Comput. Commun. Rev.* **2003**, *7*, 19–20. [CrossRef]
23. Rehman, G.U.; Haq, M.I.U.; Zubair, M.; Mahmood, Z.; Singh, M.; Singh, D. Misbehavior of nodes in IoT based vehicular delay tolerant networks VDTNs. *Multimed. Tools Appl.* **2023**, *82*, 7841–7859. [CrossRef]
24. Rehman, G.U.; Ghani, A.; Zubair, M.; Naqvi, S.H.A.; Singh, D.; Muhammad, S. IPS: Incentive and Punishment Scheme for Omitting Selfishness in the Internet of Vehicles (Iov). *IEEE Access* **2019**, *7*, 109026–109037. [CrossRef]
25. Rehman, G.U.; Zubair, M.; Qasim, I.; Badshah, A.; Mahmood, Z.; Aslam, M.; Jilani, S.F. EMS: Efficient Monitoring System to Detect Non-Cooperative Nodes in IoT-Based Vehicular Delay Tolerant Networks (VDTNs). *Sensors* **2023**, *23*, 99. [CrossRef] [PubMed]
26. Rehman, G.U.; Ghani, A.; Zubair, M.; Saeed, M.I.; Singh, D. SOS: Socially omitting selfishness in IoT for smart and connected communities. *Int. J. Commun. Syst.* **2023**, *36*, e4455. [CrossRef]
27. Scott, J.; Hui, P.; Crowcroft, J.; Diot, C. Haggle: A networking architecture designed around mobile users. In Proceedings of the Third IFIP Wireless on Demand Network Systems Conference, Les Menuires, France, 18–20 January 2006.

applied sciences

MDPI

Communication

High-Performance Microwave Photonic Transmission Enabled by an Adapter for Fundamental Mode in MMFs

Yilan Wang [1], Linbo Yang [1], Zhiqun Yang [1,*], Yaping Liu [1,*], Zhanhua Huang [1] and Lin Zhang [1,2,*]

[1] Key Laboratory of Opto-Electronic Information Technology of Ministry of Education and Tianjin Key Laboratory of Integrated Opto-Electronics Technologies and Devices, School of Precision Instruments and Opto-Electronics Engineering, Tianjin University, Tianjin 300072, China
[2] Peng Cheng Laboratory, Shenzhen 518038, China
* Correspondence: yangzhiqun@tju.edu.cn (Z.Y.); liuyp@tju.edu.cn (Y.L.); lin_zhang@tju.edu.cn (L.Z.)

Abstract: Microwave photonic links (MPLs) have long been considered as an excellent way for radio frequency (RF) transmission due to their advantages such as light weight, high bandwidth, low cost and large spurious-free dynamic range (SFDR). However, the effective mode-field area (A_{eff}) of the single-mode fiber (SMF) used in the traditional MPL is not large, so the MPL based on SMF have relatively strong nonlinearity, which limits the processing power of SMFs to a level of few milliwatts. Few-mode fibers (FMFs) have been applied in MPL as an alternative due to the larger A_{eff}, and photonic lanterns are used simultaneously to excite the high-order mode of FMFs for RF signal transmission. However, the photonic lantern could bring additional insertion loss, and the production cost of FMFs is high, so we propose an MPL based on multimode fibers (MMFs) with mode field adapters (MFAs). Since MMFs have larger A_{eff}, the nonlinearity of the link can be greatly reduced. And matched MFAs realized by reverse tapering, to excite only the fundamental mode in MMFs to reduce the crosstalk, which are very stable. As a result, the stimulated Brillouin scattering threshold and SFDR are improved by 5 dB and 14.5 dB, respectively.

Keywords: microwave photonic link; mode field adapter; reverse tapering

Citation: Wang, Y.; Yang, L.; Yang, Z.; Liu, Y.; Huang, Z.; Zhang, L. High-Performance Microwave Photonic Transmission Enabled by an Adapter for Fundamental Mode in MMFs. *Appl. Sci.* **2023**, *13*, 1794. https://doi.org/10.3390/app13031794

Academic Editor: Chi-Wai Chow

Received: 21 December 2022
Revised: 22 January 2023
Accepted: 29 January 2023
Published: 30 January 2023

1. Introduction

A microwave photonic link (MPL) works by modulating the radio frequency (RF) signal containing the baseband to the optical carrier, which is then modified by several optical devices and finally down-converted at the receiver to recover the RF signals [1]. MPLs have been increasingly applied to cable TV, military radar, radio telescope and some other fields [2–5] due to their advantages like light weight, high bandwidth, low cost and large spurious-free dynamic range (SFDR) [6–9]. SFDR is one of the very important factors to reflect the performance of MPLs. Clark et al. demonstrated a promising photonic transport and digital receiver technique for analog RF signals and proposed direct demodulation to obtain an extremely high SFDR [10]. Zheng R. et al. demonstrated a long-haul MPL achieved high SFDR by biasing the modulator at single-sideband suppressed carrier modulation and adjusting the polarization of the light entering the polarizer [1]. The stimulated Brillouin scattering (SBS) in fibers tend to limit the processing power of single-mode fibers (SMFs) to the level of milliwatts compared to high-power lasers and photodetectors that can operate with an optical power of watt-level. Therefore, the nonlinear effects of optical fibers become the bottleneck for improving the performance of MPLs. In recent years, more attention has been focused on the transmission media, that is, the fiber used in MPLs. According to fiber nonlinear theory [11], the larger the effective mode area, the weaker the nonlinear effect. Wen et al. used a higher-order mode (HOM) of few-mode fibers (FMFs) to transmit RF signals, and its large mode area helps reduce nonlinearity and improve SFDR [12], in which a photonic lantern is used to excite the corresponding modes. However, photonic lanterns are sensitive to external environment changes such as temperature or placement posture, so a specific fiber mode would not be purely excited, leading to non-negligible multi-path interference (MPI). Besides, considering the availability and cost of FMFs and mode converters, we need to seek a more stable and cost-effective solution.

Here, we propose a new MPL utilizing commercial multimode fibers (MMFs), which is featured by mode field adapters (MFAs) fabricated using reverse tapering technique and fused to both the ends of an MMF. RF signals are carried only by the fundamental mode, with greatly reduced loss and crosstalk between modes, which would significantly reduce the MPI and improve the system stability. Compared with traditional SMF-based MPLs, the SBS threshold of the MFA-based 17.6-km MMF link is improved by 5 dB, while the SFDR increases by 14.5 dB, with an input optical power of 17 dBm.

In this paper, we first introduce the background and application fields of our work and propose the novel MPL solution. Section 2 presents the principle of the proposed scheme and the fabrication of the MFAs. Section 3 shows experiments on the measurements of SBS threshold and SFDR, and the corresponding result analysis. Section 4 is the conclusion and outlook.

2. MFA Fabrication

The experimental setup for the proposed MMF-based MPL is shown in Figure 1. A light wave emitted by a 1550-nm laser is modulated by two tones (1.9 GHz and 2 GHz) that generated by arbitrary waveform generator (AWG) using an intensity modulator (MOD). The power of RF signals is attenuated by an RF attenuator (RF Att), for making the output power of the fundamental frequency (FF) signal and the third-order intermodulation (IMD3) present an approximately linear relationship with the input power. The light wave that has been modulated is amplified by an erbium-doped fiber amplifier (EDFA) then launched into a 17.6-km-long OM3 MMF via an optical circulator, and finally the Brillouin backscattering light power can be measured by the power meter. Besides, the variable optical attenuator (VOA) is used to adjust the incident optical power. Two fabricated MFAs were fused to the two ends of the MMF to excite and filter only the fundamental mode of the MMF. At the end of the link, a photodetector (PD) and an electrical spectrum analyzer (ESA) are applied to detect and analyze the output signal.

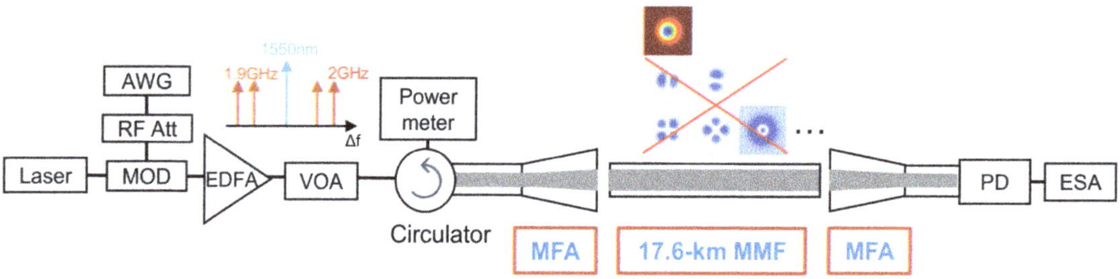

Figure 1. Schematic diagram of the microwave photonic transmission link using an MMF (OM3) with two mode field adapters, in which the large-area fundamental mode can be clearly excited.

The OM3 fibers, which are widely used in fiber communication, have a 14.5-μm mode-field diameter for the fundamental mode [13], and the A_{eff} is about 165 μm^2. As we can see, we need carefully fabricate MFAs to match the mode field of SMF and MMF. To reduce MPI, we need to expand the core of the MFA to ~14.5 μm. We plan to use reverse tapering to fabricate the MFAs by LZM-100, an optical fiber splicer produced by Fujikura. The fiber splicer uses a CO_2 laser, which makes the heating processes highly reproducible and stable. The reverse tapering process can be realized by programming, allowing any fibers to be tapered with high precision [14]. The SMF can be thickened smoothly and slowly then form a plat thickened area using the optical splicer. After that, the MFA is fabricated by cutting off at the thickened area and fusing with the MMF in alignment. We conducted six times of reverse tapering and the cross-sectional images of the MFAs are shown in

Figure 2a. The side view of the MFA obtained after the 6-th reverse tapering, which is the best match, is presented in Figure 2b. The fiber was thickened slowly from right to left. Two MFAs were fused to the two ends of the MMF, and the losses of them are 0.34 dB and 1.68 dB, respectively.

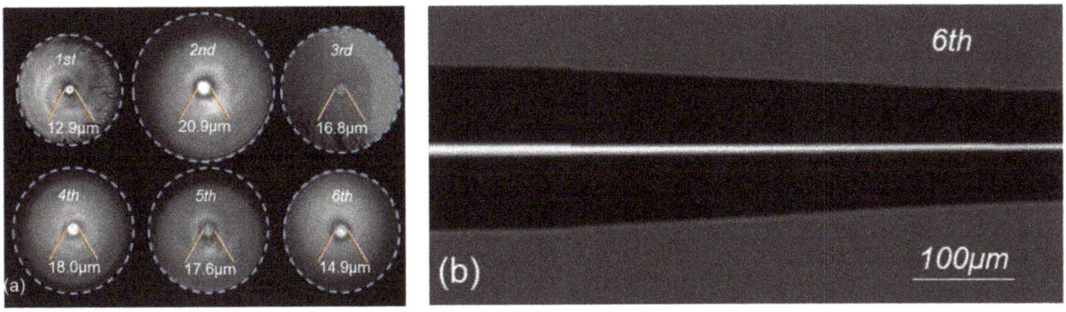

Figure 2. (**a**) The cross-sectional images of the MFAs obtained through six iterations of the reverse tapering technique, among which the 6th sample's mode field diameter can match that of the fundamental mode in the OM3 fiber. (**b**) The side view of the MFA, thickened slowly and smoothly from right to left.

3. Experimental Measurement and RF Transmission Performance

Figure 3a,b demonstrate the performances of the MMF-based MPL that we discussed above, in comparison with a G.652D SMF MPL. In order to verify whether a larger mode field area can indeed lead to lower nonlinear effects and whether the length of fiber can influence nonlinearity, we design two sets of experiments for the SBS threshold measurement. In the first set, the lengths of SMF and MMF are 10 km and 8.8 km respectively, because the SBS fluctuations are generally not significant within 2 km, while in the second set, fiber lengths are increased to 16.5 km and 17.6 km, respectively. According to the existing fibers in our lab, we chose a 17.6-km MMF for the experiment, the length of which is already in the range of typical MPLs. Besides, the characteristics of the SMF and the MMF used in the experiments are presented in Table 1 [15].

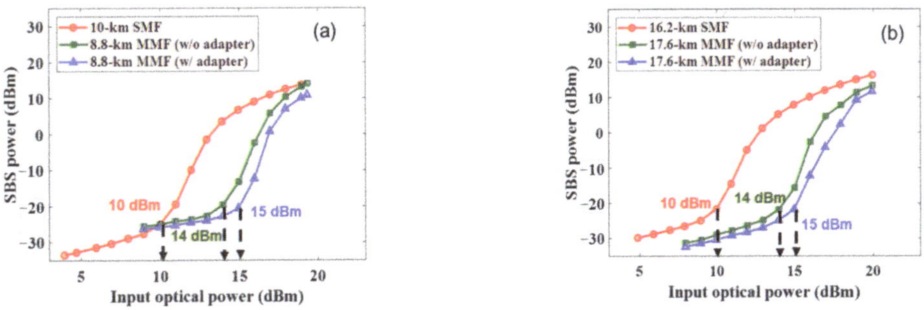

Figure 3. The back-scattered optical power versus input optical power for (**a**) a 10-km SMF and an 8.8-km MMF and (**b**) a 16.2-km SMF and 17.6-km MMF with and without adapters.

Table 1. The performance characteristics of the SMF and OM3 MMF.

Parameters	Value
α_{MMF_LP01}	0.220 dB/km
α_{SMF}	0.187 dB/km
A_{eff} of MMF_LP_{01}	165 μm^2
A_{eff} of SMF	80 μm^2

It can be seen that the SBS threshold of SMFs is only ~10 dBm, while that of MMFs is ~15 dBm. In addition, the SBS threshold without MFAs is 14 dBm, which is slightly smaller, indicating that the existence of the MFAs increases the SBS threshold by ~1 dB. We also prove that the SBS threshold of longer fibers is almost unchanged, no matter SMFs or MMFs. However, the back-scattered light power of the longer SMF becomes exactly stronger, while that of the longer MMF increases less, proving that it is feasible to use MMFs to improve the nonlinearity. The crosstalk of the link without MFAs is very high because the HOM components of the MMF could be excited in a high proportion, which means that the quality of the baseband signal would be severely affected.

Figure 4a demonstrates the relation of the output and input optical power of the link. When the input light power is 17 dBm, an increase of 3.8 dB can be obtained using MMFs compared to using SMFs. From Figure 4a, it can be seen that the transmitted power still increases after the incident optical power exceeds the SBS threshold, but not as fast as it is when below the SBS threshold, due to the presence of the Kerr effect in the fiber nonlinearity, where the power in the optical carrier is transferred to the sideband due to self-phase modulation and four-wave mixing (FWM). If the power in the sideband is below the SBS threshold, it continues to grow even though the optical carrier power is no longer growing due to exceeding the SBS threshold.

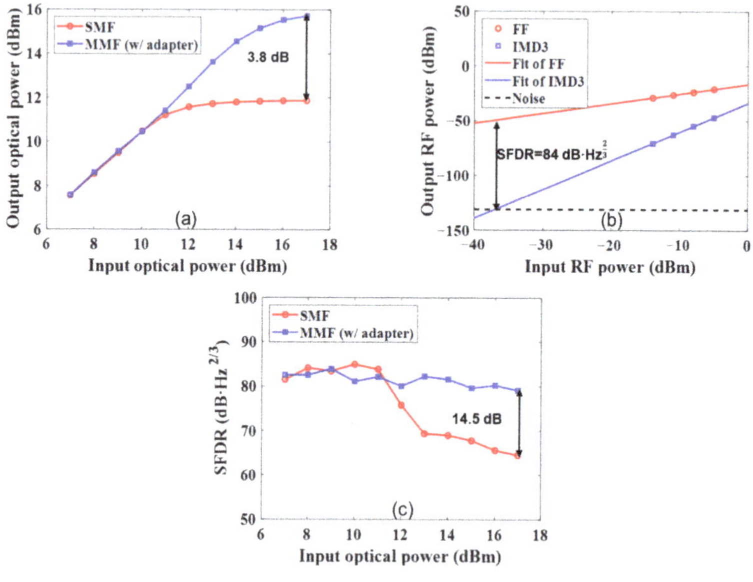

Figure 4. (**a**) The output optical power vs. input optical power. (**b**) Detected FF and IMD3 powers in SMF photonic links when input optical power is 8 dBm, as an example of measuring SFDR. (**c**) SFDR vs. input optical power.

The SFDR can be measured based on the signal output received by the ESA, which is shown in Figure 4b. The input dual-frequency RF signal can generate other frequency signals besides the fundamental frequency (FF) f_1 and f_2 due to link nonlinearity, among which the third-order intermodulation (IMD3) terms $2f_1 - f_2$ and $2f_2 - f_1$ are closest to the FF, which have a serious impact on the system. The FF and the IMD3 power values are measured a sufficient number of times at different input radio frequency (RF) signal powers and fitted to lines. The difference between the power value of FF and the noise floor is considered to be SFDR, in which the power value is obtained from the FF line based on the horizontal coordinate of the intersection of the IMD3 line and the noise floor. The formula for SFDR calculation is $\frac{2}{3}(OIP_3 - NF)$, in which OIP_3 is the intersection of the FF and IMD3 lines and NF stands for noise floor [16,17]. The IMD3 grows with the normalized optical input power and the optical spectral components that result in IMD3 include not only the optical carrier but also the modulation sidebands and FWM sidebands. Despite the large A_{eff} of the MMF and the Kerr effect is relatively weak, IMD3 in the MMF link is stronger at high input powers This is because, compared with the SMF, the MMF has a weaker SBS, which means the transmitted power can be stronger. With decreased IMD3 power and higher FF power, we obtain an improvement of 14.5-dB in SFDR based on MMFs, as presented in Figure 4c, which is measured at the input optical power of 17 dBm, a power value that is widely used in MPLs.

4. Conclusions and Discussions

We demonstrate a new method to improve the performance of microwave photonics links, that is, using MFAs fabricated by reverse tapering to match the MMF fundamental mode field, confining the signal transmitted only in the fundamental mode of MMFs. This method can effectively reduce the nonlinear effect, and the SBS threshold is 5-dB higher than that of SMFs. We also measure the SFDR of the link, and compare with the link using SMFs, there is a 14.5-dB improvement when the incident optical power is 17 dBm.

Although we have obtained the improvement on SFDR performance, the fusion of short devices like the MFA and accurate matching between the MFA and the fiber still need to be carefully improved. Besides, the MPI of the link is another subject that needs to be further investigated in the future. Even so, our work is potential to be applied in the field of high-power and long-distance RF signal transmission.

Author Contributions: Y.W. planned and wrote the paper. L.Y. participated in the experiments of the second part. Z.Y. contributed to the new idea of this work. Y.L. revised the whole manuscript and gave some valuable advice. Z.H. and L.Z. monitored the work and gave technical guidance. All co-authors participated in the review and editing job. All authors have read and agreed to the published version of the manuscript.

Funding: The work was supported by National Key R&D Program of China under grant 2019YFB2203902, and National Natural Science Foundation of China (NSFC) under grant 62105241.

Institutional Review Board Statement: Not applicable.

Informed Consent Statement: Not applicable.

Data Availability Statement: Not applicable.

Acknowledgments: The authors would like to thank the anonymous reviewers for their valuable comments and suggestions to improve this manuscript.

Conflicts of Interest: The authors declare no conflict of interest.

References

1. Marpaung, D.; Yao, J.; Capmany, J. Integrated Microwave Photonics. *Nat. Photonics* **2019**, *13*, 80–90. [CrossRef]
2. Bickers, L.; Reeve, M.; Rosher, P.; Fenning, S.; Cooper, A.; Methley, S.; Hornung, S. The analog local loop: a growing revolution in optical transmission. *J. Light. Technol.* **1989**, *7*, 1819–1824. [CrossRef]
3. Zhang, D.; Zhang, J.; Peng, X.; Lin, C.; Wo, J.; Wang, Y.; Du, P. A single sideband phase modulated radio over fiber link with spurious-free dynamic range enhancement. *Micro-Opt. MOEMS* **2021**, *12066*, 138–143. [CrossRef]
4. Zhai, W.; Wen, A.; Shan, D. Multidimensional Optimization of a Radio-Over-Fiber Link. *IEEE Trans. Microw. Theory Tech.* **2021**, *69*, 210–221. [CrossRef]
5. Urick, V.J.; Bucholtz, F.; McKinney, J.D.; Devgan, P.S.; Campillo, A.L.; Dexter, J.L.; Williams, K.J. Long-haul analog photonics. *J. Lightw. Technol.* **2011**, *29*, 1182–1205. [CrossRef]
6. Ohtsuki, T.; Aiba, T.; Matsuura, M. Simultaneous radio-frequency and baseband signal transmission over a multimode fiber. *IEEE Photonics J.* **2019**, *11*, 1–12. [CrossRef]
7. Zheng, R.; Chan, E.H.W.; Wang, X.; Feng, X.; Guan, B.-O.; Yao, J. Microwave Photonic Link with Improved Dynamic Range for Long-Haul Multi-Octave Applications. *J. Light. Technol.* **2021**, *39*, 7915–7924. [CrossRef]
8. Wen, H.; Mo, Q.; Sillard, P.; Correa, R.A.; Li, G. Transmission of RF/Microwave signals using few-mode fibers. In Proceedings of the 2016 IEEE Photonics Society Summer Topical Meeting Series (SUM), Newport Beach, CA, USA, 11–13 July 2016; IEEE: Piscataway, NJ, USA.

9. Wen, H.; Zheng, H.; Mo, Q.; Velázquez-Benítez, A.M.; Xia, C.; Huang, B.; Liu, H.; Yu, H.; Lopez, J.E.A.; Correa, R.A.; et al. Analog fiber-optic links using high-order fiber modes. In Proceedings of the 2015 European Conference on Optical Communication (ECOC), Valencia, Spain, 27 September–1 October 2015.
10. Clark, T.R.; O'Connor, S.R.; Dennis, M.L. A phase-modulation I/Q-demodulation microwave-to-digital photonic link. *IEEE Trans. Microw. Theory Tech.* **2010**, *58*, 3039–3058. [CrossRef]
11. Yang, M.; Liu, W.; Song, Y.; Wang, J.; Wei, Z.; Meng, H.; Liu, H.; Huang, Z.; Xiang, L.; Li, H.; et al. A design of dual guided modes ring-based photonic crystal fiber supporting 170 + 62 OAM modes with large effective mode field area. *Appl. Phys. B Laser Opt.* **2022**, *128*, 38. [CrossRef]
12. Wen, H.; Zheng, H.; Mo, Q.; Velázquez-Benítez, A.M.; Xia, C.; Huang, B.; Liu, H.; Yu, H.; Sillard, P.; Lopez, J.E.A.; et al. Few-mode fibre-optic microwave photonic links. *Light. Sci. Appl.* **2017**, *6*, e17021. [CrossRef] [PubMed]
13. Chen, X.; Li, K.; Wu, Q.; Clark, J.; Hurley, J.E.; Stone, J.S.; Li, M.-J. Fundamental mode transmission around 1310-nm over OM1 and OM2 multimode fibers enabled by a universal fiber modal adapter. *Opt. Fiber Technol.* **2022**, *69*, 102848. [CrossRef]
14. Yang, L.; Yang, Z.; Xu, T.; Hou, L.; Zhou, R.; Gan, L.; Cao, S.; Xiao, X.; Zhang, L. Low-loss Mode Field Adapter Using Reverse Tapering for Fundamental Mode Transmission over MMFs. In Proceedings of the Optical Fiber Communication Conference, San Diego, CA, USA, 7–9 March 2022.
15. Liu, H.; Wen, H.; Huang, B.; Li, Z.; Li, G. Low-cost and low-loss conversion of OM3 to OM4 MMFs using strong mode mixing. *Opt. Express* **2019**, *27*, 5581–5587. [CrossRef]
16. Zhang, J.; Wo, J.; Wang, A.; Luo, X.; Du, S.; Wang, D.; Wang, Y. SFDR improvement of a phase-modulated analog photonic link. *SPIE Intl. Soc. Optical Eng.* **2021**, *11763*, 1742–1747.
17. Institute of Electrical and Electronics Engineers. A 9.6 mW Low-Noise Millimeter-Wave Sub-Sampling PLL with a Divider-less Sub-Sampling Lock Detector in 65 nm CMOS. In Proceedings of the 2019 IEEE Radio Frequency Integrated Circuits Symposium (RFIC), Boston, MA, USA, 2–4 June 2019; ISBN 9781728117010.

applied sciences

MDPI

Article

Heterogeneously Integrated Multicore Fibers for Smart Oilfield Applications

Xutao Wang [1], Honglin Sun [1], Huihui Wang [1], Zhiqun Yang [1,*], Yaping Liu [1,*], Zhanhua Huang [1] and Lin Zhang [1,2,*]

[1] Key Laboratory of Opto-Electronic Information Technology of Ministry of Education and Tianjin Key Laboratory of Integrated Opto-Electronics Technologies and Devices, School of Precision Instruments and Opto-Electronics Engineering, Tianjin University, Tianjin 300072, China

[2] Peng Cheng Laboratory, Shenzhen 518038, China

* Correspondence: yangzhiqun@tju.edu.cn (Z.Y.); liuyp@tju.edu.cn (Y.L.); lin_zhang@tju.edu.cn (L.Z.)

Abstract: In the context of Industry 4.0, the smart oilfield is introduced, which relies on large-scale information exchange among various parts, and there is an urgent need for special fiber links for both increased data transmission capacity and high-sensitivity distributed sensing. Multicore fibers can be expected to play a critical role, in the parts of cores that are responsible for data transmission, while others are used for sensing. In this paper, we propose a heterogeneously integrated seven-core fiber for interconnection and awareness applications in smart oilfields, which could not only support digital and analog signal transmission but could also measure temperature and vibration. The core for digital signal transmission has a low differential mode group delay of 10 ps/km over the C-band, and the crosstalk between adjacent cores is lower than −55 dB/km at the pitch of 50 μm. A 25-Gbaud transmission over 50 km is simulated. Each core for analog signal transmission has a large effective area of 172 μm^2 to suppress the nonlinear effect due to the watt-scale input power. The proposed heterogeneously multicore fiber exhibits great potential to be applied in smart oilfields, meeting the demand for efficient and cost-effective oil production.

Keywords: space division multiplexing; multicore fiber; Industry 4.0; smart oilfield

Citation: Wang, X.; Sun, H.; Wang, H.; Yang, Z.; Liu, Y.; Huang, Z.; Zhang, L. Heterogeneously Integrated Multicore Fibers for Smart Oilfield Applications. *Appl. Sci.* **2023**, *13*, 1579. https://doi.org/10.3390/app13031579

Academic Editor: Nuno Alexandre Peixoto Silva

Received: 21 December 2022
Revised: 18 January 2023
Accepted: 24 January 2023
Published: 26 January 2023

1. Introduction

Industry 4.0 has been quickly advancing in the past few years, viewed as next-generation revolutionary technology for industrial manufacturing and production, which is featured by introduction and implementation of advanced technologies, such as artificial intelligence [1,2], cyber-physical systems, the internet of things, and cloud computing [3–6]. Large-scale machine-to-machine (M2M) communication and internet of things are integrated for increased automation, improved communication, self-monitoring, and production of smart machines [7], which could analyze and diagnose issues without the need for human involvement, enabling more efficient processes, safer working environments, and better quality and productivity.

In the context of Industry 4.0, oilfield industries call for efficient and cost-effective oil production. In the operation and management of the oilfield, a new concept called the "smart oilfield" is introduced [8], which is based on big data, the internet of things, and other technologies to automatically measure, analyze, and optimize oil production [9,10]. Recently, some companies with a strategic vision, such as Shell's NaKika Field [11], Xinjiang Oilfield [12], and Schlumberger's Haradh Smart Oilfield [13], have already started construction of smart oilfields.

Smart oilfields rely on a high-surveillance environment, via real-time data transmission, interactive business collaboration, and efficient decision-making processes [14–16]. Therefore, the large-scale information exchange among different parts is indispensable, which highly depends on the network layer (soft technologies [1–6]) and the physical layer (optical fiber links). On one hand, some wireless techniques are used in some oilfields for data communication. Because the oil will is dispersedly distributed, using remote wireless monitoring is convenient [17]; however, the communication quality is susceptible to atmospheric turbulence. In contrast, fiber communication could obtain high stability. In recent years, space division multiplexing, including mode-division multiplexing using multimode fibers or few-mode fibers and core multiplexing using multicore fibers, has been proposed and could expand the transmission capacity to overcome the single-mode fiber capacity limit. On the other hand, for fiber sensing in smart oilfields, various parameters could be monitored based on Rayleigh scattering, Raman scattering, and Brillouin scattering. To date, the data transmission and parameter monitoring are two separate parts, and it is difficult to build integrated systems, leading to the high cost of transmitters, receivers, and the process of deploying fibers. Therefore, there is an urgent need for special fiber links for both increased data transmission capacity and distributed sensing with high sensitivity [18]. Naturally, multicore optical fibers can be expected to play a critical role in smart oilfields in which parts of cores are responsible for data transmission, while other cores are used for sensing.

Multicore fibers, a significant component of space-division multiplexing, have been investigated for a few decades and are thought to be a strong candidate to meet the exponential growth of capacity demand [19,20]. On the other hand, in the area of optical sensing, there have been some reports about utilizing multicore fibers to measure multi variables, such as vibration, temperature, and strain [18,21–23]. For smart oilfields, compared with normal single-mode fibers, a unique characteristic of multicore fiber is that bending would generate local tangential strain in off-center cores, and the strain is angular-position-dependent. The cores in off-center positions lead to high bending sensitivity [24]. However, although multicore fiber design has been developed for a long time, it is still challenging to propose an integrated multicore fiber for smart oilfields applications due to the design complexity, which should take into account the performance of each core itself and the crosstalk between the adjacent cores. For digital transmission cores, the main goal is to realize low differential mode group delay (DMGD) because the computational complexity of digital signal processing (DSP) increases with that, and low DSP complexity is the basis of real-time transmission, which is limited by the processing capability of the widely used chips. As for analog transmission cores, the effective area (A_{eff}) should be enlarged to depress the nonlinear effect due to the watt-scale input power [25]. To maintain the accuracy and the sensitivity of sensing, the arrangement of different types of cores should be considered and the core pitches between neighboring cores should be optimized to keep the stability of the signal transmission. However, to date, there are few reports about multicore fiber design especially for smart oilfields, which could support real-time data transmission, device-to-device networking, and variables monitoring at the same time.

In this paper, we propose a heterogeneously integrated seven-core fiber for interconnection and awareness applications in smart oilfields, which could not only support digital and analog signal transmission but could also measure temperature and vibration through Raman effect and phase-sensitive optical time-domain reflectometry. The core arranged in the center is used for digital transmission to maintain communication stability, while others are arranged around the center for analog transmission and sensing, with equal pitches to increase sensing sensitivity. The core for digital signal transmission has a low differential mode group delay of 10 ps/km over the C-band and the crosstalk between adjacent cores is lower than −55 dB/km at the pitch of 50 μm. A 25-Gbaud transmission over 50 km is simulated. Each core for analog signal transmission has a large effective area of 172 μm^2 to suppress the nonlinear effect due to the watt-scale input power. In all, the proposed heterogeneously integrated multicore fiber exhibits great potential to be applied in smart oilfields, meeting the demand for efficient and cost-effective oil production.

2. A Case Study

To make the design process more targeted, the proposed heterogeneously integrated multicore fiber in this paper is based on the demand of a company named Beijing Perception Technology Company [26]. Figure 1 shows the schematic of the smart oilfields composed of oil production area, data distribution area, and office area.

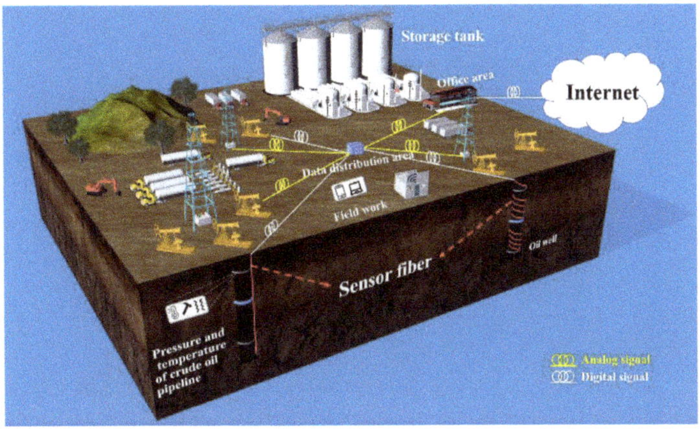

Figure 1. Schematic of the smart oilfields, which are consisted of production area, data distribution area, and office area.

The company is in urgent need of multifunctional fiber to support the construction of smart oilfields. In the production area, simultaneous distributed intrusion detection and temperature monitoring of pipelines are needed in order to achieve real-time alarm on excavation, theft, leakage, and other potential threats. This could be accomplished by implementing a Raman optical time-domain reflectometry (ROTDR) and phase-sensitive optical time-domain reflectometry (φ-OTDR) hybrid sensing system, where the ROTDR is used to monitor the temperature change and φ-OTDR is used for real-time vibration detection [18]. As for data interaction, the data throughput of the oilfield is approximately 1 Tbits per day, and it might keep increasing at a high speed in the near future. Therefore, we propose a three-mode six-polarization core for digital transmission and set the highest baud rate as 25 G with a quadrature phase-shift keying (QPSK) modulation format, which could obtain 0.3 Tbit/s and enough would be available for the situation, even for an instantaneous sharp increase of data. To realize the real-time transmission, a big challenge is the DSP complexity due to the limitation of the processing capacity of field programmable gate arrays (FPGAs). Although some companies, such as XILINX and Altera [27,28], have made a breakthrough in high-performance FPGAs, these products are not cost-effective and cannot be widely used at present. We set the filter taps as 32, which means a common FPGA could meet the demand. Moreover, the oilfields are mostly located in remote areas with weak signals. In this sense, it is essential to transmit analog signals to build a temporary network for connecting the devices and correspondence. Compared with long-haul transmission, fibers in smart oilfields might be used for short-distance transmission, so the maximum transmission distance is set as 50 km.

3. Heterogeneously Integrated Multicore Fibers

The proposed heterogeneously integrated multicore fiber could simultaneously support digital signal transmission, analog signal transmission, and temperature and vibration measurement. The schematic of the designed fiber is shown in Figure 2. It can be seen that different cores are responsible for different functions. To maintain the stability of digital signals and depress the bending loss, cores for digital transmission are set in the center of the cladding, while cores for sensing and analog transmission are arranged alternately in the outside circle. The numbers and types of different cores are determined by the demand of data rates and application scenarios. In addition, the pitch between adjacent cores is another key parameter needing to be optimized due to the trade-off between crosstalk and bending loss with a limited diameter of cladding. The cladding diameter should not be larger than 200 μm to guarantee mechanical reliability [29]. To maintain a crosstalk lower than −50 dB/km, the core pitch is set as 50 μm. To reduce micro bending loss, the outer cladding thickness is set as 40 μm. Therefore, the cladding diameter is 180 μm.

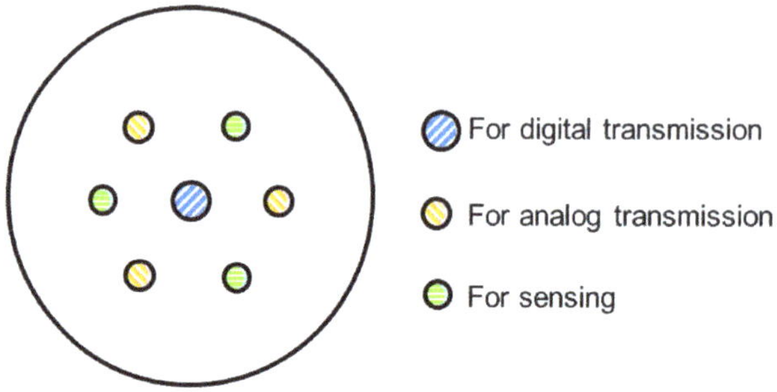

Figure 2. Schematic of the designed heterogeneous 7-core multicore fiber.

In general, there are three types of fibers: step-index (SI) fibers, grade-index (GI) fibers, and trench-assisted (TA) fibers. From the perspective of the manufacturing process, SI fibers and GI fibers are easier to fabricate than TA fibers due to their relatively simple structures. However, TA fibers could be designed by optimizing the width and depth of the trench and other parameters to obtain high performance, such as low DMGD or large A_{eff}. Figure 3 reveals the refractive index profiles of various types of fibers. r_{core}, w_1, and w_2 are the core radius, the distance between the core and the trench, and the width of the trench, respectively. α is the gradient parameter determined by the shape of the refractive index profile.

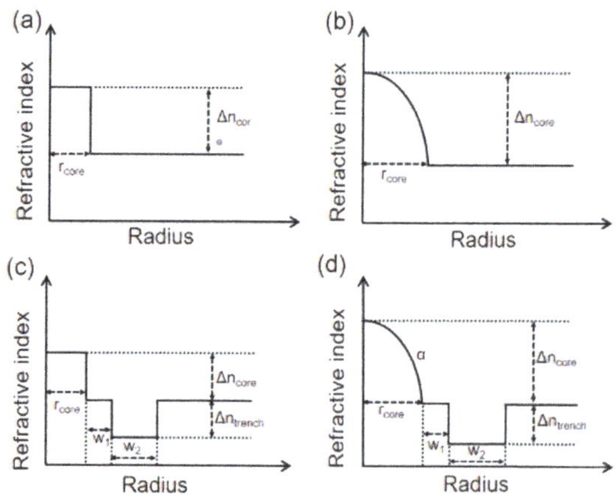

Figure 3. Refractive index profiles of various types of fibers. (**a**) step-index fiber, (**b**) graded-index fiber, (**c**) step-index trench-assisted fiber, (**d**) graded-index trench-assisted fiber.

3.1. Design of Digital Transmission Cores

For fiber design, the choice of structure is dependent on the applications. The key consideration for digital signal transmission is DMGD because the computational complexity of DSP increases with it, and the level of DSP complexity determines whether the real-time transmission could be realized. We sweep the parameters, including r_{core}, α, and

Δn_{core}, and the setting of ranges follows two principles. One is the ranges should be as large as possible and the other is the designed fiber should only support three guide modes. We calculate the DMGD over the C-band based on full vector finite element analysis. The group delay of any propagation mode in FMF is given by [30]

$$\tau_g = \frac{z(n_{eff} - \lambda\frac{dn_{eff}}{d\lambda})}{c} \tag{1}$$

where z represents transmission distance, c is the speed of light in vacuum, λ is the wavelength and n_{eff} is the effective refraction index. Therefore, the DMGD between LP_{mn} mode and LP_{01} mode is given by:

$$\mathrm{DMGD} = \frac{\left(n_{eff}^{LP_{mn}} - \lambda\frac{dn_{eff}^{LP_{mn}}}{d\lambda}\right)}{c} - \frac{\left(n_{eff}^{LP_{01}} - \lambda\frac{dn_{eff}^{LP_{01}}}{d\lambda}\right)}{c} \tag{2}$$

The DMGD results are shown in Figure 4.

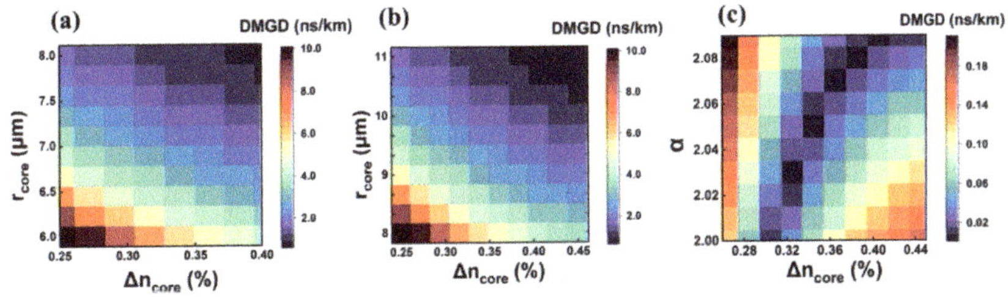

Figure 4. The DMGD variation of digital transmission cores with the setting parameters. (**a**) sweeping r_{core} and Δn_{core}, (**b**) sweeping r_{core} and Δn_{core}, and the α is 2, (**c**) sweeping α and Δn_{core}, and the r_{core} is 10.5 μm.

Within the sweeping ranges, the DMGD of GI fibers and TA-GI fibers have inflection points, while that of SI fibers have a gradual decrease with the increment of r_{core}, and Δn_{core}. We list the parameters of the three types of fibers at their lowest DMGD levels in Table 1, and the lowest DMGD of the TA-GI fiber is only 10 ps/km, which is 1/18 of the DMGD of SI fiber (183 ps/km) and one percent of that of the GI fiber (1000 ps/km). Although low DMGD could be obtained in the TA-GI fiber, the manufacturing process is more complex than that of SI fibers. In a word, it is not possible to use the GI fiber as the digital transmission core in this scene, and the choice of the SI fiber or the TA-GI fiber is dependent on the DSP complexity tolerance, which is positively correlated with DMGD. In addition, bending loss is considered and calculated by the finite element method. The bending radius is defined as the radius when the loss of the highest guided mode is 0.5 dB/turn, and the bending radius is calculated by the method in [31]. For the SI fiber, GI fiber, and TA-GI fiber, the bending radii are 70 mm, 30 mm and 20 mm. A smaller bending radius means better anti-bending performance. Considering the applied conditions in smart oilfields, the bending radius is m-scale or even larger, the anti-bending performance of the designed fiber is up to standard. A phrase.

Table 1. The parameters of various types of fibers applied for digital transmission.

Type	r_{core} (μm)	Δn_{core} (%)	Δn_{trench} (%)	α	w_1 (μm)	w_2 (μm)
SI	7.1	0.27	\	\	\	\
GI	10.5	0.40	\	2	\	\
TA-GI	10.5	0.345	0.34	2.05	2	9.3

3.2. Design of Analog Transmission Cores

As for the design of the analog transmission cores, A_{eff} should be enlarged to depress the nonlinear effect. For analog signal transmission, the input power is watt-scale (~10 dBm), which is 10 times larger than that of digital transmission (~0 dBm) [32]. The A_{eff} of standard G.654E fiber is approximately 120 μm^2. To obtain a larger A_{eff}, we only excite the fundamental mode in a three-mode fiber. In smart oilfields, the transmission distance is usually less than 50 km, and there are few intra-link splices. The multipath interference (MPI) in this scenario is approximately −40 dB, which sightly limits transmission performance [33]. As for the connection of transmitters and receivers, the MPI could be induced by the mismatch between mode fields. Low-loss mode field adapters produced by stepwise reverse tapering technique and thermally expanded core technique could further effectively reduce the MPI [34]. Referring to the design of cores for digital transmission, the results are shown in Figure 5 and Table 2. It shows that the A_{eff} variation of the TA-SI fiber is smaller than that of the other two types within the sweeping ranges, which means that TA-SI fibers could maintain the stability of large A_{eff} in a larger range. When the A_{eff} is set as 172 μm^2, the bending radii of the SI-fiber, GI-fiber and TA-SI fiber are 50 mm, 80 mm, and 40 mm, respectively.

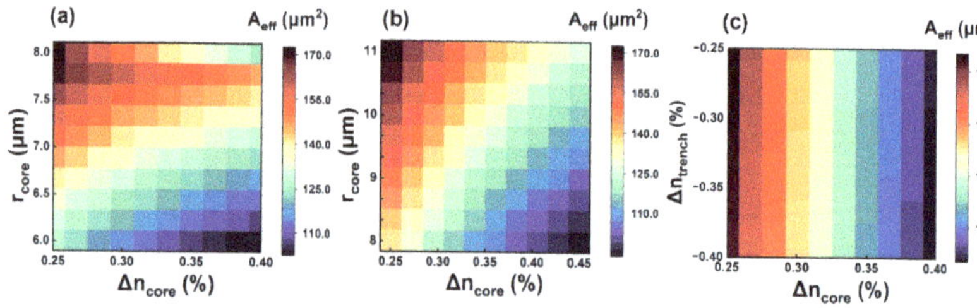

Figure 5. The A_{eff} variation of analog transmission cores with the setting parameters. (**a**) sweeping r_{core} and Δn_{core}, (**b**) sweeping r_{core} and Δn_{core}, and α is 2, (**c**) sweeping Δn_{trench} and Δn_{core}, and r_{core} is 8 μm.

Table 2. The parameters of various types of fibers applied for analog transmission.

Type	r_{core} (μm)	Δn_{core} (%)	Δn_{trench} (%)	α	w_1 (μm)	w_2 (μm)
SI	8.0	0.27	\	\	\	\
GI	11.0	0.25	\	2	\	\
TA-SI	8.0	0.25	0.25	\	4	5

3.3. The Analysis of the Crosstalk

To maintain the signal quality, the crosstalk between adjacent cores, which is calculated by the finite element method to evaluate the performance of the multicore fiber, should be lower than −55 dB/km [35]. As we mentioned above, the cores for digital transmission and analog transmission support three modes while the core for analog transmission only excites the fundamental mode. For sensing cores, we use the standard G.652 fiber. We know that the distance between the edges of cores influences crosstalk greatly and it is a negative correlation with the crosstalk. To simplify the calculation, we suppose that the crosstalk between two three-mode cores is the crosstalk of the whole multicore fiber because the core radius of G.652 fiber (3 μm) is much smaller than the other two three-mode fibers, which leads to lower crosstalk when the core pitch is fixed. The parameters of cores for digital transmission and analog transmission are listed in Tables 1 and 2. The bending radius is assumed as one meter according to the diameter of the pipeline. The crosstalk

(XT) is obtained by using coupled-power theory [36]. The mode-coupling coefficient could be shown as:

$$\kappa_{pq} = \frac{\omega\varepsilon_0 \iint_{-\infty}^{+\infty} \left(N^2 - N_q^2\right) E_p^* \cdot E_q dxdy}{\iint_{-\infty}^{+\infty} u_z \cdot \left(E_p^* \times H_p + E_p \times H_p^*\right) dxdy} \tag{3}$$

where ω is an angular frequency of the sinusoidally varying electromagnetic fields, ε_0 is the permittivity of the medium, and u_z means the outward-directed unit vector. E_p and E_q represent the electric field distribution of core inside the range of core p, and the electric field distribution of core inside the range of core q, respectively. The crosstalk (XT) between the neighboring cores with length L is estimated as:

$$XT = \tanh\left(\overline{h}_{pq} \mathrm{L}\right) \tag{4}$$

$$h_{pq}(z) = \frac{2\mathrm{K}_{pq}^2 d}{1 + \left(\Delta\beta'_{pq} d\right)^2} \tag{5}$$

where K_{pq} is the average value of κ_{pq} and κ_{qp}. $\Delta\beta'_{pq}$ is the difference of equivalent propagation constant between two cores. d means the correlation length [37]. Here, d is assumed to be 0.05 m.

Figure 6 shows the crosstalk as a function of the pitch between neighboring cores. It could be seen that the crosstalk reduces as the core pitch increases. To compare all the combinations, we set the core pitch as 50 μm. For each row, the third combination performs better than the other two, which means that we could obtain a better performance if the SI-TA fiber is applied in the analog-transmission core. For example, in the first row, the core for digital transmission is the SI fiber and that for analog transmission is the SI fiber, the GI fiber, and the SI-TA fiber, respectively. The maximum crosstalk of the third combination is −63.4 dB/km, while the other two combinations' maximum crosstalk values are approximately −55.4 dB/km and −34.1 dB/km. Furthermore, comparing the results by column is another dimension. In the third column, when the SI-TA structure is used for analog transmission, there is a slight difference among the three combinations, but the maximum crosstalk values of them are all lower than −55 dB/km, which satisfies the communication requirements of crosstalk. Above all, we suppose that the core for analog transmission should use the SI-TA fiber and that for digital transmission could be chosen from the SI fiber or the GI-TA fiber, depending on the processing capability of chips.

3.4. Transmission System Demonstration

To further investigate the performance of the multicore fiber in actual applicable conditions, we model the signal transmission process in VPItransmissionMaker Optical Systems. A 100 Gb/s quadrature phase-shift keying (QPSK) signal is generated at 1550 nm, and the input powers of the digital signal and the analog signal are 0 dBm and 13 dBm, respectively. The core pitch is set as 50 μm. The number of splices is 2 and the splice loss is 0.4 dB. The loss of LP_{01} and LP_{11} is 0.20 dB/km and 0.22 dB/km, respectively. The mode crosstalk is −60 dB/km. In the simulation, the fan-in and fan-out devices and (de)multiplexers are somewhat ideal devices. The crosstalk of these two types of devices is −50 dB, and the insertion loss is 1 dB and 3 dB, respectively. Taking into account the processing capacity of widely used chips, the number of taps is set as 32 to maintain the possibility of real-time transmission. The DSP algorithm is referred to in [38], which could adjust the distribution of filter taps and enhance the utilization to efficiently recover signal. The SI-TA fiber is set as the core for analog transmission. The Q^2-factor performances versus different transmission distances with the digital signal transmission cores of the SI fiber, the GI-TA fiber, and the fabricated GI-TA fiber are shown in Figure 7. Because the designed fiber would be applied in smart oilfields, the maximum transmission distance is set as 50 km. The Q^2-factor threshold of forward error correction (FEC) is 6.5 dB, which is recommended by IEEE Standard 802.3 [39]. We could see that when the core for digital transmission is the optimized low-DMGD three-mode TA-GI fiber, the Q^2-factors of two guided modes decrease slightly as the length increases and those are over 15 dB within 50 km. Compared with this, if the SI fiber is set as the core for digital transmission, the Q^2-factor of each guided mode decreases more sharply from approximately 15 dB to 6 dB. The effective transmission distance is approximately 43 km, which is smaller than 50 km. It shows that the performance of the TA-GI fiber is more available and could be applied in this environment. As for analog transmission performance, due to the 13 dBm input power of analog signal transmission, it is almost 20 times larger than the input power of digital signal transmission. For such an asymmetry channel, the performance of the analog single transmission is almost the same as that in a single-core transmission system.

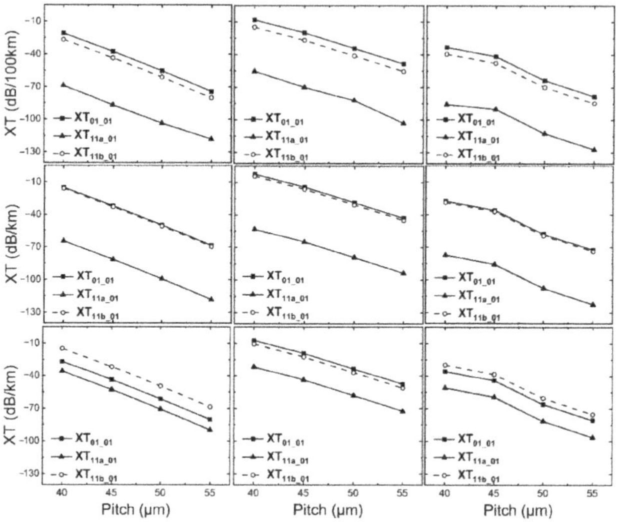

Figure 6. The relationship between the crosstalk of guided modes and the core pitches. In each row, the cores for digital transmission are the SI fiber, the GI fiber, and the GI-TA fiber. In each column, the cores for analog transmission are the SI fiber, the GI fiber, and the SI-TA fiber, respectively.

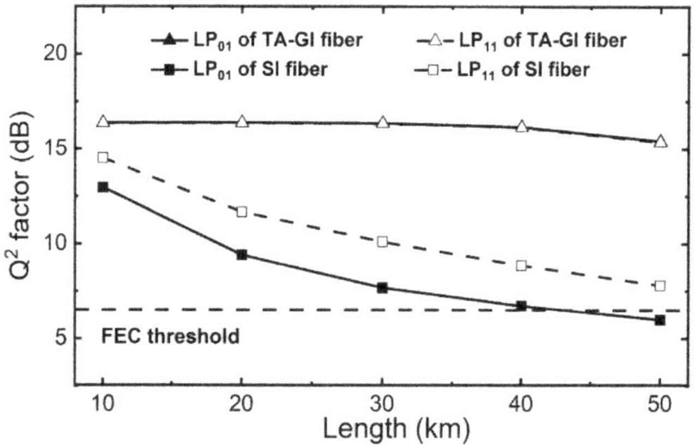

Figure 7. Q^2 factor performances versus transmission distances with different types of fibers for digital transmission.

4. Conclusions

In this paper, we propose a heterogeneously integrated seven-core fiber for smart oilfields, which could not only support digital and analog signal transmission but could also measure temperature and vibration through Raman effect and phase-sensitive optical time-domain reflectometry. The core arranged in the center is used for digital transmission to maintain communication stability, while others are arranged around the center for analog transmission and sensing, with equal pitches to increase sensing sensitivity. The core for digital signal transmission has a low differential mode group delay of 10 ps/km over the C-band. Each core for analog signal transmission has a large effective area of 172 μm^2 to suppress the nonlinear effect. The crosstalk between adjacent cores is lower than −55 dB/km, which could support 25-Gbaud real-time transmission over 50 km, based on widely used chips for data processing. We believe that the proposed multicore fiber could support and enhance the construction of smart oilfields.

Author Contributions: X.W. planned and wrote the paper. H.S. contributed to the plotting of Figure 3. H.W. contributed to the writing of the fourth part. Y.L. took the overall responsibility for managing the manuscript. Z.Y. and Z.H. revised the whole article and gave advice. L.Z. supervised the work and provided technical leadership. All co-authors contributed to the final version with suggestions and critical comments. All authors have read and agreed to the published version of the manuscript.

Funding: This research was funded by the National Key R&D Program of China under grant 2019YFB2203902, the National Natural Science Foundation of China under grant 62105241, and the Postgraduate Research Program of Tianjin under grant 2021YJSB129.

Institutional Review Board Statement: Not applicable.

Informed Consent Statement: Not applicable.

Data Availability Statement: Not applicable.

Acknowledgments: The authors would like to thank the anonymous reviewers for their valuable comments and suggestions to improve this manuscript.

Conflicts of Interest: The authors declare no conflict of interest.

References

1. Nilsson, N.J. *Principles of Artificial Intelligence*; Morgan Kaufmann: San Mateo, CA, USA, 1982.
2. Jordan, M.I.; Mitchell, T.M. Machine learning: Trends, perspectives, and prospects. *Science* **2015**, *349*, 255–260. [CrossRef] [PubMed]
3. Lee, E.A. Cyber physical systems: Design challenges. In Proceedings of the IEEE Symposium on Object/Component/Service-Oriented Real-Time Distributed Computing, Orlando, FL, USA, 5–7 May 2008; pp. 363–369.
4. Armbrust, M.; Fox, A.; Griffith, R.; Joseph, A.D.; Katz, R.; Konwinski, A.; Lee, G.; Patterson, D.; Rabkin, A.; Stoica, I.; et al. A view of cloud computing. *Commun. ACM* **2010**, *53*, 50–58. [CrossRef]
5. Atzori, L.; Iera, A.; Morabito, G. The Internet of Things: A survey. *Comput. Netw.* **2010**, *54*, 2787–2805. [CrossRef]
6. Wang, S.; Wan, J.; Li, D.; Zhang, C. Implementing smart factory of industrie 4.0: An outlook. *Int. J. Distrib. Sens. Netw.* **2016**, *12*, 3159805. [CrossRef]
7. Weyrich, M.; Schmidt, J.-P.; Ebert, C. Machine-to-machine communication. *IEEE Softw.* **2014**, *31*, 19–23. [CrossRef]
8. Du, D.; Zhang, X.; Guo, Q.; Zhang, B.; Zhang, G. Smart oilfield technology. In Proceedings of the International Field Exploration and Development Conference, Xi'an, China, 18–20 September 2018; pp. 685–694.
9. Ahmad, I.; Pothuganti, K. Smart field monitoring using ToxTrac: A cyber-physical system approach in agriculture. In Proceedings of the International Conference on Smart Electronics and Communication, Trichy, India, 10–12 September 2020; pp. 723–727.
10. Eifert, T.; Eisen, K.; Maiwald, M.; Herwig, C. Current and future requirements to industrial analytical infrastructure—Part 2: Smart sensors. *Anal. Bioanal. Chem.* **2020**, *412*, 2037–2045. [CrossRef]
11. Available online: https://www.shell.com (accessed on 21 November 2022).
12. Available online: https://www.quantum.com/zh-cn/resources/customer-success/xinjiang-oilfield-company/ (accessed on 21 November 2022).
13. Available online: https://www.slb.com (accessed on 21 November 2022).
14. Temizel, C.; Canbaz, C.H.; Palabiyik, Y.; Putra, D.; Asena, A.; Ranjith, R.; Jongkittinarukorn, K. A comprehensive review of smart/intelligent oilfield technologies and applications in the oil and gas industry. In Proceedings of the SPE Middle East Oil and Gas Show and Conference, Manama, Bahrain, 18–21 March 2019.

15. Al-Subaiei, D.; Al-Hamer, M.; Al-Zaidan, A.; Nawaz, M.S. Smart production surveillance: Production monitoring and optimization using integrated digital oil field. In Proceedings of the SPE Kuwait Oil and Gas Show and Conference, Mishref, Kuwait, 13–16 October 2019.
16. Carvajal, G.; Maucec, M.; Cullick, S. Components of artificial intelligence and data analytics. In *Intelligent Digital Oil and Gas Fields*, 1st ed.; Gulf Professional Publishing: Boston, MA, USA, 2018; pp. 101–148.
17. Hussain, R.F.; Salehi, M.A.; Kovalenko, A.; Feng, Y.; Semiari, O. Federated Edge Computing for Disaster Management in Remote Smart Oil Fields. In Proceedings of the 2019 IEEE 21st International Conference on High Performance Computing and Communications; IEEE 17th International Conference on Smart City; IEEE 5th International Conference on Data Science and Systems (HPCC/SmartCity/DSS), Zhangjiajie, China, 10–12 August 2019; pp. 929–936.
18. Lu, P.; Lalam, N.; Badar, M.; Liu, B.; Chorpening, B.T.; Buric, M.P.; Ohodnicki, P.R. Distributed optical fiber sensing: Review and perspective. *Appl. Phys. Rev.* **2019**, *6*, 041302. [CrossRef]
19. Li, G.; Bai, N.; Zhao, N.; Xia, C. Space-division multiplexing: The next frontier in optical communication. *Adv. Opt. Photon.* **2014**, *6*, 413–487. [CrossRef]
20. Saitoh, K.; Matsuo, S. Multicore Fiber Technology. *J. Light. Technol.* **2016**, *34*, 55–66. [CrossRef]
21. Dang, Y.; Zhao, Z.; Wang, X.; Liao, R.; Lu, C. Simultaneous distributed vibration and temperature sensing using multicore fiber. *IEEE Access* **2019**, *7*, 151818–151826. [CrossRef]
22. Zhao, Z.; Dang, Y.; Tang, M.; Li, B.; Gan, L.; Fu, S.; Wei, H.; Tong, W.; Shum, P.; Liu, D. Spatial-division multiplexed Brillouin distributed sensing based on a heterogeneous multicore fiber. *Opt. Lett.* **2017**, *42*, 171–174. [CrossRef] [PubMed]
23. Yan, B.; Li, J.; Zhang, M.; Xu, Y.; Yu, T.; Zhang, J.; Qiao, L.; Wang, T. Temperature accuracy and resolution improvement for a Raman distributed fiber-optics sensor by using the Rayleigh noise suppression method. *Appl. Opt.* **2020**, *59*, 22–27. [CrossRef] [PubMed]
24. Zhao, Z.Y.; Tang, M.; Lu, C. Distributed multicore fiber sensors. *Opto-Electron. Adv* **2020**, *3*, 190024. [CrossRef]
25. Zhang, L.; Agarwal, A.M.; Kimerling, L.C.; Michel, J. Nonlinear Group IV photonics based on silicon and germanium: From near-infrared to mid-infrared. *Nanophotonics* **2014**, *3*, 247–268. [CrossRef]
26. Available online: http://www.zjofs.com (accessed on 21 November 2022).
27. Available online: https://www.xilinx.com (accessed on 21 November 2022).
28. Available online: https://www.intel.com/content/www/us/en/products/programmable.html (accessed on 21 November 2022).
29. Matsui, T.; Nakajima, K.; Fukai, C. Applicability of Photonic Crystal Fiber with Uniform Air-Hole Structure to High-Speed and Wide-Band Transmission Over Conventional Telecommunication Bands. *J. Light. Technol.* **2009**, *27*, 5410–5416. [CrossRef]
30. Zhang, H.; Zhao, J.; Yang, Z.Q.; Peng, G.J.; Di, Z.X. Low-DMGD, Large-Effective-Area and Low-Bending-Loss 12-LP-Mode Fiber for Mode-Division-Multiplexing. *IEEE Photonics J.* **2019**, *11*, 1–8. [CrossRef]
31. Tu, J.; Saitoh, K.; Koshiba, M.; Takenaga, K.; Matsuo, S. Design and analysis of large-effective-area heterogeneous trench-assisted multi-core fiber. *Opt. Express* **2012**, *20*, 15157–15170. [CrossRef]
32. Wen, H.; Zheng, H.; Mo, Q.; Velázquez-Benítez, A.M.; Xia, C.; Huang, B.; Liu, H.; Yu, H.; Sillard, P.; Lopez, J.E.A.; et al. Few-mode fibre-optic microwave photonic links. *Light Sci. Appl.* **2017**, *6*, e17021. [CrossRef]
33. Downie, J.D.; Mlejnek, M.; Roudas, I.; Wood, W.A.; Zakharian, A.; Hurley, J.E.; Mishra, S.; Yaman, F.; Zhang, S.; Ip, E.; et al. Quasi-Single-Mode Fiber Transmission for Optical Communications. *IEEE J. Sel. Top. Quantum Electron.* **2017**, *23*, 31–42. [CrossRef]
34. Yang, L.; Yang, Z.; Xu, T.; Hou, L.; Zhou, R.; Gan, L.; Cao, S.; Xiao, X.; Zhang, L. Low-loss Mode Field Adapter Using Reverse Tapering for Fundamental Mode Transmission over MMFs. In Proceedings of the Optical Fiber Communication Conference (OFC) 2022, San Diego, CA, USA, 6 March 2022; p. M4E.7.
35. Xie, Y.; Pei, L.; Zheng, J.; Zhao, Q.; Ning, T.; Li, J. Low-DMD and low-crosstalk few-mode multi-core fiber with air-trench/holes assisted graded-index profile. *Opt. Commun.* **2020**, *474*, 126155. [CrossRef]
36. Marcuse, D. Derivation of Coupled Power Equations. *Bell Syst. Tech. J.* **1972**, *51*, 229–237. [CrossRef]
37. Tu, J.; Saitoh, K.; Koshiba, M.; Takenaga, K.; Matsuo, S. Optimized Design Method for Bend-Insensitive Heterogeneous Trench-Assisted Multi-Core Fiber with Ultra-Low Crosstalk and High Core Density. *J. Light. Technol.* **2013**, *31*, 2590–2598. [CrossRef]
38. Arık, S.Ö.; Askarov, D.; Kahn, J.M. Adaptive Frequency-Domain Equalization in Mode-Division Multiplexing Systems. *J. Light. Technol.* **2014**, *32*, 1841–1852. [CrossRef]
39. IEEE. *IEEE Std 802.3bt-2018 (Amendment to IEEE Std 802.3-2018 as amended by IEEE Std 802.3cb-2018)*; IEEE Standard for Ethernet Amendment 2: Physical Layer and Management Parameters for Power over Ethernet over 4 Pairs. IEEE: Piscataway Township, NJ, USA, 2019; pp. 1–291. [CrossRef]

applied sciences

MDPI

Article

Reliability Study for Communication System: A Case Study of an Underground Mine

Batzorig Bazargur [1,*], Otgonbayar Bataa [1] and Uuganbayar Budjav [2]

1 Department of Communications Engineering Technology, School of Information and Telecommunication Technology, Mongolian University of Science and Technology, Ulaanbaatar 14191, Mongolia
2 Department of Mathematics, School of Applied Sciences, Mongolian University of Science and Technology, Ulaanbaatar 14191, Mongolia
* Correspondence: batzorigbazargur@must.edu.mn

Abstract: After summarizing and evaluating works on system reliability, various models and results for predicting and evaluating system reliability have been introduced. However, we have not seen a study conducted to assess the reliability of communication systems in an underground mine. Underground mining operations are normally dependent on communication system reliability. The main purpose of this work is to study the failure of a theoretical underground mine communication system, propose a method to improve its reliability, and predict the results of the suggested method using system dynamic modeling. This study contributes to improving the reliability of communication systems in underground mines. In the case of a single nonredundant system, three options were implemented: doubling corrective maintenance, increasing preventative maintenance by 50%, and combining both measures. These three options were modeled by combining Markov modelling with system dynamic modelling methodology and were confirmed by experiments and simulation results. This combination of modelling constitutes the novelty of this study. In this paper, actual system component failure data was used for simulation for the single nonredundant system, after doubling corrective maintenance, increasing preventative maintenance by 50%, and combining in the case of implementing simultaneously both maintenance changes, but not for developing single and dual standby system models. Therefore, these models should be suitable for practical use, as they are based on actual working systems. Modelling confirmed that placing a communication system in each shaft of the theoretical underground mine increases the reliability of the communication system. The degree of availability of the communication system with single standby device shows the result of 62.38% while the communication system with two standby devices and three parallel communication systems' availability rating shows a result of 85.18%.

Keywords: underground mine; communication system; reliability; availability; system dynamic; results

Citation: Bazargur, B.; Bataa, O.; Budjav, U. Reliability Study for Communication System: A Case Study of an Underground Mine. *Appl. Sci.* **2023**, *13*, 821. https://doi.org/10.3390/app13020821

Academic Editor: Christos Bouras

Received: 9 December 2022
Revised: 27 December 2022
Accepted: 30 December 2022
Published: 6 January 2023

1. Introduction and Literature Review

Since an underground mine is a very dangerous work environment, the safety of workers can only be ensured with the help of reliable radio communication and information technology systems. It is important to develop and calculate the reliable operation of these systems on a scientific basis [1]. This study is an important part of a doctoral dissertation on the reliability of underground communication systems, which analyzes the exponential distribution of seismic data using statistics on the reliability of underground communication systems to determine the probability of an earthquake in an underground mine. An earthquake is one of the largest natural risk factors for a mine. The probability of an earthquake was determined using seismic data from 1957 to 2017 in Khanbogd soum, Umnugovi aimag, kept by the Institute of Astronomy and Geophysics of the Mongolian Academy of Sciences [2].

Another important part of our doctoral study is improving the reliability of communication systems and the modeling of standby communication systems [1]. We studied the complex structure of underground mines and tunnel distribution [3]. The results show that the isolation of RF repeaters and the distribution of radiation cables within the tunnel have been studied in accordance with the complex structure of an underground mine. In addition, in the event of an accident at an underground mine, radio communication services were relocated from the area where radio communication services were disrupted to a place where radio communication services were normal, and miners were quickly evacuated. We developed an emergency analysis using graph theory and Dijkstra's algorithm developed in the MATLAB program and developed an algorithm [4].

The communication system for underground mining is one of the most important parts of a mine. The main objective of this study is to evaluate the possibility of improving the reliability of underground communication systems. Relationship between key elements was established using reliability causality diagram based on maintenance by system dynamics method and even there are advantages as we have recommended preventive and corrective maintenance to improve system reliability in the study, but the study was not addressed to standby systems [5].

In addition, the aging and maintenance times of single-source and dual parallel systems are considered in the Weibull distribution law. The system's mathematical model was modeled using the Markov model, but the model was not studied for dual parallel and triple parallel systems, only for Weibull distribution [6]. A current and future reliability analysis of a smart grid consisting of an electric power distribution system and an integrated communications network based on Monte Carlo simulation was developed and tested in this study [7], whereas in another study [8], a reliability model and analysis were carried out using Poisson distribution to improve the reliability of the distributed network by reducing the energy consumption of the data center.

However, in studies [9,10], the researchers reviewed these studies about the reliability of systems due to the challenges of the 4th industrial revolution and the requirements set. In other words, it is concluded that new solutions using artificial intelligence, machine learning and big data mean the goal of sustainable production to achieve a balance between environmental, social, and economic dimensions. In addition, it is emphasized that machine learning is the most important tool for assessing the reliability of a system, and in doing so, it is emphasized that using a combination of different methods will help to better analyze the reliability.

To increase the reliability of underground communication systems, this study identified new models for increasing and testing corrective and preventative maintenance, as well as developing an availability rating for single and dual standby systems. The entrances and exits to underground mines are called shafts, and the placement of only one communication device in one shaft limits reliable communication in the event of various incidents in a multi-shaft underground mine.

The following were included in this study for improvement the reliable performance of underground mine communication systems:

- Checking the availability of a three-shaft underground mine communication system;
- Options for improving maintenance;
- Suggesting the overall approach to keep the resource system ready and calculating the results using system dynamic modeling.

Markov analysis is used to calculate the reliability of production systems that can be maintained, and the quality of maintenance can be improved, which we have used to calculate. In this study, the communication system of the Oyu Tolgoi underground mine was taken as an example, and the reliability of the communication system of the mine was analyzed separately for each part of the communication system (in the example of the site link) and also for the overall communication system.

Repairable and non-repairable parts on Komatsu dump trucks for open pit copper mining were analyzed and it was revealed that the wheels are as the most important parts of the dump trucks. According to the analysis, results were introduced, showing results on using reliability-centered maintenance, such as reducing the downtime of heavy equipment, increasing productivity, and reducing operating costs, pre-saving important necessary parts and optimally planning the interval between preventive maintenance [11].

Therefore, the installation of a communication device in each shaft is a key factor in influencing the system's reliability, since the studies conducted on communication systems, including underground mine communication systems, are very rare, and we have considered the studies made on the examples of other systems other than the communication system.

To prolong the life of power systems and reduce the cost of maintenance, data were collected using environmental monitoring technology. Based on that information, various causes of damage to power system components are analyzed. Although a repair strategy is proposed in the paper by Roengchai based on the results of the system's dynamic design to prevent major failures, shortcomings were noted, such as the model being too general and not modeled for each power system detail [12]. The authors introduced how the system dynamics were used in the study methodology by simulating the reliability of the actual system [13]. There are not many works that have analyzed the reliability of a system using system dynamics modeling. A mixed model of Markov system dynamics is used as an example of resource system maintenance. Based on results assessment of other studies in this field, it was found that most of the studies considered exponential distribution a single-standby system. The methodological variables used in these works contain many differential equations. These differential equations are difficult to calculate. Therefore, this study uses system dynamics modeling to analyze the reliability of standby systems [14]. In this paper [15], Markovian processes are modeled using exponential distributions and calculated analytically. The model in this study was described as from $P_t[1,0]$ to $P_t[0,1]$, from $P_t[1,0]$ to $P_t[1,1]$, and from $P_t[0,1]$ to $P_t[1,1]$. The model was taken without transition between processes, and it was different from our case. However, principally it is not very different from our model.

Then, in their paper, [16] proposed a hybrid approach called the Markov system dynamics (MSD) approach, which combines the Markov model with system dynamics simulation for time-dependent availability analysis. A difference from our model was the addition of a reduced state to the model described in this work. In the next part of this paper, the direction of the study is determined, the methodology is developed, and the results are discussed. In other words, the developed model is explained in more detail.

2. Materials and Methods

Availability is the probability that the system will be operational at a given time. It combines aspects of reliability, maintainability, and maintenance support, and implies that the system is either in active operation or able to operate if required. This study determines the probability of the reliable operational availability of an underground communication system and considers its possibility for improvement through system maintenance and the use of a system with high resource [6]. In our study, we determine the availability probability of reliable operation of an underground mine communication system and consider the possibility of its improvement in terms of system maintenance and its standby system. To improve the reliability of the system, a cost-effective preventive and corrective maintenance plan was proposed [5], and in our paper, we discuss how the reliability of the system can be improved by using a standby system in the example of an underground mine communication system and present the study results.

Maintenance improvement options:

- Doubling the amount of money, manpower, and time spent on maintenance (during corrective maintenance);
- Carrying out regular preventative maintenance to reach 50 percent of total performance (during the scheduled repair);
- Using a mixed method in which abovementioned 2 options are taken simultaneously. The results are considered for one communication system device and for the entire communication system.

Another way to increase the reliability of a communication system is to have a standby system available. Weibull, Poisson, exponential distribution laws are used to show the degree of availability or reliability of a single and dual standby (connected in parallel) system. To show how the system availability rate improves by taking the measures of the first two options for improving the system availability rate, system dynamics modeling is used to simulate the probability of availability in the Vensim program. The result confirms that when any system undergoes high-quality maintenance regularly and has a resource system, then its availability rate improves.

The general structure of the underground mine communication system with three shafts is shown in Figure 1 below. An underground mine can have one or more shafts, and they are characterized by varying distances from each other. A shaft is an excavated engineering structure for the access of people and materials from the surface to the underground mine, and for the return of people and ore bodies from the underground mine to the surface. In other words, this shaft can be compared to a lift in a high-rise building. In terms of purpose, shafts are divided into those for access only by people, and only for ventilation systems and for transportation. For our study, we developed a reliability model for a two-reserve or three-parallel communication system for a three-shaft underground mine. Core systems, base stations and optical cable devices are located near the shaft and on the surface. The optical cable from the optical cable device, called the site link, is connected to the optical repeater located in the underground mine through the shaft tunnel, and the radiation cable from the optical repeater is attached to the side wall along the underground mining tunnel. In underground mine tunnels where radiation cables are laid, users or miners communicate with each other and with control rooms and people of other teams on the surface using radios. Factors that affect the probability of reliable operation of the communication system shown in Figure 1 below, or the causes of faults, using the information of the communication system of Oyu Tolgoi mine, if we determine each component (see Supplementary Materials):

Fiber optic cable:

- Failure of electricity;
- Fiber optic cable breakage;
- Damage to the power supply battery.

Core system:

- Database jams and crashes due to an outdated core system;
- Hard disk damage due to outdated core system hardware.

Base station:

- The device shuts down due to power failure;
- The base station overheats, jams, and shuts down due to excessive room temperature;
- Damaging of internal parts;
- Turning off the base station's connection to the device;
- Disruption of the site link;
- Damage to the antenna cable.

Transmission:

- Switching off the transmission due to a power failure;
- Overheating, jamming, or shutting down the connection due to excessive room temperature;
- Loss of transmission device configuration;
- Ignition of the transmission device power supply;
- Deterioration of connection quality due to contamination of the fiber optic cable connection module;
- Disconnection during fiber optic cable arrangement;
- Cutting the fiber optic cable during unpermitted excavation work.

Repeater device:

- Burning out the device due to a power supply voltage difference;
- Fiber optic cable breakage;
- Radiation cable breakage;
- Damage to internal parts.

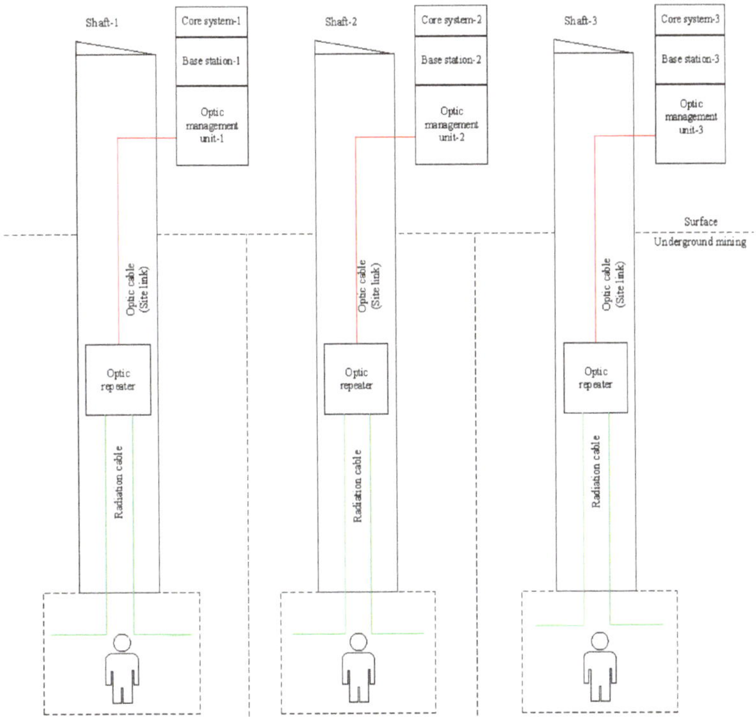

Figure 1. The general structure of an underground mine communication system with three shafts.

The causes of failure and the number of failures per communication system device at the Oyu Tolgoi mine in 2014–2018 were included in the statistical data, as shown in Table 1 (see Supplementary Materials).

Table 1. Number of failures of communication system components.

Duration of Failure:	Device Name, Number of Failures:	Duration of Failure by Hour
November 2014–December 2018	Site link (transmission line)—107 Base stations—151 Repeater—19 OMU (Optic management unit)—15 Core system—13	Site link (transmission line)—18 Base station—21 Repeater—4 OMU—3 Core system—24.5

According to these statistics, base stations failed 151 times, connections failed 107 times, site link failed 19 times, the OMU failed 15 times, and failure of the core system occurred 13 times. However, according to the total hours of failure, core system failure was 24.5 h, failure of the base stations was 21 h, and connection line failure was 18 h. Considering the reasons for these failures, the main reason for the longest failure of the core system was that the system was too outdated in terms of hardware and software regarding system resources. According to the manufacturer's design, the core system of the Motorola Dimetra system is manufactured with a backup core system. Therefore, the standby core system was operating during the longest core system failure, and the communication system service was unaffected. In terms of the base stations, the total duration of the Oyu Tolgoi mine communication system's six base station failures was 21 h, and the base stations were mostly out of order due to power failure. Base stations have a standby device in their internal structure, just like the core system, but they can be interrupted by a power failure, no matter how many standby devices they have. From the general structure of the underground mine communication system shown in Figure 1, the reliable operation of the system is highly dependent on operational site link. The reason is that the core system of the underground mine communication system, the base station and the optical cable device, are located on the surface, and they are connected to the repeater devices located underground only using the optical cable connection line; therefore, the site link is a very important part of the underground mine communication system. It is important to install a communication system base station (whether it is located on the surface or underground), a site link, an OMU connected to the site link, and a repeater for each underground mining shaft. This study concludes that the installation of these system components can improve an underground mine's communication system reliability.

The best way to improve the reliability of an underground mine is to install additional communication equipment at each underground mining shaft. The study is based on the design of the core system of the underground mining communication system Motorola Dimetra, the base station, and the repeaters located inside the underground mine. In this paper, a model was developed based on the core system of the Motorola Dimetra underground mine communication system, the base station and the repeater devices located inside the underground mine and using the reliability availability information of each device. Moreover, the effect of corrective and preventive maintenance on each device of the communication system and on the entire system; the rating of standby systems availability for underground mine communication systems with one shaft, two shafts and three shafts were tested by simulation; and the simulation results were compared.

Reliability or probability of communication system availability: The probability of a system's working state is determined by the probability of its availability. According to the definition of a repaired system and Markov analysis, we consider the system as being in one of two states: working or broken. According to Markov analysis, the state of the system is shown in Figure 2. In Figure 2, the system switches from the working state (1) to the broken state (0) and back to the working state due to maintenance. This was considered by utilizing the Markov process. Here, λ_{10} is the failure state and μ_{01} is the repair state.

The probability $P_j(t)$ for each state (j) in the given time (t) is expressed by the following differential Equations as (1) and (2):

$$\frac{dP_0(t)}{dt} = -\mu_{01}P_0(t) + \lambda_{10}P_1(t) \tag{1}$$

$$\frac{dP_1(t)}{dt} = -\lambda_{10}P_1(t) + \mu_{01}P_0(t) \tag{2}$$

where $P_1(t)$ is availability and $P_1(t) = 1 - P_0(t)$. Moreover, the initial condition of the system is $P_1(t) = 1,\ P_0(t) = 0$. The causal loop diagram of this process is shown in Figure 2.

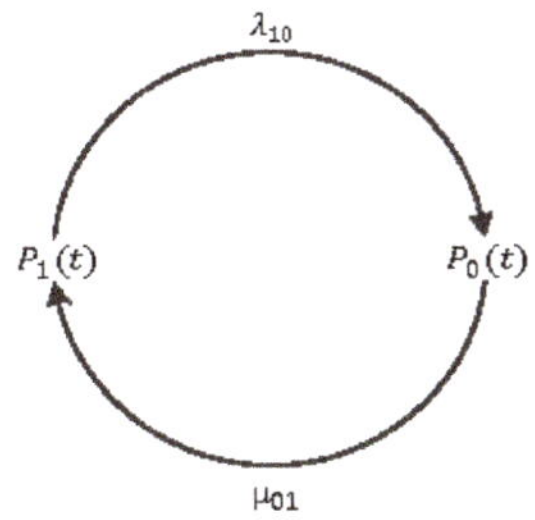

Figure 2. System transmission state.

As shown in Figure 3, there are two different causal loops, which are normal and fault operation on system availability, and each has negative and positive effects. The failure and normal operation of system availability have both negative and positive effects and that there are two types of causal loops. The loop between normal operation and readiness is positive, while the loop between failure and availability is negative. As the failure and repair functions increase, the failure and normal operation states increase, respectively [5]. The probability of the reliable operation of each underground communication system device is modeled for the first core system device with a single standby core system device and two parallel core system devices.

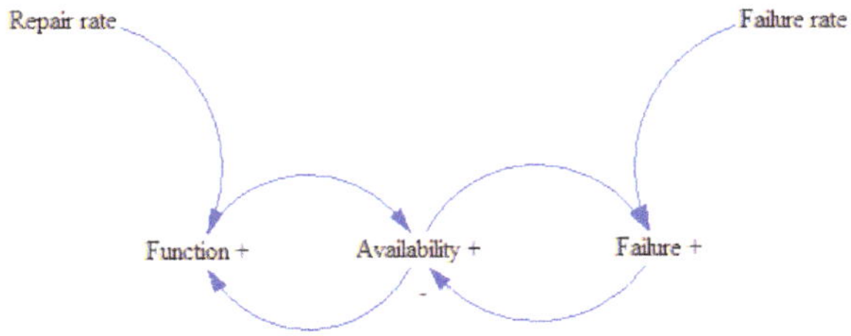

Figure 3. Casual loop diagram of availability.

To determine the probability of system availability, this study was developed through the following methodology steps (Table 2 and Figure 4).

Table 2. Research methodology.

1.	Identifying the problem	How to ensure reliable operation of an underground mine communication system?
2.	Choosing the research objective:	To improve the reliability of communication systems in an underground mine.
3.	Literature review and defining variables:	The values of the variables were determined based on the results of the last four years of failure and maintenance data of the Oyu Tolgoi mining communication system.
4.	Revealing space and time:	The results of some measures to improve reliability were revealed based on 4 years of data (2014–2018) and used to develop a forecast for the next 10 years. Moreover, a 10-year period was chosen for the simulation test since the average service life of communication system devices is 10 years.
5.	Defining the problem's dynamics characteristics:	Statistics on the failure and repairs of communication system devices at Oyu Tolgoi over four years were developed and the probability of dynamic reliability was determined.
6.	Formulating a dynamic hypothesis:	One of the ways and opportunities to improve the reliability of the current communication system is to increase the instances and quality of maintenance and install standby devices.

Table 2. *Cont.*

7.	Formulating the simulation model:	A cause-and-effect diagram of the communication system's key components' failures and maintenance was drawn, and a resource flow diagram was developed in the Vensim program using Markov analysis and system dynamics modeling. The effects of corrective and preventative maintenance on each communication system device and the system, as well as the probability of a system's reliable operation with single or dual standby systems, were simulated and tested.
8.	Findings:	The simulation model of communication system availability was tested, and the results increased the initial communication system availability rate and improved the reliability of the overall communication system. Some sensitivities of the failure and repair parameters were checked, and the upper and lower variable boundaries were set.

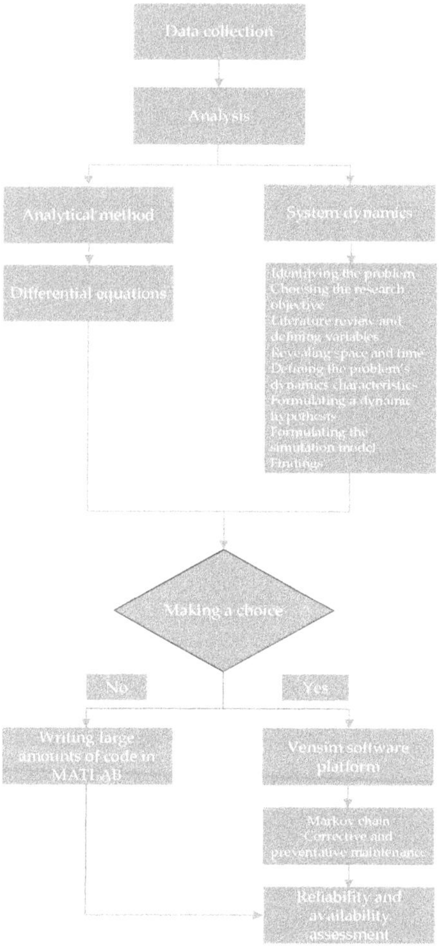

Figure 4. Schematic diagram of research methodology.

Figure 5 shows that the connection line reliability of the three-shaft underground mine communication system shown in the previous Figure 1 (see Supplementary Materials) is modeled only for the site link of the first shaft.

Recommended algorithm:

Step 1: The values of variables (μ, λ) are taken as inputs. These were based on data from four years (2014–2018) of data on failures of the Oyu Tolgoi underground mining communication system. The total time (T) is taken as input, i.e., the time for which the system must be simulated.

Step 2: $P_0[0,0] = 1$ is taken as the initial value is equal to one.

Step 3: When $Pt[i,j] = 0$, $i \neq 0$, $j \neq 0$ is taken as the value being 0.

Step 4: A conditional loop is formed with the condition $t > T$.

Step 5: In each execution of the loop, the time is increased by dt, i.e., $t = t + dt$. The loop continues until time (T) with each step taken at time difference dt.

Step 6: As assumed, $P_1(t)$ initially has probability unity, and the system fails when it equals 0. When $P_1(t) > 0$ is satisfied, repetition is executed.

Step 7: At each repetition, all variables are calculated according to Formulas (3)–(6).

Step 8: All the required values are displayed, and graphs can be drawn using these values. Various experiments can be simulated.

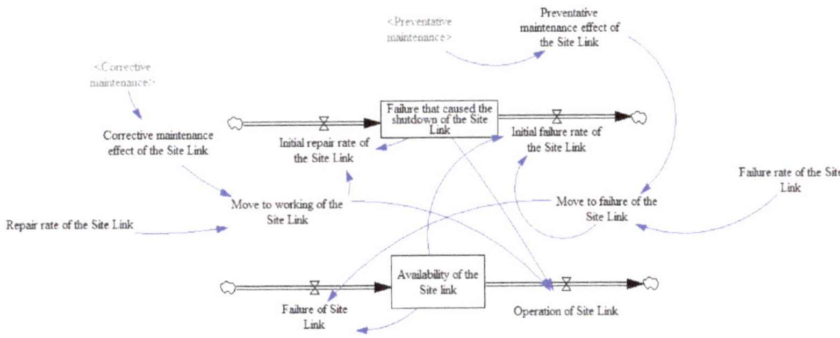

Figure 5. Site link model.

The comparison results of the site link model shown in Figure 5 were shown in Figure 9 (see Supplementary Materials) in the Section 3.

The overall availability of a communication system without standby in a one-shaft underground mine is based on Markov analysis and system dynamics modeling, using data from the availability ratings provided by the manufacturer if the equipment was not damaged during this period. Availability model of the communication system without standby in a one- shaft underground mine which was shown in Figure 6 (see Supplementary Materials) and comparing results of the model were shown in Figure 11 (see Supplementary Materials) in the Section 3.

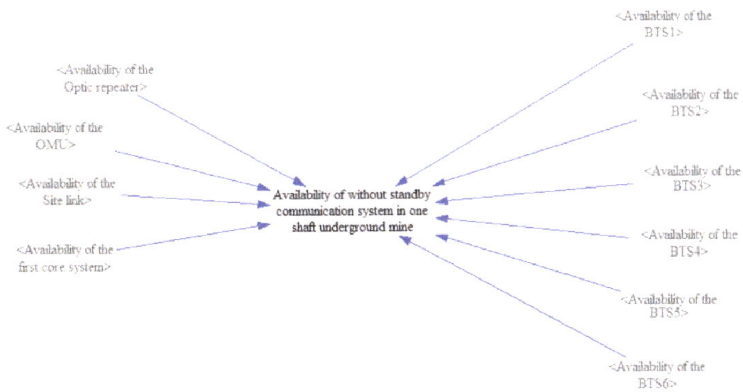

Figure 6. Availability of a communication system without standby in a one-shaft underground mine.

2.1. Increasing the Probability of Communication System Availability or the Reliability of a Single Standby System

The standby system consists of two independently operating communication devices with parallel connections. It is believed that one device failure does not affect the other device. The system is defined by four states: (0, 0), (1, 0), and (0, 1) are operating states, and (1, 1) is a failure state. Figure 7 shows all possible states and transitions between them.

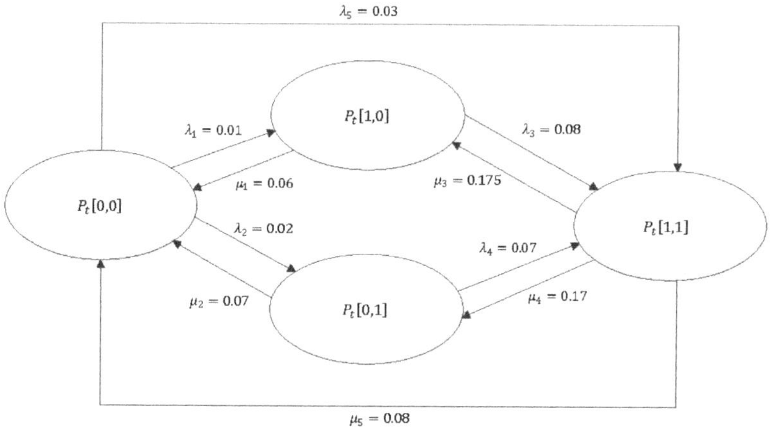

Figure 7. Reliability model of a communication system with a single standby device.

Description of probability is shown in Figure 7:

$P_t[0,0]$ is the probability that the system will run at time (t).

$P_t[1,0]$ is the probability of the second device functioning and failure of the first device at time (t).

$P_t[0,1]$ is the probability of the first device functioning and failure of the second device at time (t).

$P_t[1,1]$ is the probability that the system will not function at time (t).

This system is modeled as the following system of four ordinary differential equations based on Markov analysis:

$$\frac{dP_t[0,0]}{dt} = -[\lambda_1(t) + \lambda_2(t) + \lambda_5(t)]P_t[0,0] + \mu_1(t)P_t[1,0] + \mu_2(t)P_t[0,1] + \mu_5(t)P_t \quad (3)$$

$$\frac{dP_t[1,0]}{dt} = -[\lambda_3(t) + \mu_1(t)]P_t[1,0] + \lambda_1(t)P_t[0,0] + \mu_3(t)P_t[1,1] \quad (4)$$

$$\frac{dP_t[0,1]}{dt} = -[\lambda_4(t) + \mu_2(t)]P_t[0,1] + \lambda_2(t)P_t[0,0] + \mu_1(t)P_t[1,1] \quad (5)$$

$$\frac{dP_t[1,1]}{dt} = -[\mu_4(t) + \mu_3 + \mu_5(t)]P_t[1,1] + \lambda_5(t)P_t[0,0] + \lambda_3(t)P_t[1,0] + \lambda_4(t)P_t[0,1] \quad (6)$$

Initial condition of the system is: $P_0[0,0] = 1$, $P_0[1,0] = 0$, $P_0[0,1] = 0$, $P_0[1,1] = 0$

To calculate the equation above, a distribution law to model the failure and repair parameters must be chosen. As the law of exponential distribution has been used often in research papers on device reliability [6], this study has been developed on exponential, Weibull, and Poisson distribution. The development results were almost identical and were considered only in the case of the Weibull distribution law. The density function of Weibull distribution is determined using the following formula:

$$f(t, \alpha, \beta) = \frac{\alpha}{\beta^{\alpha}} t^{\alpha-1} e^{-(t/\beta)^{\alpha}} \quad (7)$$

2.2. Model for a Dual Standby Communication System

The system consists of three communication devices with parallel connections, which operate independently.

The failure of one device was not considered to affect other devices. The system is divided into eight simple states based on Markov analysis: (0, 0, 0), (1, 0, 0), (0, 1, 0), (0, 0, 1), (1, 1, 0), (1, 0, 1), and (0, 1, 1) being operational states, and (1, 1, 1) a failure state. All possible states between them also transition, with changes shown in Figure 8.

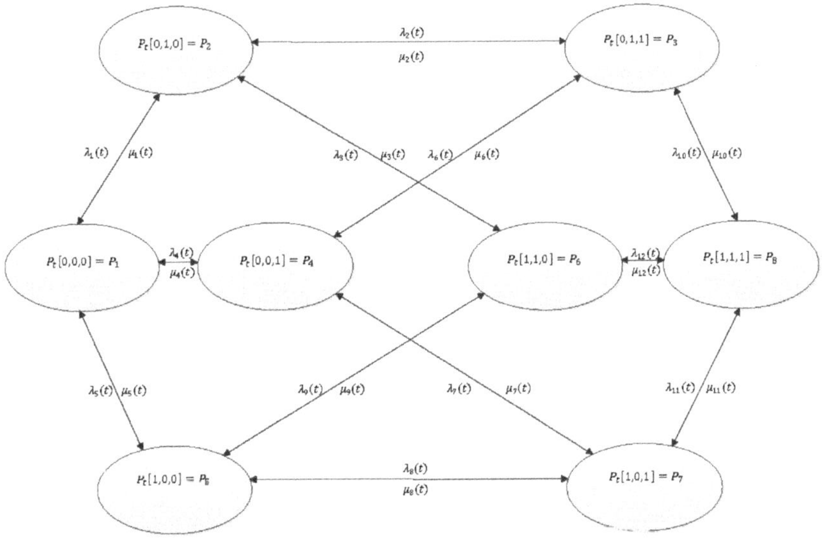

Figure 8. Reliability model of a communication system with dual standby devices.

The "normal" mode of operating is when all devices are operating without interruption; one can fail, and when one fails, the second device can fail. Two devices cannot be in a failure state simultaneously; if the first one fails, and the second one fails, all three will fail. The repair parameters are the same as the failure parameters.

Description of probability is shown in Figure 8:

$P_t[0,0,0]$ is the probability of system operation at time (t).

$P_t[1,0,0]$ is the probability of the first device failing and operation of the second and third devices at time (t).

$P_t[0,1,0]$ is the probability of the first and third devices being operational and the probability of the second device failing at time (t).

$P_t[0,0,1]$ is the probability of the first and second devices being operational and the third device failing at time (t).

$P_t[1,1,0]$ is the probability of the first and second devices failing and the third device being operational at time (t).

$P_t[1,0,1]$ is the probability of the first and third devices failing and the second device being operational at time (t).

$P_t[0,1,1]$ is the probability of the first device being operational and the second and third devices failing at time (t).

$P_t[1,1,1]$ is the probability of the system not operating at time (t).

$\lambda_1[t]$ is the failure function when the operational status shifts from all three devices being in a normal state to the second device failing.

$\lambda_2[t]$ is the transition failure function when the operational status shifts from the first and third devices being in a normal state and the second device failing to the first device being in a normal state and the second and third devices failing.

$\lambda_3[t]$ is the transition failure function when the operational status shifts from the first and third devices being in a normal state and the second device failing to the third device being in a normal state and the first and the second devices failing.

$\lambda_4[t]$ is the transition failure function when the operational status shifts from the three devices being in a normal state to the first and second devices being in a normal state and the third device failing.

$\lambda_5[t]$ is the transition failure function when the operational status shifts from the three devices all being in a normal state to the second and third devices being in a normal state and the first device failing.

$\lambda_6[t]$ is the transition failure function when the operational status shifts from the first and second devices being in a normal state and the third device failing to the first device being in a normal state and the second and third devices failing.

$\lambda_7[t]$ is the transition failure function when the operational status shifts from the first and second devices being in a normal state and the third device failing to the second device being in a normal state and the first and third devices failing.

$\lambda_8[t]$ is the transition failure function when the operational status shifts from the second and the third devices being in a normal state and the first device failing to the second device being in a normal state and the first and third devices failing.

$\lambda_9[t]$ is the transition failure function when the operational status shifts from the second and third devices being in a normal state and the first device is failing to the third device being in a normal state and the first and second devices failing.

$\lambda_{10}[t]$ is the transition failure function when the operational status shifts from the first device being in a normal state and the second and third devices failing to all three devices failing.

$\lambda_{11}[t]$ is the transition failure function when the operational status shifts from the second device being in a normal state and the first and third devices failing to all three devices failing.

$\lambda_{12}[t]$ is the transition failure function when the operational status shifts from the third device being in a normal state and the first and second devices failing to all three devices failing.

The repair functions $\mu_i[t], i = \overline{1,10}$ were defined in the same way as the matching failure functions.

The system is modeled with the following system of ordinary differential equations:

$$\frac{dP1}{dt} = -[\lambda_1(t) + \lambda_4(t) + \lambda_5(t)]P1 + \mu_1(t)P2 + \mu_4(t)P4 + \mu_5(t)P5 \tag{8}$$

$$\frac{dP2}{dt} = -[\lambda_2(t) + \lambda_3(t) + \mu_1(t)]P2 + \mu_2(t)P3 + \mu_3(t)P6 + \lambda_1(t)P1 \tag{9}$$

$$\frac{dP3}{dt} = -[\lambda_2(t) + \lambda_6(t) + \mu_{10}(t)]P3 + \mu_2(t)P2 + \mu_6(t)P4 + \lambda_{10}(t)P8 \tag{10}$$

$$\frac{dP4}{dt} = -[\lambda_6(t) + \lambda_7(t) + \mu_4(t)]P4 + \lambda_4(t)P1 + \mu_7(t)P7 + \mu_6(t)P3 \tag{11}$$

$$\frac{dP5}{dt} = -[\mu_5(t) + \lambda_9(t) + \lambda_8(t)]P5 + \lambda_5(t)P1 + \mu_9(t)P6 + \mu_8(t)P7 \tag{12}$$

$$\frac{dP6}{dt} = -[\lambda_{12}(t) + \mu_3(t) + \mu_9(t)]P6 + \lambda_3(t)P2 + \lambda_9(t)P5 + \mu_{12}(t)P8 \tag{13}$$

$$\frac{dP7}{dt} = -[\lambda_{11}(t) + \mu_7(t) + \mu_8(t)]P7 + \mu_{11}(t)P8 + \lambda_8(t)P5 + \lambda_7(t)P4 \tag{14}$$

$$\frac{dP8}{dt} = -[\mu_{10}(t) + \mu_{11}(t) + \mu_{12}(t)]P8 + \lambda_{11}(t)P3 + \lambda_{11}(t)P7 + \lambda_{12}(t)P6 \tag{15}$$

The initial condition of the system was defined as:

$$P1 = 0;\ P2 = 0;\ P3 = 0;\ P4 = 1;\ P5 = 0; P6 = 0;\ P7 = 0;\ P8 = 0$$

The availability function can be expressed as:

$$A(t) = P1 + P2 + P3 + P4 + P5 + P6 + P7$$

Reliability operation probability can be expressed as the formula:

$$A = \lim_{t \to \infty} A(t) \quad (16)$$

We used the case of Weibull distribution in this paper:

$$\lambda_i(t) = \beta \lambda_i t^{\beta - 1},\ i = \overline{1,12} \quad (17)$$

$$\mu_i(t) = \beta \mu_i t^{\beta - 1},\ i = \overline{1,12} \quad (18)$$

3. Results and Discussion

The general structure of the three-shaft underground mine communication system shown in Figure 1 is illustrated for the Motorola Dimetra (TETRA) communication system, for the first one-shaft underground mine communication system, Oyutolgoi mine communication system failure data (see Supplementary Materials) was used and modeled using availability information (see Supplementary Materials) provided by the manufacturer of the Motorola Dimetra (TETRA) communication system to develop availability probabilities for two- and three-shaft underground mine communication systems. Figure 9 graphically shows the availability simulation results for one-, two-, and three-shaft underground mine communication systems, separately and together. According to the results of the probabilities of these communication systems availability (shown in Table 3) the availability probability of the three-shaft underground mine communication system shows the highest result of 85.18%, compared to the availability probability results of the one- and two-shaft underground mine communication system. In other words, the probability of communication system availability after an average of 10 years of operation in underground mines is 81.26% for the one-shaft communication system, 62.38% for the two-shaft underground mine communication system or single-standby communication system, and the three-shaft showed results of 85.18% for an underground mine communication system or dual-standby communication system. Moreover, the three-shaft communication system can work with 100% availability probability for the first three years of operation. While the two-shaft underground mine communication system can work with 100% availability probability for the first two years in single-shaft underground mine communication systems or for communication systems without standby, the results show that it is possible to operate with 100% availability probability in the first year.

For a three-shaft underground mine, according to Motorola Dimetra (TETRA) system specifications, each shaft has one core system, six base stations, one site link, one optical repeater and one optical cable management device, (a total of 3 core systems, 18 base stations, 3 site links, 3 optical repeaters and 3 optical cable management devices) which were modeled and developed for each shaft and for in general (Figure 9) (see Supplementary Materials).

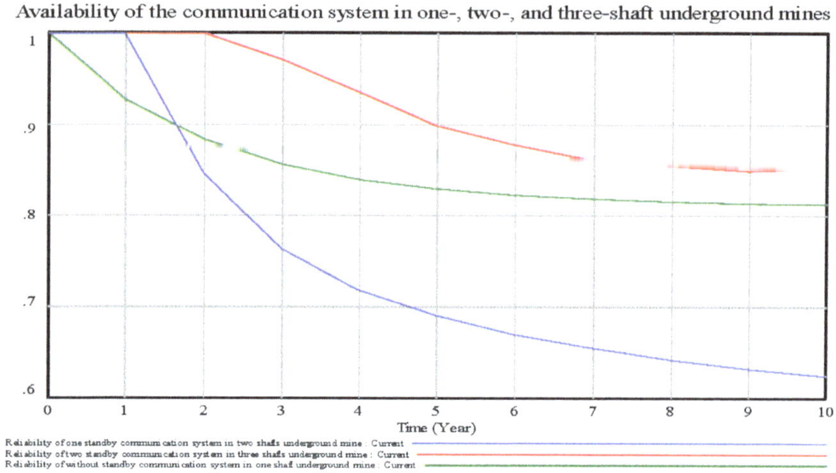

Figure 9. Availability rate of the communication system with a one-, two-, and three-shafts mine.

Table 3. Simulation results of one-, two-, and three-shaft underground mine communication system availability.

Availability Rate of the Communication System with One-, Two-, and Three-Shaft mine/years	0	1	2	3	4	5	6	7	8	9	10
Availability of communication systemwithout standbyin one-shaft underground mine	100	92.72	88.32	85.62	83.94	82.88	82.22	81.8	81.53	81.36	81.26
Availability of one standby communication system in two-shaftunderground mine	100	100	84.55	76.42	71.88	69.02	67	65.46	64.22	63.21	62.38
Availability of two-standby communication system in three-shaftunderground mine	100	100	100	97.12	93.51	89.87	87.75	86.06	85.5	84.86	85.18

Figure 10 and Table 4 (see Supplementary Materials), on the other hand, shows the simulation results of combined corrective and preventative maintenance, results after corrective maintenance was doubled, and when preventative maintenance was increased by 50%. In the current situation, there is a single site link that has no recourse site link. According to the results shown in this graph, the current availability rate of the site link was 78.27%. When preventative maintenance was increased by 50%, the availability rate of the site link increased by 84.41% and by 6.14% compared to the status. After doubling corrective maintenance, the rate increased by 87.77% and by 9.5% compared to the status, and after combining preventative and corrective maintenance, availability increased by 91.5% and by 13.23% compared to the status. Simulations were developed for the existing and without standby communication system, after doubling corrective maintenance and a 50% increase in preventive maintenance, as well as combining both corrective and preventive maintenance.

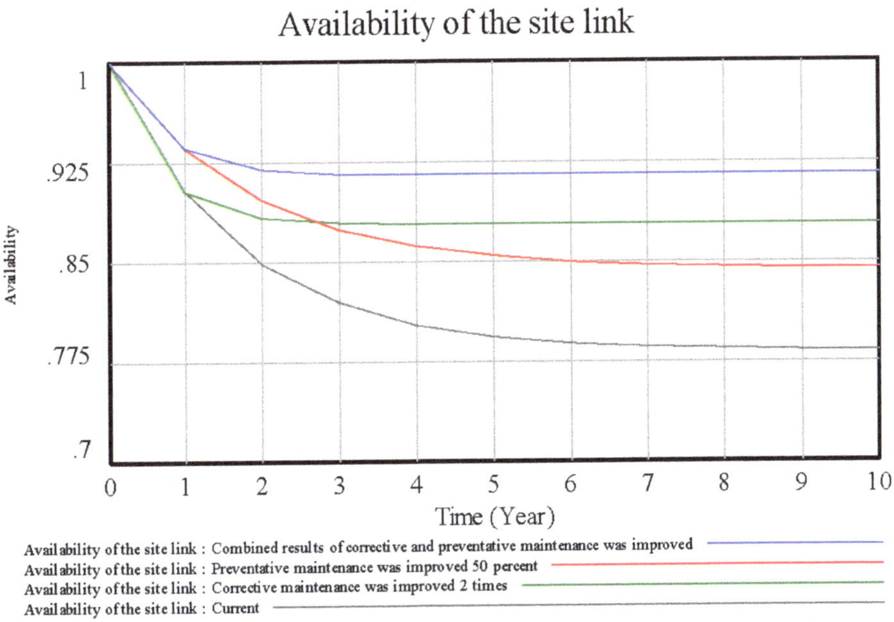

Figure 10. Results of the compared simulation of the site link.

Table 4. Simulation results of site link availability.

Site Link Availability Model (years)	0	1	2	3	4	5	6	7	8	9	10
Combined results of improved corrective and preventative maintenance	100	93.5	92	91.6	91.5	91.5	91.5	91.5	91.5	91.5	91.5
Preventative maintenance improved by 50%	100	93.5	89.7	87.5	86.2	85.4	85	84.7	84.6	84.5	84.4
Corrective maintenance doubled	100	90.3	88.3	87.9	87.8	87.8	87.8	87.8	87.8	87.8	87.8
Current	100	90.3	84.9	81.9	80.2	79.3	78.8	78.6	78.4	78.3	78.3

The simulation was performed by doubling the current and corrective maintenance to the without standby communication system in one shaft underground mine and by combining corrective and preventative maintenance after a 50% increase in preventative maintenance. In the communication system without standby in the one-shaft underground mine, the current availability rate was 73.95%, and after increasing preventative maintenance by 50%, the availability rate was 81.26%, and 9.8% compared to the status. After doubling corrective maintenance, it was 85.27% and 15.6% compared to the status, and after combined corrective and preventative maintenance, it was 89.75% and 21.3% compared to the status (Figure 11 and Table 5) (see Supplementary Materials).

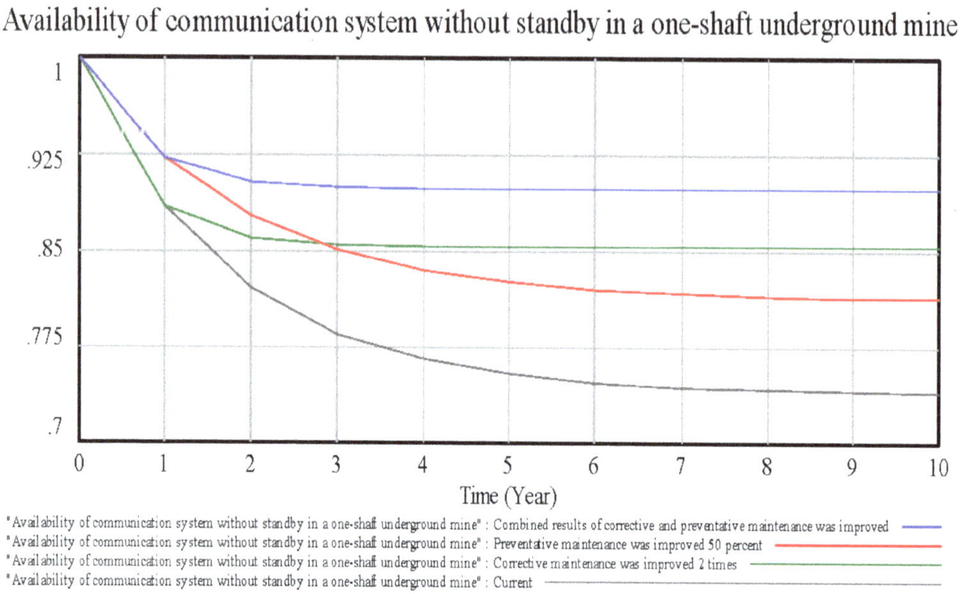

Figure 11. Simulation results of the without standby communication system in one shaft underground mine.

To improve the reliability of the communication system, a standby system is necessary. In the last part of the study, the results of the development using system dynamics modeling show how the reliability of the single- and dual-standby system is improved.

Table 5. Simulation results of single-shaft underground mine communication system reliability without standby system (included preventive and corrective maintenance analysis).

Communication System Availability model/years	0	1	2	3	4	5	6	7	8	9	10
Combined results of improved corrective and preventative maintenance	100	92.23	90.36	89.9	89.79	89.76	89.75	89.75	89.75	89.75	89.75
Preventative maintenance was improved by 50%	100	92.23	87.7	85.04	83.47	82.54	81.98	81.65	81.45	81.33	81.26
Corrective maintenance doubled	100	88.41	85.97	85.43	85.31	85.28	85.27	85.27	85.27	85.27	85.27
Current	100	88.41	82.04	78.5	76.5	75.37	74.73	74.36	74.15	74.03	73.95

According to the system dynamic model—which was developed during the study—and results of the simulations, the models fully achieved the originally set goals of improving the communication system's reliability. The results of this research are compared to previous studies. Our research dealt with the standby system and is based on the example of underground mine communication systems. This study is innovative in the way it is modeled and developed for each device and the system as a whole [5]. In addition, it studied single standby parallel systems and dual standby three-parallel systems. Weibull, Poisson, and exponential distribution were uniquely developed to model the system dynamics of a communication system [6].

4. Implications and Limitations of the Study

The main contribution of our study is to develop a traditional differential equation system and system dynamic modeling method and suggested a methodology for calculating the reliability of the communication system using a combined method. The data about Oyu Tolgoi communication system failure was used in development this method. The most important result is that the interdependence model of the devices included in the communication system was established based on their conditional probabilities (Figure 11) on the example of the one-shaft underground mine communication system and it was possible to test the availability degree of each device and the entire system by simulating with the help of Oyu Tolgoi mine communication system failure and maintenance functions and availability information provided by Motorola Dimetra (TETRA) communication system manufacturer. The results in Figure 9 were results of comparing the availability degrees of the three-shaft underground mine communication system shown in Figure 1. But Figure 10 was the results for the model shown in Figure 5 and when there was not any condition and when doubling the preventive maintenance and after increasing the preventive maintenance by 50% and was considered a feature by simulating both corrective and preventive maintenance. Also, single- and dual standby system models are suggested in Figures 7 and 8. As for the single- and dual standby system the models shown in Figures 7 and 8 have been tested in MATLAB program only, given random data for the availability degree for the Exponential, Poisson, and Weibull distributions. In other words, the failure and maintenance functions of the Oyu Tolgoi mine communication system and the availability information provided by the Motorola Dimetra (TETRA) communication system manufacturer were not used in both models. Thus, the models shown in Figures 5 and 6 and the results shown in Figures 9–11 were varied from the models shown in Figures 7 and 8 by processing using TETRA communication system devices and their failure and manufacturer's availability data. The results for the Exponential, Poisson, and Weibull distributions of the models shown in Figures 7 and 8 varied according to the time, due to the availability degree was unstable across the three distributions. In this regard, the results of the models shown in Figures 7 and 8 were not included in this paper, and the problems faced during the study of this model were included detailly in this section. As for the methodology of the study, the system dynamic modeling methodology was not considered as the main research methodology, and we considered the system dynamic modeling methodology as a way of calculating the formulated calculation. In other words, we use a causal loop diagram, which is the basis of the system dynamic modeling methodology, formulated the basic model in the form of an ordinary differential equation without defining the feedback links, and performed its analysis using the Vensim program. This is another feature of this study. The work in [5] was based on the idea that one way to improve the reliability of any equipment in the research work is to increase the maintenance, using the failure data of Oyu Tolgoi underground mine communication system in 2014–2018, the probability of reliable operation was calculated for each device of the communication system in the mine. This is one of the distinguishing features of the work [5]—calculating the probability of reliable operation of the entire communication system by expanding it. The authors of [6] showed in a research paper that one way to improve the reliability of any equipment is to have a resource system available, and it is formulated as an ordinary differential equation for a system with single resource, by Weibull's distribution law. By using it, the system with single- and two-resources considering the reliability of an underground mine communication system is formulated in ordinary differential form by Weibull distribution law and was calculated using the system dynamic modeling methodology in Vensim software, and the results were compared with each other and with the maintenance improvement method. This was an advantage of the study. Oyu Tolgoi communication system failure data were collected only between 2014 and 2018, and it was difficult to collect data after 2018, because the underground mine communication system was in the construction phase, and this process is still ongoing. In addition, since the monitoring system at that time was in the process of being updated from 2018, the failure data after 2018 were not recorded correctly until 2021, which were the main limitations that occurred in the time and data collection process of our study.

The following problems occurred when the study was conducted (Figures 7 and 8):

- Simulating and testing in MATLAB requires a lot of programming code, and it was relatively easy to simulate and test in Vensim. One of the main advantages of system dynamic modeling was the ease of conducting experiments with different parameter values.
- When the failure function (λ) and repair function (μ) values of three parallel communication systems (communication system in a one-, two-, and three-shaft mine) were modeled by the Weibull distribution law, the reliability was not converged to a fixed number, that is, it was not stabilized. Therefore, for the case of three parallel communication systems, the Weibull distribution was left.
- Moreover, for dual- and triple-parallel communication systems, we entered the failure function (λ) and correction function (μ) mixed, or exponent on the failure function (λ), Poisson on the correction function (μ), or failure function (λ) or Weibull on the correction function (μ), exponential distribution on the correction function (μ), etc. But this idea was not implemented because the reliability index was not stabilized.
- While calculating reliability model of a communication system with a single standby device (Figure 7) and reliability model of a communication system with dual standby devices (Figure 8), we tried using the dsolve command in MATLAB, but it took a long time to receive the results. Therefore, the dsolve command was replaced by the ode23 command to complete the calculation.
- When the values of failure function (λ) and repair function (μ) were assumed to be governed by Poisson distribution law, the factorial of t (t!) was included our formula, which made it difficult to calculate differential equations. Therefore, the factorial of t (t!) was replaced by Stirling formula.
- In the design of reliability model of a communication system with dual standby devices considering all options, some cases were left because differential equation systems were computationally and conceptually difficult. For example: it was left out that these three systems are working and all three stop at the same time, i.e., one device starts to stop, then the second device stops, and finally the third device stops, so all three parallel systems are modeled in such way that all of them stop.

5. Conclusions

As a result of this study, the impact of increasing corrective and preventative maintenance to improve the reliability of underground mine communication systems is made evident. In the case of the communication system without standby in the one-shaft underground mine studied, the current availability rate was 73.95%, and after increasing preventative maintenance by 50%, the availability rate was 81.26% and 9.8% compared to the status. After doubling corrective maintenance, it was 85.27% and 15.6%, compared to the status and after combined corrective and preventative maintenance, it was 89.75% and 21.3% compared to the status. The most significant result was a model of the interdependence of devices included in the communication system without standby in the one-shaft underground mine, built based on their conditional probabilities (Figure 11). The model makes it possible to simulate the availability rate of each device and the availability of the communication system without standby in the one-shaft underground mine using failure and maintenance functions. The new model is also suggested for developing the rate probability of the availability of underground mine communication system with one, two, and three shafts. According to the original objectives of our study, a methodology was developed that allows for simulation development of the communication system without standby in the one-shaft underground mine, which shows the impact of doubling the current and corrective maintenance without standby after increasing preventative maintenance by 50% and increasing corrective and preventative maintenance simultaneously. The results of our study show that using system dynamics modeling instead of traditional methods provides a simple simulation model that can be used to study the reliability and availability of any communication system. In this paper, real-time failure data of a system was used for simulation development. System failure data was not used to develop a model for single and dual standby systems. By being based on a real operating system, the models that were developed are suitable for practical use.

The artificial intelligence-based neural system dynamic research methodology can be further developed based on our model. Moreover, based on the source in [17] it seems that it is possible to develop a diagnostic model based on Vensim software, not only on MATLAB (LabVIEW). Regarding future work on communication systems with single- and dual standby system:

- Conduct experiment by replacing variable values (μ, λ) instead of real system values (system failure values);
- In the dual standby communication system, it is expected that no two devices will be in failure state at the same time, and if one is in failure, then the second will be in failure state, and then all three will be in failure state. Moreover, the maintenance parameters are considered to be exactly the same as the failure parameters.
- Conducting an experiment in the case of if two devices fail simultaneously, the third device does not fail, and vice versa; when the third device fails, the first and second devices do not fail.
- The reason why these conditions were not considered in this paper is because the system of differential equations was very large and complicated to calculate.

6. Patents

The part included in this paper that uses the system dynamic model, under the name of the work "Analysis of the reliability of underground mine radio communication system using system dynamic modeling" received copyright certificate No. 12,388 from the Mongolian Intellectual Property Authority on 25 January 2021.

Supplementary Materials: The following supporting information can be downloaded at: https://www.mdpi.com/article/10.3390/app13020821/s1.

Author Contributions: Conceptualization, B.B. and U.B.; methodology, B.B.; software, U.B.; validation, B.B., O.B. and U.B.; formal analysis, B.B.; investigation, B.B.; resources, B.B.; data curation, B.B.; writing—original draft preparation, B.B.; writing—review and editing, B.B.; visualization, B.B.; supervision, B.B.; project administration, U.B.; funding acquisition, B.B. All authors have read and agreed to the published version of the manuscript.

Funding: This research received no external funding.

Institutional Review Board Statement: Not applicable.

Informed Consent Statement: Not applicable.

Data Availability Statement: All data are contained within the article and the Supplementary materials.

Conflicts of Interest: The authors declare no conflict of interest.

Abbreviations

The following abbreviations are used in this manuscript:

Dimetra	Digital Motorola Enhanced Trunked Radio
TETRA	Terrestrial Trunked Radio
MSD	Markov system dynamics
OMU	Optic management unit
BTS	Base transceiver station
RF	Radio frequency

References

1. Bazargur, B.; Bataa, O. A study of Mining Communication System Reliability Model. In Proceedings of the 13th International Forum on Strategic Technology (IFOST 2018), Harbin, China, 31 May–1 June 2018.
2. Bazargur, B.; Bataa, O.; Budjav, U. Underground Mining Radio Communication System's Risk and Reliability. *Batzorig Bazargur J. Eng. Res. Appl.* **2018**, *8 Pt II*, 36–39.
3. Bazargur, B., Bataa, O.; Khurelbaatar, Z.; Battseren, B. Experimental Measurements and Antenna Isolation for TETRA Communication System in Underground Mining and decline. In Proceedings of the ETTC-2018, Conference Proceeding, Europian Test and Telemetry Conference, Nuremberg, Germany, 26–28 June 2018.
4. Bazargur, B.; Bataa, O.; Budjav, U.; Gantumur, G. Analysis of Radio Communication Networks in Underground Mining for Emergencies. In Proceedings of the 2019 21st International Conference on Advanced Communication Technology (ICACT), Pyeongchang, Republic of Korea, 17–20 February 2019.
5. AKhorshidi, H.; Indra Gunawan, M.; Ibrahim, Y. Reliability Centered Maintenance Using System Dynamics Approach. In Proceedings of the IEEE International Conference on Industrial Technology (ICIT), Seville, Spain, 10–12 March 2021.
6. Temraz, N.S. Availability and Reliability Analysis for System with Bivariate Weibull Lifetime Distribution. *Int. J. Sci. Eng. Res.* **2017**, *8*, 376–389.
7. Parol, M.; Wasilewski, J.; Wojtowicz, T.; Arendarski, B.; Komarnicki, P. Reliability Analysis of MV Electric Distribution Networks Including Distributed Generation and ICT Infrastructure. *Energies* **2022**, *15*, 5311. [CrossRef]
8. Wei, G. System Reliability Modeling and Analysis of Distributed Networks, Hindawi. *Adv. Multimed.* **2022**, *2022*, 9719427. [CrossRef]
9. Antosz, K.; Machado, J.; Mazurkiewicz, D.; Antonelli, D.; Soares, F. Systems Engineering: Availability and Reliability. *Appl. Sci.* **2022**, *12*, 2504. [CrossRef]
10. Odeyar, P.; Apel, D.B.; Hall, R.; Zon, B.; Skrzypkowski, K. A Review of Reliability and Fault Analysis Methods for Heavy Equipment and Their Components Used in Mining. *Energies* **2022**, *15*, 6263. [CrossRef]
11. Morad, A.M.; Pourgol-Mohammad, M.; Sattarvand, J. Application of reliability-centered maintenance for productivity improvement of open pit mining equipment:Case study of Sungun Copper Mine. *J. Cent. South Univ.* **2014**, *21*, 2372–2382. [CrossRef]
12. Chumai, R. *System Dynamic Modeling of Plant Maintenance Strategy in Thailand*; Asian University: Chon Buri, Thailand, 2009.
13. Rully, M.; Hendrawana, A.; Aristioa, A.P.; Bulialib, J.L.; Yuniartoc, M.N. Testing Methods on System Dynamics: A Model of Reliability, Average Reliability, and Demand of Service. ScienceDirect. *Procedia Comput. Sci.* **2019**, *161*, 968–975.
14. Rao, M.S.; Naikan, V.N.A. Reliability analysis of repairable systems using system dynamics modeling and simulation. *J. Ind. Eng. Int.* **2014**, *10*, 69. [CrossRef]
15. Tamaloussi, N.; Bouzaouit, A. Study of Reliability in a Repairable System by Markov Chains. *Acta Univ. Sapientiae Electr. Mech. Eng.* **2020**, *12*, 66–76. [CrossRef]
16. Rao, M.S.; Naikan, V.A. A hybrid Markov system dynamics approach for availability analysis of degraded systems. In Proceedings of the 2011 International Conference on Industrial Engineering and Operations Management, Kuala Lumpur, Malaysia, 22–24 January 2011.
17. Liu, Y. Artificial Intelligence–Based Neural Network for the Diagnosis of Diabetes: Model Development. *JMIR Med. Inform.* **2020**, *8*, e18682. [CrossRef] [PubMed]

applied sciences

MDPI

Article

A Controller-Driven Approach for Opportunistic Networking

MariaCarmen de Toro *, Carlos Borrego and Sergi Robles

Department of Information and Communications Engineering (dEIC), Universitat Autònoma de Barcelona, 08193 Bellaterra, Spain
* Correspondence: mariacarmen.detoro@uab.cat

Abstract: Opportunistic networks (OppNets) leverage opportunistic contacts to flow data across an infrastructure-free network. As of yet, OppNets' performance depends on applying the most suitable forwarding strategy based on the OppNet typology. On the other hand, software-defined networking (SDN) is a paradigm for wired networks that decouples the control and data planes. The control plane oversees the network to configure the data plane optimally. Our proposal uses SDN-like controllers to build a partial overview of the opportunistic network. The forwarding strategy uses this context information to achieve better network performance. As a use case of our proposal, in the context of an OppNet quota-based forwarding algorithm, we present a controller-driven architecture to tackle the congestion problem. Particularly, the controller-driven architecture uses the context information on the congestion of the network to dynamically determine the message replication limit used by the forwarding algorithm. A simulation based on real and synthetic mobility traces shows that using context information provided by the controller to configure the forwarding protocol increments the delivery ratio and keeps a good latency average and a low overhead compared with the baseline forwarding protocols based on message replication. These results strengthen the benefits of using supervised context information in the forwarding strategy in OppNets.

Keywords: congestion control; data forwarding; intermittently connected networks; opportunistic networks

Citation: de Toro, M.; Borrego, C.; Robles, S. A Controller-Driven Approach for Opportunistic Networking. *Appl. Sci.* **2022**, *12*, 12479. https://doi.org/10.3390/app122312479

Academic Editor: Runzhou Zhang

Received: 2 November 2022
Accepted: 1 December 2022
Published: 6 December 2022

1. Introduction

The Internet is growing fast with ever-increasing machine-to-machine communication (M2M), the expansion of micro and proximity services, the use of cloud services, and the continuous increase in the number of connected devices, among other reasons. The International Telecommunication Union report for the years 2020 to 2030 [1] estimates that there will be 100 billion connected devices by the year 2030. The fastest-growing mobile device category is M2M, followed by smartphones and other smart devices.

Improving the legacy infrastructure to cope with the demand in terms of bandwidth, coverage, quality of service (QoS), and specific requirements for emerging applications is costly and, therefore, not the ultimate solution. Thus, offloading traffic from core networks is a concern. For that matter, device-to-device (D2D) communication [2] performs direct transmissions between peers in range without needing a base station.

A type of network based on D2D communication is opportunistic networks (OppNets). OppNets are characterized by the mobility of their nodes, which leads to an undefined network topology that hinders contemporaneous end-to-end connectivity. Therefore, in OppNets, communication is led by the contact opportunity between peers. This paradigm is very convenient in networks as vehicular ad hoc networks [3], mobile wireless sensor networks [4], pocket-switched networks [5], people-centric networks [6], and mesh networks [7], among others.

OppNets, due to the mobility of their nodes, are prone to frequent disconnections, segmentation, and long delay paths. Hence, traditional routing schemes based on end-to-end connectivity are not applicable. Therefore, OppNets nodes function as routers which use the store-carry-and-forward (SCF) principle [8] to forward data from the source to the destination on a hop-to-hop basis. Moreover, in OppNets, routing efficiency directly affects the network performance and is highly coupled with the type of application running over the OppNet [9]. In this environment, routing orchestration is a challenge.

In this regard, software-defined networking (SDN) [10] is a network paradigm applied to connected networks that programmatically orchestrates the traffic routing and network configuration by using a program called a controller. The controller has an overview of the whole network and makes routing decisions based on this information. Although SDN protocols are based on TCP and, therefore, are not straightforwardly applicable to OppNets, we apply the concept of having a controller to orchestrate the OppNet traffic routing and agents receiving/sending information to the controller.

Hence, the goal of this proposal consists of using the SDN building blocks to build a context-aware system over an OppNet. This context-aware system leverages context information to dynamically configure the OppNet's forwarding policy parameters aiming to achieve better performance. In the proposed context-aware system, some OppNet's nodes perform as SDN-like controllers (controllers) and the rest as SDN-like agents (nodes). The nodes gather context measurements, specifically device measurements, and send them opportunistically on a hop-to-hop basis to the controllers, which will use this information to dynamically tune the forwarding algorithm parameters. The nodes extend the SCF paradigm to (1) gather device context measurements, (2) aggregate the gathered measurements, (3) forward them to the controllers, and (4) apply the received policies generated by the controllers. In this manuscript, we name this paradigm GAFA. We refer to the OppNet extended with the control system and the GAFA functionalities as *controller-driven* OppNet. In developing this proposal, we make the following contributions:

- Design of the architecture of a novel context-aware system for OppNets inspired by the SDN paradigm where nodes operate based on the GAFA paradigm to feed controllers with device-context information and apply the controllers' policies. The controllers use that information to tune the forwarding algorithm parameters through configuration policies emitted to the nodes to obtain a better network performance.
- Use of the controller-driven OppNet architecture for tackling the congestion in an OppNet characterized by a high unpredictability of the nodes' mobility and a multi-copy replication forwarding strategy. Specifically, the OppNet's controllers orchestrate the value of the replication limit of the forwarding algorithm based on the buffer occupancy readings gathered by the nodes.
- An evaluation of the performance and benefits of the controller-driven OppNet for the use case of congestion control. To evaluate our proposal, we have simulated diverse network scenarios based on real and synthetic mobility traces using different message generation distributions. We have evaluated the controller-driven OppNet on the basis of the standard performance metrics for OppNets. Furthermore, we have run the aforementioned simulations over an OppNet without the control layer (context oblivious) using an epidemic and a quota-based forwarding protocol. We have compared and evaluated the performance of both configurations. We have proven that a controller-driven OppNet performs better than a context-oblivious one.

The rest of the paper is structured as follows. Section 2 introduces the related work in the field of opportunistic networks, focusing on data forwarding and congestion control. Section 3 presents the controller-driven OppNet architecture. Next, in Section 4, the controller-driven OppNet architecture is used to manage congestion. The paper follows with Section 5, where through simulation-based experimentation, we evaluate the performance of the controller-driven OppNet congestion use case, and we compare these results with the performance obtained by a non-controlled OppNet. Finally, Section 6 contains the conclusions drawn from this work.

2. Related Work

First, this section describes the current state of the field of opportunistic networks. This proposal's targeted OppNet uses a multi-copy forwarding strategy prone to congestion, so this section accosts congestion control mechanisms for OppNets. Finally, this section develops on the related work in the scope of context-based routing because this proposal's goal consists of using context knowledge to tune the OppNet's multi-copy-based forwarding algorithm.

2.1. Opportunistic Networks Overview

An OppNet is a structureless multi-hop network built upon fixed and mobile nodes via wireless links. Due to the mobility of the nodes, OppNets are prone to disruptions, segmentation, and long delay paths. Under these conditions, where there is no guarantee of an end-to-end path between the source and destination at a specific instant in time, the TCP/IP protocol suite is not effective. Hence, the communication in an OppNet is driven by the direct contact opportunity between peers in range, and data flow is achieved by exploiting the pairwise contact opportunity provided by the nodes' mobility. OppNets have been conceived as a complement to connected networks to provide connectivity under specific conditions. Therefore, they are targeted for well-defined practical use case applications characterized by having a limited or even nonexistent infrastructure [8]. Trifunovic et al. [11] stated that the emerging technologies providing global Internet connectivity are not the ultimate solution to settle the classical target OppNet applications domain yet. Moreover, despite the fact that numerous OppNet proposals are formulated as a prototype, there are several commercial solutions [11] and several realistic prototypes, such as [12,13], among others.

2.2. Context-Based Routing in Opportunistic Networks

OppNets are prone to long delay paths, disconnections, and segmentation, hindering end-to-end connectivity. Hence, traditional routing based on contemporaneous end-to-end connectivity is not feasible. Therefore, in OppNets, data forwarding is driven hop-to-hop using the store-carry-and-forward (SCF) paradigm originally designed for DTNs [14]. This forwarding paradigm consists of the node *storing* the data and *carrying* the data along the network according to the node's mobility until a contact opportunity occurs and then *forwarding* the data to the contacted node.

Data forwarding is principal in OppNets as application deployment relies on the forwarding as a guarantee of their particular QoS requirements [14]. Seeking forwarding efficiency, Jain et al. [14] state that context information helps to make a more efficient forwarding. CC et al. [15] classified data forwarding algorithms into two main categories depending on the context information used to make routing decisions: social-based routing and pure opportunistic routing. The latter mainly considers device context information.

Under the pure opportunistic routing category, sound forwarding strategies have been proposed in the literature. Those proposals fit well under determined network conditions and application requirements. In this regard, flooding-based strategies, consisting of message replication, have proven to maximize the delivery ratio with a low latency when the OppNet is characterized by the unpredictable nodes' movement [9]. Under the aegis of multi-copy forwarding, the epidemic flooding approach proposed by Vahdat et al. [16] is a context-oblivious strategy prone to suffer from the congestion derived from the replication overhead. Spyropoulos et al. [17] addressed the congestion overhead by establishing a static configured replication quota. Context-aware strategies aim to reduce the effects of a naive replication by calculating the utility of a relay based on historical information. CC et al. [15] highlighted the most relevant routing proposals in this category.

Finally, Boldrini et al. [18] pointed out the relevance of a contextual middleware to manage context information in an OppNet. In this regard, but in the context of wired networks, SDNs [10] decouple the control from the network devices in a control plane where a software named controller gathers network information from the data plane to build an overview of the network. The controller uses this information to orchestrate the network data flows and resources optimally. To our knowledge, SDN and OppNets have been converged by Li et al. [19]. The authors applied the SDN paradigm in OppNets to implement a mobile crowdsensing system. Nevertheless, southbound communication relies on a cellular network.

2.3. Congestion Control in OppNets

In OppNets, traditional mechanisms based on contemporaneous end-to-end connectivity to provide feedback regarding congestion are unsuitable. Furthermore, due to the node's mobility, congestion at a link level is very rare; thus, buffer overload is the main issue. OppNets use the SCF paradigm to forward data across the network. In this paradigm, if a node is affected by congestion, meaning that the buffer is overloaded, the node will need to reallocate, drop queued messages, or reject incoming ones. Either case is highly undesirable as losing messages caused by the node congestion could lead to a delivery failure. Therefore, congestion control is especially critical for OppNets.

Buffer management is a strategy for congestion control applied on the relayed node [20]. Buffer management determines which messages must be dropped in case an incoming one needs to be fitted in. The basic buffer management policies are based on the local's node information such as message priority, lifetime, size, or delivery probability. Krifa et al. [21] stated that basic buffer management policies as drop-tail, drop-head, etc., are suboptimal. They proposed an optimal policy associating a utility function to each queued message, producing a marginal value for a selected optimization metric (delay or delivery ratio). They used statistical learning about encounters to approximate the global knowledge of the network. Pan et al. [22] proposed a mechanism that integrates all the aspects of buffer management.

Another congestion control strategy applied by the sending node is congestion avoidance [20]. Under this category, Goudar et al. [23] stated that basic congestion measures based on buffer management, such as message drops, are not accurate in detecting congestion. They stated that under the inherent characteristics of a mobile OppNet, congestion may occur before buffers are overwhelmed. They proposed an analytical model where the forwarder node calculates the instantaneous forwarding probability of a relay, and they found that this probability decreases dramatically beyond a certain buffer occupancy (buffer occupancy threshold). The node is considered to be congested when it reaches this occupancy threshold.

Furthermore, Lakkakorpi et al. [24] stated that an effective congestion control system should not be based on the network conditions at the time the message was created. Instead, the congestion control system should consider the current network conditions before relaying the message. They proposed a mechanism where each node, upon a contact, shares its buffer availability. Nodes use this information to determine if a relay node has enough buffer resources to custody the message. Thomson et al. [25], measured the congestion as the ratio of drops over message replication per node. They used this information to adjust the replication limit of the messages. Goudar et al. [26] proposed a probabilistic model using an estimator to predict the average buffer occupancy of the nodes in the network. They used this information to discard relay nodes without enough storage to hold the messages to be relayed. Similarly, Batabyal et al. [27] derived a steady-state probability distribution for buffer occupancy.

3. Control Layer Architecture

This proposal uses the controller concept from the SDN architecture. We have designed a control layer running on top of the convergence layer of the nodes as a context-aware system. Some of these nodes, the ones selected to be controllers, run the controller module of the control layer. Similar to the SDN controller, the proposed controllers keep an overview of the network by gathering network measurements from the data plane. On the other hand, in OppNets, the control and data planes are coupled in the node. Thereby, the controller-driven OppNet nodes perform the GAFA functionality introduced in Section 1 for gathering, aggregating, and disseminating network measurements.

Any node could potentially be a controller. Whether a node functions as a controller depends on the nature of the network. In a vehicular network (VANET), the controllers could be the roadside units; in an information-centric network, they could be well-connected nodes. For this particular work, we consider a generic OppNet. Thus, the generic criteria of selecting the more central nodes, i.e., the ones with more contacts, has been applied.

Figure 1 shows a summary of how the control layer implements the node's GAFA and the controller functionalities. Firstly, the nodes in the OppNet sense local context measurements (Figure 1a). Detailed information can be found in Section 3.1. The nodes disseminate an aggregation of context measurements upon a contact (Figure 1b). Section 3.3 develops the aggregation methodology and Section 3.2 shows how the dissemination of context information is performed. Finally, Figure 1c shows how controllers process the received context information to obtain a prediction of a network indicator at a future time (Section 3.5). From this prediction, the controller determines an action to be performed by the nodes consisting of the modification of a forwarding algorithm parameter (Section 3.6). The controller disseminates this action upon a contact with another node (Section 3.7).

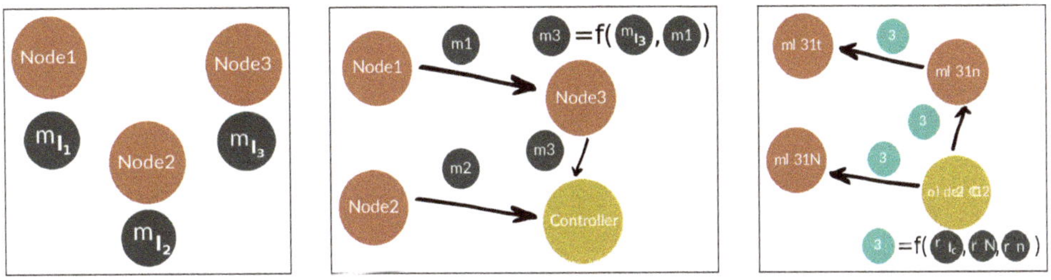

(**a**) Taking network measurements. (**b**) Measurements aggregation/dissemination. (**c**) Directive creation and dissemination.

Figure 1. Node's GAFA functionality: (**a**) gathering measurements, (**b**) aggregating and forwarding them, (**c**) applying the directive created by the controller.

A list of abbreviations is provided in Abbreviations section to help read the rest of the document.

3.1. Control Metadata

The control metadata used in the control layer are context measurements and control directives. When two nodes come into range, they disseminate control metadata before delivering or relaying buffered messages. The following two sections describe the aforementioned concepts.

3.1.1. Context Indicator Measurement

A *context indicator measurement* is the local reading of a context indicator taken at a time by the nodes and the controllers (Figure 1a). The controller receives those measurements upon opportunistic contacts with nodes (Figure 1b) and builds an overview of the network based on this information (Figure 1c). In an OppNet, the contact time, bandwidth, and nodes' energy are limited resources, so we have opted to aggregate those measurements (see Section 3.3). Therefore, the context information we consider is this aggregation.

3.1.2. Control Directive

A *control directive* is an action to be performed by the contacted node to modify a forwarding algorithm parameter. A directive is represented as the tuple: $\delta_s = (Id_s, \vartheta_s)$, where Id_s identifies a forwarding algorithm setting from the list of settings managed by the controller, and ϑ_s is the value for this setting. The controller generates the directive based on context information (Figure 1c) aiming to improve the forwarding performance. The controller disseminates this directive to the nodes. When a node receives the directive, it applies it by modifying the node's forwarding setting Id_s with the new value ϑ_s. As examples of forwarding settings, we could consider the message TTL, weights, and thresholds intrinsic to the forwarding algorithm, among others.

3.2. Context Measurements Dissemination

The node that receives a context measurement stores it in an indexed list. Considering $\{n_1, \ldots, n_z\}$ as the set of nodes in the network at time t, the indexed list of received context measurements for the node n_i for $1 < i \leq z$ is represented as $M_i = [m_j \mid 1 \leq j \leq z]$ where $M_i(j) = m_j$.

Figure 2 shows how a node (n_1) disseminates its context measurement when comes into range with another node (n_2). The control metadata are shared bidirectionally, hence, n_2 will follow the same flow when it is its turn to do the dissemination.

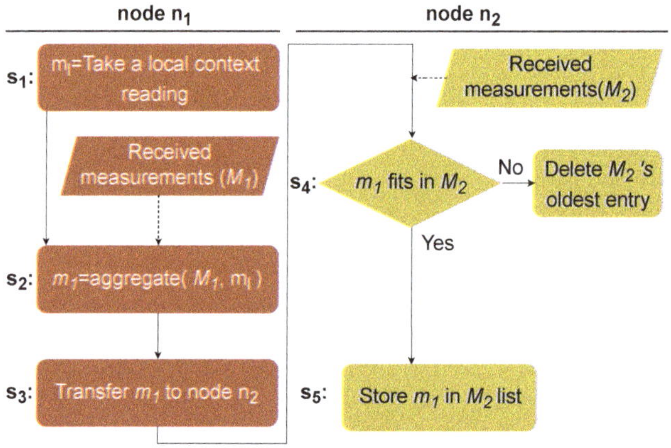

Figure 2. Context measurements dissemination when two nodes are in range.

In Figure 2, *step* s_1, when node n_1 contacts node n_2, n_1 takes a local context indicator reading, m_l. In *step* s_2, n_1 aggregates this context reading along with all the context measurements received by contacted nodes stored in the list M_1. The aggregation process is described in Section 3.3. In *step* s_3, the aggregated measurement, m_1, is shared with the contacted node n_2 which stores it in its context measurements list M_2 (*steps* s_4 and s_5). At this point, n_2 follows the same flow to calculate m_2 with the slight difference that it will not use the entry $M_2[n_1]$. This entry contains the received measurement m_1 built out of the information provided by n_1. Using m_1 would interfere with the network perception n_2 is about to share with n_1.

3.3. Context Measurements Aggregation

Considering $\{n_1, \ldots, n_z\}$ as the set of nodes in the network at time t, we represent the aggregated context measurement generated by node n_i for $1 \leq i \leq z$ as the tuple $m_i = (v_i, \eta_i, t_{c_i})$, where v_i is the aggregated result, η_i is the number of measurements used for the calculation of v_i, and t_{c_i} is the time when the calculation was made. The node n_i's local context measurement is represented as $m_{l_i} = (v_{l_i}, 1, t_{c_i})$, where $\eta_i = 1$ as v_{l_i} is a straight reading, not the result of an aggregation.

The process that performs the aggregation is *getAggregatedContextMeasurement* (*line 28*) in Algorithm 1. First of all, the node n_i gets the local context reading, v_l (*line 29*). The node uses a two-factor weighted average to aggregate all the context measurements stored in

M_i and its own context reading v_l. Each of the context measurements (m_j) stored in M_i is weighted by two factors: (1) the number of aggregations used to create it (η_j)—in the case of the local measurement, this value would be 1—and (2) its decay (d_j) at the current time t. In the case of the local measurement, the decay would be 1 (no decay).

The following logistic function calculates the decay of a measurement at time t:

$$d(t) = \frac{1}{1 + r^2 t} \tag{1}$$

where r is the reduction factor or decay degree to apply to obtain a certain decay. By isolating this variable we obtain:

$$r(t) = \left(\frac{1-d}{td}\right)^{\frac{1}{2}}. \tag{2}$$

This resulting equation is used to obtain the necessary decay degree (r) to be used in (1) to obtain the desired decay at a particular time t. The decay is inversely proportional to time (t), and it ranges from 0 to 1, where a decay of 1 means no decay. With a decay of 1, the congestion reading is not lowered when aggregated. In contrast, the older the reading is, the higher the decay (lower value), hence the more diminished the reading is when the controller aggregates it along with the other readings received during the aggregation interval.

The aforementioned two-factor weighted average used to aggregate the context measurements in M_i along with the perceived local context reading v_l (*lines 33–41*) is:

$$v_i = \sum_{j=1}^{k} v_j\left(\alpha \frac{\eta_j}{\sum_{p=1}^{k} \eta_p} + (1-\alpha)\frac{d_j}{\sum_{p=1}^{k} d_p}\right) \tag{3}$$

where v_i is the value resulting from this aggregation; k is the size of M_i plus one (to include the local context reading), v_j is the *value* of the measurement being aggregated ($m_j = M_i[j] = (v_j, \eta_j, t_{c_j})$); η_j is the number of aggregations used to generate m_j, d_j is the decay of m_j at the current time, and α is the weight factor, specified as a control setting (see Section 5.4.2), used to weigh the two factors of the weighted average. The function *getSumOfNrofAggrs* at *line 17*, calculates the normalizing factor applied in (3) over the number of aggregations the context measurement being processed is formed by ($\sum_{p=1}^{k} \eta_p$). Similarly, the function *getSumOfDecays* (*line 1*) calculates the normalization factor to be applied over the decay of a measurement ($\sum_{p=1}^{k} d_p$).

Finally, the node n_i creates the aggregated measurement $m_i = (v_i, \eta_i, t_{c_i})$ where v_i is the calculated aggregation value, η_i is the number of entries in M_i that have been aggregated plus the node's own reading (*lines 32, 39*), and t_{c_i} is the current time. At this point, *step 3* in the flowchart in Figure 2, the calculated aggregated measurement m_i is transferred to the contacted node.

3.4. Controller Architecture

The goal of a controller is to have an overview of its nearby part of the network, considering that the mobility nature of the nodes keeps changing the network's topology. This mobility brings on a network "segmentation" in terms of groups of nodes that eventually are connected between them. Ideally, some controllers would be required to cover all the possible network segments.

The controller operates opportunistically, i.e., its actuation is triggered by contacting another node or controller. When that happens, the controller shares both its aggregated context measurement and a directive with the contacted node.

Algorithm 1 Context measurements aggregation algorithm.

```
▷ M_i: List of context measurements received from contacted nodes.
▷ hostID: Identifier of a host.
▷ excludeHost: Id of the host whose measurement in M_i will not be used.
▷ sumDecays: Sum of the decays of all the measurements in M_i.
▷ M_i_keys: Measurement list indexes.
▷ m: A received context measurement represented by the tuple (v, η, t)
▷ threshold: Decay threshold under which the measurement is considered to be expired.
▷ sumNrAggr: Sum of the # aggregations a measurement is made of.
▷ #aggrEntries: # of elements in M_i that have been aggregated.
▷ aContextReading: A local context measurement.
▷ v_i: Aggregated measurement value.
function GETSUMOFDECAYS(M_i,excludeHost)
    sumDecays ← 0
    for all hostID ∈ M_i_keys do
        if hostID != excludeHost then
            m ← M_i[hostID]
            d ← decay(current_time − m[t])
            if d < threshold then
                M_i.remove(hostID)
            else
                sumDecays += d
            end if
        end if
    end for
    sumDecays ++                    ▷ Adding the decay of my own reading.
    return sumDecays
end function
function GETSUMOFNROFAGGRS(M_i, excludeHost)
    sumNrAggr ← 0
    for all hostID ∈ M_i_keys do
        if hostID != excludeHost then
            m ← M_i[hostID]
            sumNrAggr += m[η]
        end if
    end for
    sumNrAggr ++                    ▷ Considering its own reading.
    return sumNrAggr
end function
function GETAGGREGATEDCONTEXTMEASUREMENT(M_i, excludeHost)
    v_l ← aContextReading
    decays ← getSumOfDecays(M_i, excludeHost)
    aggrs ← getSumOfNrofAggrs(M_i, excludeHost)
    #aggrEntries ← 1                ▷ Considering its own reading.
    v_i ← v_l((α · 1/aggrs) + ((1 − α) · 1/decays))
    for all hostID ∈ M_i_keys do
        if hostID != excludeHost then
            m ← M_i[hostID]
            d ← decay(current_time − m[t])
            v_i += m[v]((α · m[η]/aggrs) + ((1 − α) · d/decays))
            #aggrEntries ++
        end if
    end for
    return m_i = (v_i, #aggrEntries, t)
end function
```

To generate a directive, the controller implements the closed-loop control system, also known as feedback control system [28], showed in Figure 3. A closed-loop control system is a control system that maintains a constant relation between the output of the system ($\mathfrak{c}$: controlled variable) and the desired value ($\mathfrak{r}$: the reference input) by subtracting one from the other as a measure of control.

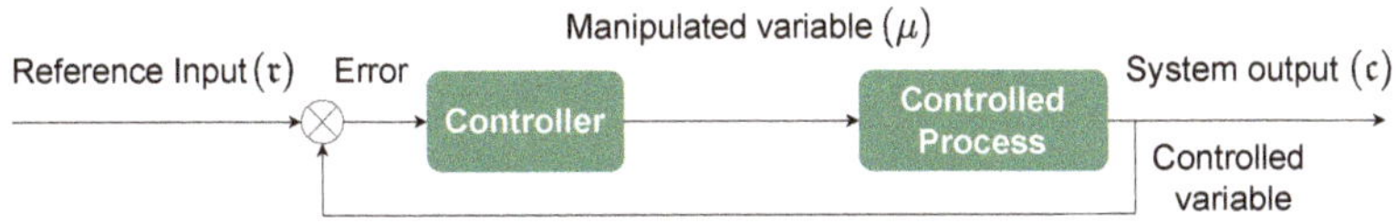

Figure 3. Closed-loop control system for congestion control.

In the proposed control system, the controller feeds from incoming aggregated context measurements (m_i) and determines the value of the manipulated variable μ out of those measurements and the system's reference input $\mathfrak{r}$. The resulting manipulated variable μ is encapsulated in an outgoing directive and shared with any node that might come into contact.

Additionally, Goudar et al. [26] stated that considering the current condition of the OppNet to trigger an actuation is not effective as it would be too late, given the variability of the network. Following their considerations, proactively, we decide the actions to be taken based on predicting the network situation. By this time, presumably, the directive would have been propagated and applied throughout the network. Indeed, if the control system would directly inject the raw aggregated context measurements received from other nodes to the controller for generating a directive containing μ, when eventually this directive would reach the nodes, the network situation might have changed and, possibly, the directive would be no longer adequate for the current situation. The following section shows how we predict the value of a context indicator.

3.5. Context Indicator Prediction

The controller anticipates the context indicator value using a two-step strategy. Firstly, instead of directly considering the received context measurements, the controller aggregates them for a configurable time ($\hat{t}$) using (3). This step results in two lists: $\check{M}$ and $\check{T}$. $\check{M}$ contains the aggregation of the measurements received for periods of $\hat{t}$ seconds. This list is represented as $\check{M} = [\check{m}_j \mid 1 \leq j \leq \check{z}]$, where $\check{z}$ is the max size of $\check{M}$. $\check{T}$ contains the time when each entry in $\check{M}$ was calculated and is represented as $\check{T} = [\check{t}_j \mid 1 \leq j \leq \check{z}]$. Secondly, once several aggregated samples of context measurements are available, these are used as an input of a linear regression (ρ) to calculate the prediction of this context indicator value for time t_{t+n}:

$$m_{t+n} = \rho(t_{t+n}, \check{M}, \check{T}) \tag{4}$$

where $\check{M}$ and $\check{T}$ are sliding lists of size $\check{z}$, t_{t+n} is the time ahead for the context indicator prediction and it is calculated as $t = t + n$ where n is an offset, and m_{t+n} is the resulting predicted value of the context indicator value at time t_{t+n}.

Algorithm 2 shows how the controller calculates the prediction of a context indicator. The *addContextMeasurement* procedure is executed by the controller when a contacted node shares its aggregated context measurement (m). This measurement is stored in the controller's received measurements list M_i (*line* 2). If the window time for receiving measurements from contacted nodes has expired (*line* 3), the controller aggregates all the received aggregated context measurements in the list M_i and its local context measurement, by using the procedure *getAggregatedContextMeasurement* (*line* 5) defined in Algorithm 1. This aggregation result ($\check{m}$) is stored in the list $\check{M}$ along with the current time (*lines 6–7* of Algorithm 2). The controller uses the entries in the above lists to calculate a prediction of the value of the context indicator (m_{t+m}) using a linear regression function ρ (*line 10*). The controller maintains the size of the aggregation calculations list $\check{M}$ at the constant value $\check{z}$ by using a FIFO discarding policy (*lines 12–13*). Notice that at least two inputs are needed to use the linear regression function ρ (*lines 8* and *9*). If this condition is not fulfilled, the predicted measurement (m_{t+n}) assumes the value of the $\check{m}$ calculated at *line 5* (*line 15*). Once m_{t+n} is calculated, M_i is emptied, ready to receive new context measurements from other contacts (*line 19*). A new aggregation window period ($aggrTimeout$) is configured, so all the measurement gathering and the prediction process starts over (*line 20*).

Algorithm 2 Controller's context indicator prediction.

▷ M_i: List of context measurements received from contacted nodes.
▷ m: Received context measurement (controlled variable).
▷ t: Current time.
▷ $aggrTimeout$: Timeout for aggregating incoming context measurements.
▷ $\hat{t}$: Time period for aggregating incoming context measurements.
▷ $\check{m}$: Aggregation of the received context measurements during $\hat{t}$ s.
▷ $\check{M}$: Sliding list of the aggregations so far.
▷ $\check{T}$: Sliding list of the times when the aggregations were performed.
▷ t_{t+n}: Prediction time.
▷ m_{t+n} Predicted value of the context indicator at t_{t+n}.
▷ $directive$: Manipulated variable (μ) encapsulated in a directive

1: **function** ADDCONTEXTMEASUREMENT(m)
2: $\quad M_i.add(m)$
3: $\quad$ **if** $t >= aggrTimeout$ **then**
4: $\quad\quad$ //The window time for receiving context measurements has finished.
5: $\quad\quad \check{m} \leftarrow getAggregatedContextMeasurement(M_i, NULL)$
6: $\quad\quad \check{M}.add(\check{m})$
7: $\quad\quad \check{T}.add(t)$
8: $\quad\quad$ **if** $\check{M}.size() > 0$ **then**
9: $\quad\quad\quad$ **if** $\check{M}.size() > 1$ **then**
10: $\quad\quad\quad\quad m_{t+n} = \rho(t_{t+n}, \check{M}, \check{T})$
11: $\quad\quad\quad\quad$ //Just keep $\check{z}$ values.
12: $\quad\quad\quad\quad \check{M}.removeEldestN()$ ▷ Sliding the list.
13: $\quad\quad\quad\quad \check{T}.removeEldestN()$
14: $\quad\quad\quad$ **else**
15: $\quad\quad\quad\quad m_{t+n} = \check{m}$
16: $\quad\quad\quad$ **end if**
17: $\quad\quad\quad directive \leftarrow createDirective(m_{t+n})$
18: $\quad\quad$ **end if**
19: $\quad\quad M_i.clear()$
20: $\quad\quad aggrTimeout \leftarrow t + \hat{t}$
21: $\quad$ **end if**
22: **end function**

3.6. Directive Generation

The function *createDirective*(m_{t+n}) (*line 17*, Algorithm 2) is called to generate a directive encapsulating the context indicator prediction (m_{t+n}). The former function, defined in Algorithm 3, calculates the manipulated variable's value (μ') from of the context indicator prediction and the reference value (*line 4*). Next, μ' is encapsulated in a directive: $\delta_s = (Id_s, \mu')$ (*line 6*).

As previously mentioned, the reception of a context measurement after contacting a node triggers the generation of a directive. Nevertheless, if the time window for receiving context measurements is set to a high value, it would take a controller a long time to generate a directive. Hence, the nodes would not receive any directive to adjust their initial configured manipulated variable (μ) based on the current network condition, and they would have the perception that there is no controller nearby. Therefore, to prevent this situation, the controller is configured to generate a directive periodically, provided no directive has been generated opportunistically during this period. This directive encapsulates the last calculated manipulated variable (μ), and it acts as a beacon announcing the presence of a nearby controller. Algorithm 3 describes the above behaviour.

Algorithm 3 Directive creation algorithm.

▷ m_{t+n}: Predicted measurement.
▷ opp: Execution mode's flag.
▷ μ: Manipulated variable's current value.
▷ μ': Manipulated variable's new value.
▷ τ Controller system's reference input.
1: **function** CREATEDIRECTIVE(m_{t+n}, opp = FALSE)
2: $\quad\mu' = \mu$
3: $\quad$**if** opp == *false* **then**
4: $\quad\quad\mu' \leftarrow$ apply controller_adjustment(τ, m_{t+n})
5: $\quad$**end if**
6: $\quad$return new Directive(μ')
7: **end function**

Notice that the function *createDirective* in Algorithm 3, receives the parameter *opp* which indicates whether the function is executed periodically, as described above or opportunistically after receiving a context measurement from a contacted node (see Algorithm 2). In the case of a periodic execution of the function *createDirective*, the new value for the manipulated variable (μ') is directly the current one (μ) (*lines* 2 and 6 in Algorithm 3). In the case of an opportunistic execution (*line* 3), the controller calculates the new value for the manipulated variable (*line* 4).

3.7. Directive Dissemination

Although it is a controller that generates a directive, it is stored, carried, and forwarded by any node in the network that receives it. Figure 4 shows this behaviour. When a node receives a directive (dir_{n_1}) (*step* s_4), if the directive is newer than the one the node might be carrying (*steps* s_5–s_6), the node discards the old one (*step* s_7), executes the action encapsulated in the new directive (*step* s_8), and stores, carries, and forwards the new directive (*step* s_9).

Discarding the older directive is a measure to deal with possible inconsistent directives as several controllers are allowed in the network. The newer a directive is, the closer the node is to the controller and, therefore, the more appropriate the directive is. If a node receives several directives, if they are consistent, it means that the controllers are also nearby and are mainly sensing the same context. Conversely, if the received directives are inconsistent, it indicates that the controllers are sensing different parts of the network. In this case, the node likely belongs to the network "segment" controlled by the controller generating the newer directive.

In the following section, we will describe how to apply the control layer for the specific use case of congestion control.

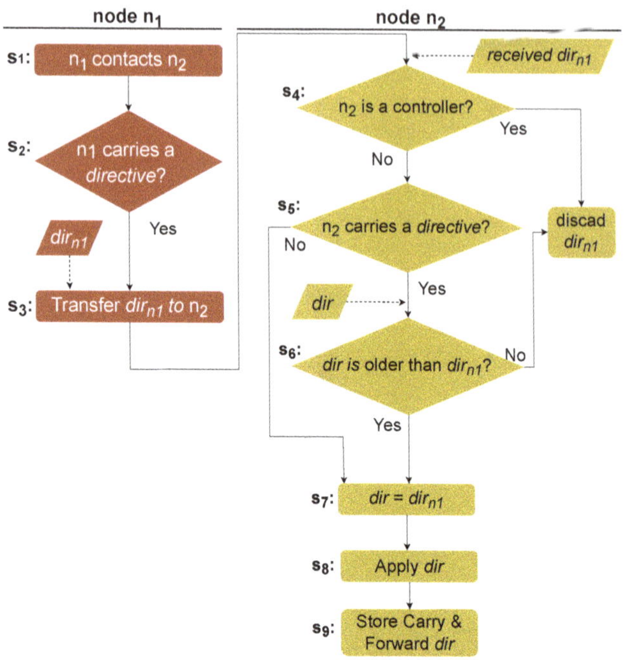

Figure 4. Disseminating a directive when two nodes are in range.

4. Use of the Controller-Driven OppNet Architecture to Manage Congestion

This section describes how the control layer, presented in Section 3, is used over an OppNet characterized by a multi-copy forwarding strategy to manage the congestion intrinsic to replication [20]. Specifically, the controller-driven OppNet will approximate how congested the network is and, out of this knowledge, will adjust the replication limit of the newly created messages and the ones being forwarded.

In the following sections, we specialize for the use case of congestion control: (1) the data message representation the control layer will work with, (2) the adaptation of the controller architecture (presented in Section 3.4) for this particular use case, (3) the context indicator prediction (introduced in Section 3.5), and (4) the directive generation (Section 3.6) and the directive dissemination (Section 3.7). Finally, we summarize the buffer management techniques used by the control layer as a congestion control strategy.

4.1. Control Layer Data Message

In the scope of forwarding algorithms based on message replication, one of the forwarding algorithm parameters is the replication limit of the message. The replication limit determines the total number of copies of the message allowed to exist in the network. As seen in Section 2.2, multiple replication strategies exist. For this particular use case, we consider a forwarding algorithm that uses a binary replication scheme consisting in relaying a copy of the message to the contacted node and reducing by half the replication limit of both the node's message and the relayed copy of it.

The control layer encapsulates the data messages generated by the application layer in the tuple $g = (\mathfrak{a}, l, \varrho, \varphi)$. From this tuple, $\mathfrak{a}$ is the data message generated at the application layer. l is the replication limit of the message (in the case of a binary replication scheme, the message can be relayed $\lceil l/2 \rceil$ times). ϱ is the number of times this particularly message copy has been relayed. This field is incremented each time the message is relayed to the next hop. φ is the *alive* flag. This flag is set to false to indicate that this message is marked to be deleted in case buffer space is required. This message is not deleted straightforwardly as it could happen that the next contact could be the message destination. Furthermore, it could occur that applying another directive would update this message's replication limit to a value equal to or higher than one, providing the message with more chances to be delivered. With this congestion measure, the messages with the field φ set to *false* are reactively removed in case of need, but also, in a proactive way, the message is given a chance to be carried along the network while there is no need for buffer space. This strategy requires the node to not consider the messages with the φ field set to *false* when calculating its buffer occupancy measurement.

4.2. Control Layer Tailored for Congestion Control

The control layer proposed in Section 3.4 has been adapted to use a congestion policy to efficiently limit the number of messages in the network to limit congestion. This congestion policy dynamically determines the replication limit for a newly created message based on the network congestion perception of a controller.

As a network congestion measure, we use the node's buffer occupancy rate (o):

$$o = \frac{\sum_{i=1}^{n} sizeof(g_i)}{b} \tag{5}$$

where n is the number of buffered data messages, *sizeof* is the function that returns a message size in bytes, g_i is a buffered message, and b is the buffer capacity in bytes.

The generic closed-loop control system in Figure 3, for the use case of congestion control, is specialized in Figure 5 with the slight difference that the controller assumes calculating the difference between the reference input ($\mathfrak{r}$) and the controlled variable ($\mathfrak{c}$). The controller's reference input ($\mathfrak{r}$) is an optimal buffer occupancy congestion interval. This interval consists of a range of buffer occupancy rates, defined by a lower and upper bound, o_{min} and o_{max}, respectively. The buffer occupancy is considered optimal when the buffer is neither underused nor close to its maximum capacity.

The controller feeds from the buffer occupancy measurements (o) received from contacted nodes and uses those measurements to calculate a buffer occupancy prediction (o_{t+n}) following Algorithm 2.

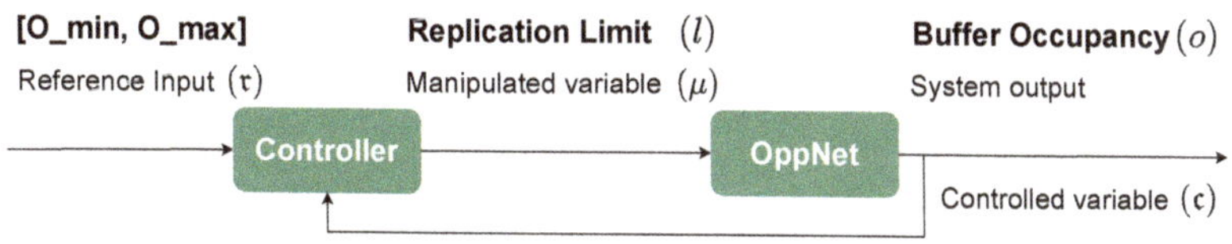

Figure 5. Closed-loop control system for congestion control.

4.3. Network Congestion Prediction

The controller uses the optimal congestion interval along with the calculated o_{t+n} to determine its prediction of the network's congestion. The network's congestion status falls into three states: *UNDER_USED*, *OPTIMAL*, and *CONGESTED*. The controller uses the following function to sentence which is its prediction of the network's congestion state (c_{t+n}):

$$c_{t+n}(o_{t+n}) = \begin{cases} \text{UNDER\_USED} & \text{if } o_{t+n} < o_{min} \\ \text{OPTIMAL} & \text{if } o_{min} \leq o_{t+n} < o_{max} \\ \text{CONGESTED} & \text{if } o_{t+n} \geq o_{max}. \end{cases} \tag{6}$$

4.4. Directive Generation for Congestion Management

Based on its congestion prediction (c_{t+n}), the controller calculates the maximum number of copies of a message allowed to be in the network (l). For that matter, this proposal considers a proportional controller (P-controller) [29] using an additive increase and multiplicative decrease (AIMD) factor to adjust the manipulated variable (l). The considered

AIMD factor uses k_1 as a multiplicative decrease factor and k_2 as an additive increase factor. Out of the predicted network congestion, the P-controller calculates the new replication limit l'.

$$l'(c_{t+n}) = \begin{cases} l & \text{if } c_{t+n} = \text{OPTIMAL} \\ l \cdot k_1 & \text{if } c_{t+n} = \text{CONGESTED} \\ l + k_2 & \text{if } c_{t+n} = \text{UNDER\_USED} \end{cases} \quad (7)$$

With this control function, if the predicted network's congestion state is *OPTIMAL*, it is not necessary neither to increase nor decrease the replication limit of a message. When the predicted network's congestion state is *CONGESTED* or *UNDER_USED*, the controller decreases or increases the message's replication limit, respectively.

The controller encapsulates the calculated l' in a directive: $\delta_l \rightarrow (Id_l, l')$, where Id_l is the identifier of the message replication limit setting and l' is the calculated replication limit. Following the flowchart described in Figure 4, when the controller contacts a node, it forwards the former directive along with the calculation of its buffer occupancy.

4.5. Applying a Directive

Applying a directive $\delta_l = (Id_l, l')$ entails two actions: (1) using the encapsulated replication limit l' when creating new data messages and (2) updating all the buffered messages according to the new l'. The first action consists of using the new replication limit l' when an application creates a new message ($g = (\alpha, l', \varrho, \varphi)$). The second action consists of updating the message's field l of the buffered messages, considering that some are already copies.

Indeed, it would not be accurate to update the buffered messages with the new replication limit l' received through δ_l, as some of these messages are already copies. Therefore, the number of times this message has been relayed (message's ϱ field) is considered. This information is used to calculate the current replication limit of the buffered message, considering that: (1) the message was created with the replication limit set by the last received δ_l and (2) $\lceil l/2 \rceil$ copies of the message are forwarded at each encounter:

$$l'' = \frac{l'}{2^{\varrho}} \quad (8)$$

where l'' is the remaining replication limit after relaying the message ϱ times.

Algorithm 4 shows the procedure for modifying the replication limit of the buffered messages. For each of the buffered messages (*line 2*), provided the replication limit it was assigned when created was l' (from the received directive $\delta_l = (Id_l, l')$) and taking into account the number of times that the message has been relayed (ϱ), the remaining replication limit (l'') is calculated using (8) (*line 3*). In case the calculated l'' is less than one, it would mean that by starting with the replication limit specified in the received directive, the message would not have any copies left to disseminate at this point and, therefore, the message would not reach the current node. If this is the case, the message's field φ is set to *false* (*line 5*), indicating that this message is marked to be deleted in case buffer resources are required.

4.6. Buffer Management

The control layer applies a congestion control mechanism based on buffer management. Besides the congestion policy limiting the number of message copies in the network, the control layer applies a hybrid drop policy.

The control layer foresees the network congestion (Section 4.3). Based on the predicted congestion, it limits the number of message copies (Section 4.4) and updates the copies left of the queued messages (Section 4.5). Proactively, if this update results in the message having no copies remaining, the message is marked to be deleted by setting the message's flag φ to false. Reactively, in case of buffer overflow, the messages marked to be deleted are dropped. The control layer applies basic drop policies if more buffer space is required. It first applies a *drop-oldest* policy based on the message's remaining TTL. Next, if necessary, it applies a *drop-head* policy removing the oldest ones.

Algorithm 4 Procedure to update the buffered messages with the l set by a directive.

```
▷ δ_l, Received directive (δ_l = (Id_l, l'))
▷ l': The replication limit in the received directive (δ_l = (Id_l, l')).
▷ g: A message (g = (α, l, ϱ, φ)).
▷ B: A buffer to store messages.
function APPLYDIRECTIVETOBUFFEREDMESSAGES(δ_l)
    for all g ∈ B do
        l'' = l' / 2^ϱ
        g[l] ← l''                    ▷ Setting the new rep. limit in the message's field l
        g[φ] ← (l'' < 1) ? false : true
    end for
end function
```

5. Experimentation

The controller-driven OppNet, which uses a quota-based multi-copy forwarding with a dynamic message replication limit, is named the control configuration (Control). The control configuration is compared with two *No-Control* multi-copy baseline forwarding algorithms: epidemic (EP) and a quota-based algorithm (Static), both presented in Section 2.2. In epidemic routing, nodes forward messages to every encountered node to achieve maximum network coverage. Static routing sets a static upper bound of the number of message replicas in the network. It distributes half of these copies to each contact (provided the contact does not carry copies of the message yet) until the node has only one copy left, which will carry up to the destination. Both no-control approaches are multi-copy baseline forwarding algorithms considered for benchmarking in opportunistic networking research [30].

This section describes how the control configuration performs. The experimentation methodology follows the guidelines pointed out by Dede et al. [31], and Kuppusamy et al. [32]: (1) appropriate mobility models to favour different congestion degree situations are designed; (2) our proposal is compared with benchmark multi-copy forwarding algorithms; (3) several performance metrics, listed in Section 5.1, are evaluated over the Control and No-Control configurations; moreover, the network performance is also evaluated for the Control configuration for different values of its configuration settings listed in Section 5.4.2; (4) the experimentation setup is detailed and well documented to be reproduced for benchmark purposes; and (5) the simulator provides the link model and the physical aspects are not considered.

5.1. Performance Metrics

We use the standard metrics to measure the performance of the Control and No-Control configurations [33]:

Delivery Ratio (σ): measures the ratio of created messages that are delivered to the final destination:

$$\sigma = \frac{\#g_d}{\#g_c} \tag{9}$$

where $\#g_d$ is the number of delivered messages, and $\#g_c$ is the number of created messages.

Latency average ($\overline{\lambda}$): the average time it takes for the created messages to get delivered to their final destination:

$$\overline{\lambda} = \frac{\sum_{i=1}^{w} \lambda_i}{w} \tag{10}$$

where w is the number of messages delivered to the destination, and λ_i is the elapsed time from the message creation to its delivery.

Overhead ratio (θ): measures the average of the message copies needed to deliver the message to its final destination:

$$\theta = \frac{\#g_r - \#g_d}{\#g_d} \tag{11}$$

where $\#g_r$ is the number of relayed copies, and $\#g_d$ is the number of delivered messages.

5.2. Scenarios

We use four scenarios which use different mobility patterns representing different network conditions. These scenarios are classified into two groups: (1) the scenarios based on real-world mobility traces, available at the Crawdad database [34] and (2) the synthetic scenarios generated by a random waypoint model (RWP) in a grid with reflective barriers where the nodes move at a configured speed for a configured distance. The nodes keep changing direction each time they cover that distance. The scenarios based on real mobility traces are very convenient for evaluating this proposal under real network conditions. In contrast, the synthetic scenarios help us to recreate particular network conditions as emergencies, not yet covered with real mobility traces samples. The considered scenarios are:

Taxis: tracks 304 Yellow Cab taxis in the San Francisco Bay area for one week. The traces are available at [35].

Info5: tracks the movement activity of 41 students attending the Infocom conference in 2005 over three days [36].

Campus: a synthetic map-based scenario that simulates the mobility activity of 80 students at Autonomous University of Barcelona campus, covering an area of 4.5 km $\times$ 3.4 km with defined points of interest (POI) corresponding to eight faculties and the railway station. The students walk throughout the campus arriving/leaving their faculty and the railway station with a certain probability.

Emergency: a synthetic RWP-based scenario with 100 nodes randomly walking in an area of one square kilometre. In this scenario, the pattern of message generation changes abruptly in the second half of the simulation, aiming to roughly simulate network conditions under the disrupting events of an emergency.

Each scenario has different node densities. A sparse nature in a scenario entails fewer contact opportunities among the nodes, whereas dense natures are prone to more contacts between nodes. The Taxis scenario is a sparse scenario, and the Campus is a sparse scenario with POI promoting a temporary high density of nodes. Info5 and Emergency are both dense scenarios.

5.3. Message Generation Distribution

Depending on the scenario, for the data message generation, we will be using either a constant bit rate (CBR) distribution or an inverted smoothed top hat distribution (ISTH) [37]. Next, we formalize the ISTH for the specific case of a 24 h working day.

5.3.1. Inverted Smoothed Top Hat Distribution (ISTH)

We aim to mimic the network traffic in a working day with the ISTH distribution. The flat region (FR) of the ISTH represents the working hours where the network traffic is the heaviest, i.e., messages will be generated at a higher rate. The transient regions (TR) are shaped in two ways: (1) to logarithmically increase the rate of message generation up to reach the peak rate of the FR; and (2) to logarithmically decrease the rate of the message creation from the peak rate in the FR to a shallow rate when approaching the 24th hour of the day.

The ISTH function is built as a composition of a descendant logistic functions, an ascendant logistic function, and a linear function. Both logistic functions determine the message generation frequency for a time:

- Descendant logistic function (exponential growth rate$(k) > 0$):

$$f(x) = \frac{L_2}{1 + ae^{k(x-x_0)}} \tag{12}$$

- Ascendant logistic function ($k < 0$):

$$g(x) = \frac{L_2}{1 + ae^{-k(x-x_0)}}. \tag{13}$$

These functions are bounded by two limits: L_2 and L_1. L_2 corresponds to the lowest message generation rate (highest value) used to generate very low traffic. L_1 corresponds to the highest message generation rate (lowest value) used to generate the highest network traffic. $x - x_0$ is the flexible horizontal translation. No horizontal translation is considered: $x_0 = 0$. k is the exponential growth rate.

From 00:00 a.m. to 09:00 a.m., the descendant function defined in (12) needs to descend from L_2 down to L_1, i.e., $f(0) = L_2$ and $f(9) = L_1$. Both L_2 and L_1 limits are specified through the control configuration settings depending on the scenario. Hence, from (12), the only unknown variable (a) is isolated:

$$a = \frac{(L_2 - L_1)}{L_1 e^{k9}}.$$

The ascendant logistic function in (13) follows the same process to ascend from L_1 at the end of the working hours of the day (17:00 h) up to L_2 at (00:00 h): $g(17) = L_1$ and $g(0) = L_2$.

Finally, to build the ISTH function it is necessary to combine the descendant and ascendant logistic functions with a flat linear function that covers the eight working hours of the day (from 09:00 a.m. to 17:00 p.m.). During this time window, messages are generated at L_1 rate:

$$h(x) = \begin{cases} 0 \leq x < 9: & f(x) \\ 9 \leq x < 17: & L1 \\ 17 \leq x < 24: & g(x) \end{cases}. \tag{14}$$

5.3.2. Scenarios' Message Generation Distribution

Table 1 summarizes the message creation distribution for the different scenarios. The Taxis scenario uses an ISTH distribution that replicates two working days, where the messages are generated each 10 to 60 s uniformly distributed for each working day during eight peak hours. With all the scenario specifics, it is considered a low- to medium-congestion scenario.

The Info5 scenario uses the aforementioned message generation distribution. Given the message generation distribution and the fact that the nodes congregate around the events of the congress, it is considered a medium- to high-congestion scenario.

The Campus scenario uses an ISTH distribution resembling a workday where, during the peak hours, messages are created every 10 s. Hence, this scenario is considered a high-congestion scenario.

For the Emergency scenario, during the first eight hours, messages are generated at the high rate of each 10 s using a CBR distribution followed by eight hours of low-rate message generation (every 80 s). Therefore, Emergency is a variable-congestion scenario.

Table 1. Message generation distribution per scenario.

Settings	Taxis	Info5	Campus	Emergency
Simulation time	69 h	70 h	34 h	23 h
Message distribution	ISTH	ISTH	ISTH	CBR
Message generation frequency	FR: [10–60] s; TR: 1800 s	FR: [10–60] s; TR: 900 s	FR: 10 s; TR: 1800 s	0–8 h: 10 s; 8–16 h: 80 s
Congestion level	Low-Medium	Medium-High	High	Medium-High

5.4. *Environment Setup*

For the experimentation, the Opportunistic Network Environment (ONE) simulator [38] was used, which is designed specifically to simulate OppNets. Recent works show that it is the most used simulator for OppNets [32]. The ONE has proven easy to configure and provides an extensive set of mobility, traffic models, and propagation protocols [31]. The control layer has been developed on top of the simulator's network layer and is available through a public repository (https://github.com/MCarmen/the-one/tree/control, accessed on 3 December 2022).

The simulations over the synthetic scenarios use a traces file with node contacts generated with the built-in RWP model of the simulator to preserve the same scenario over different simulation rounds. Next, we will describe the configuration settings common to all the scenarios and the specific settings by scenario.

5.4.1. Common Configuration Settings

Table 2 lists the common simulation configuration settings. For all the scenarios, the nodes are configured with a WiFi interface with a transmission speed of 100 Mbps and a transmission range of 60 m as an approximation of the WiFi 5 (802.11ac) standard.

In each simulation cycle, any random node in the network creates a message to a randomly selected node to approximate a real-world communication model. For the simulations, it is considered that when two nodes are in range, they have enough time to exchange the control protocol data, the messages to be delivered to the contacted node, and the messages to be relayed.

Table 2. Summary of the common simulation settings for all the scenarios.

Setting	Value	Setting	Value
Network interface	Wi-Fi	Battery	none
Transmission speed	100 Mbps	Type of nodes	pedestrians
Transmission range	60 m	User behaviour	none
Interference	none	Application	Single destination
Power consumption	none		

5.4.2. Scenario and Control Configuration Settings

The nodes' buffer size, the messages' TTL, the messages' size, and message generation frequency, which directly affect the network congestion, vary for each scenario to create different congestion conditions. The first part of Table 3 lists the values for the above scenarios' settings. In this table, the intervals specifying the value for the settings: message generation frequency, message size and walk speed denote a uniform distribution between the two interval limits. Notice that for all the scenarios, the nodes' buffer size is set to 10 M to favour congested situations, mainly when messages are generated at a high rate.

Also, the control layer can be customized through the settings listed in the second part of Table 3. Next, we describe the customisable control settings:

Number of controllers: indicates the number of nodes that will act as controllers in the network.

Optimal congestion interval ($o_{min} - o_{max}$): as presented in Section 4.2, this setting specifies the optimal range of the node's buffer occupancy.

Additive increase (k_2) and multiplicative decrease (k_1) (AIMD): denote the factor to be added to (k_2) and the factor to multiply by (k_1) the current replication limit (l) to update the former replication limit based on the network congestion status through (7).

Aggregation interval ($\hat{t}$): time while the controller gathers congestion measurements (see Section 3.5).

LR nrof inputs ($\check{z}$): maximum size of the congestion readings aggregation list $\check{M}$. This list is used as an input for the congestion prediction function. $\check{M}$ works as a sliding list of size $\check{z}$ to consider a recent history of readings for the prediction.

Prediction time factor (ϕ): the multiplicative factor applied over the aggregation interval setting ($\hat{t}$) to determine the time (t_{n+t}) for a congestion prediction (Section 3.5, Algorithm 2, *line 10*). ϕ is calculated by the equation:

$$t_{t+n} = t + \hat{t}\phi \tag{15}$$

where t is the current time, and $\hat{t}$ is the time interval for aggregating congestion readings.

Directive generation frequency: determines the periodicity of the automatic directive generation triggered in case the controller does not receive any congestion reading for this period of time (see Section 3.6).

Reduction factor for a specific decay (r): the reduction factor to apply to get a certain decay. Used in (1) (Section 3.3).

Decay threshold: when a controller receives a congestion measurement with a decay lower than *decay threshold*, it is discarded, and hence, it is not used to estimate the congestion.

Number of aggregations weight (α): weight factor applied over the number of aggregations a congestion measurement is built on. Used in (3) (Section 3.3).

5.5. Results

This section shows and evaluates our proposal (Control) and the No-Control configurations, introduced in Section 5, for different scenarios, in terms of (1) the buffer occupancy, (2) the performance metrics listed in Section 5.1, and (3) the delivery ratio for different values of the controller settings listed in Section 5.4.2. Before delving into the comparison between the Control and No-Control configurations, for the Control configurations, we analyze the replication limit (l) it tends toward. Along this section, the Control configuration will be compared to the Static configurations with the same l that the Control tends toward (Static*).

For the No-Control Static routing policy, simulations have been run with different replication limits to show the tendency of the metric's value. We have narrowed the replication limit to the scenario's number of nodes. We consider that having as many copies of the message as nodes are in the network is an approximation of epidemically flooding the network.

Finally, we provide all the obtained results, the ONE configuration files for all the scenarios, the script files to run the simulations, and the data traces for the reproducibility of these results (https://deic.uab.cat/~mcdetoro/controller-driven_OppNet_results.zip, accessed on 3 December 2022).

Table 3. Summary of the specific simulation settings per scenario.

Scenario Setting	**Taxis**	**Info5**	**Campus**	**Emergency**
Simulation time	69 h	70 h	34 h	23 h
Simulation area	San Francisco Bay	hotel	4.5 × 3.4 km	1 km^2
Mobility model	Contact Traces	Contact Traces	Map-Based + POI	RWP
# Nodes	304	41	80	100
# Contacts	69,412	22,459	168,442	38,013
TTL (s)	10,000	10,000	10,000	4000
Buffer size	10 M	10 M	10 M	10 M
Message generation distribution	ISTH	ISTH	ISTH	CBR
Message generation frequency (s)	FR: [10–60] s; TR: 1800 s	FR: [10–60] s; TR: 900 s	FR: 10 s; TR: 1800 s	0–8 h: 10 s; 8 h–end: 80 s
Message size	[10–500] k	[10–500] k	[10–500] k	0–8 h: 500 k; 8 h–end: 10 k
Walk speed	N/A	N/A	0.5 m/s	[0.5–1] m/s
Control Settings	**Taxis**	**Info5**	**Campus**	**Emergency**
Nrof controllers	10	2	4	20
$[o_{min} - o_{max}]$	[0.6–0.7]%	[0.3–0.5]%	[0.6–0.7]%	[0.7–0.9]%
Additive increase (k_2)	1	1	1	1
Multiplicative decrease (k_1)	0.25	0.25	0.25	0.25
LR nrof inputs ($\check{z}$)	6	10	6	6
Aggregation interval ($\hat{t}$)	60 s	30 s	300 s	120 s
Prediction time factor (ϕ)	2	5	2	2
Directive generation frequency	900 s	900 s	900 s	900 s
Reduction factor for a Decay (r)	0.103 (d of 5% at 1800 s)	0.3 (d of 1% at 300 s)	0.058 (d of 5% at 300 s)	0.3 (d of 1% at 300 s)
Decay Threshold	0.1	0.1	0.1	0.1
nrofAggregations weight (α)	0.2	0.2	0.2	0.2

5.5.1. Replication Limit Tendency for the Control Configuration

Figure 6 depicts the tendency of the calculated Control replication limit along the simulation. We observe that the calculated l tendency is inverse to the filling up of the buffer for each scenario (Figure 7), i.e., the calculated l values are initially high, corresponding to the period where buffers are still not overwhelmed but tend to decrease along the simulation depending on the congestion readings. Specifically, the l values for Taxis tends toward 8, Info5 tends toward 2, Campus tends toward 2, and Emergency tends toward 4.

The controller's goal is keeping a high l, aiming for a higher delivery ratio and lower latency while the buffers are not stressed, and lowering l to prevent this stress from happening. Following this behaviour, for the less congested scenario (Taxis), where the controller can keep a higher l value, we observe that the controller decreases l slowly. Conversely, for higher congested scenarios (Info5 and Campus), where high l values would rapidly overwhelm the buffers, we see that the controller decreases l much faster. Specifically, we observe that the controller reduces the l in the Info5 scenario faster than in the Campus scenario. Indeed, for the Info5 scenario for all the No-Control configurations, the buffer is similarly overwhelmed, whereas, for the Campus scenario, it depends on the No-Control replication limit configuration. More precisely, we can observe the controller's capacity to adjust the l in the Emergency scenario, where in the first half of the simulation (the more congested phase) the controller decreases the l and increases it in the second half of the simulation (the less congested phase).

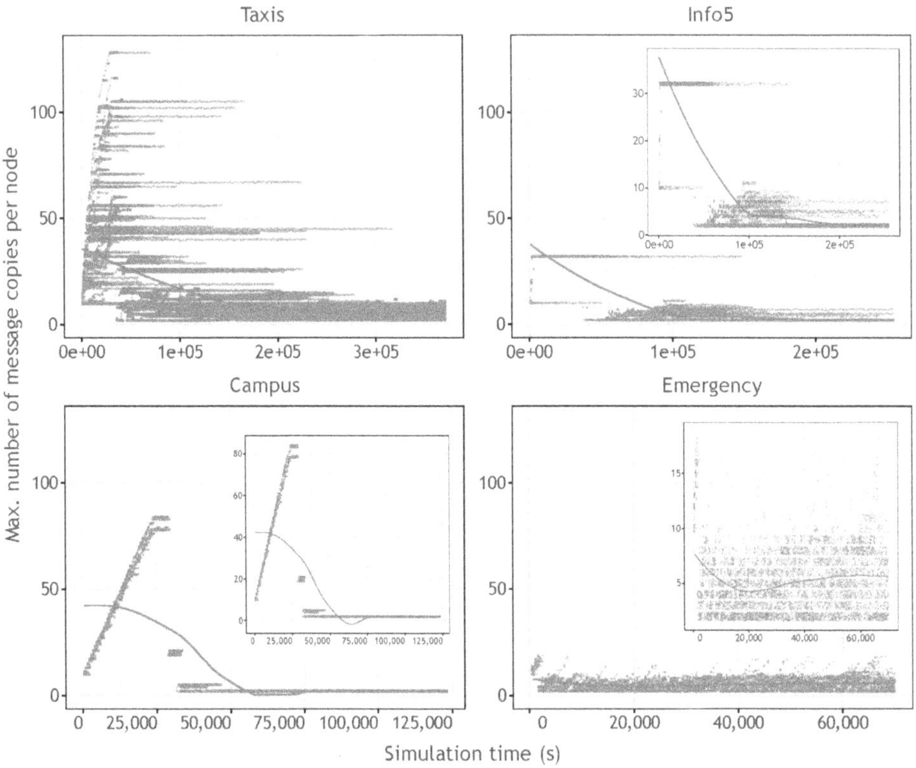

Figure 6. For the Control configuration, progression of the replication limit used by each node at each time unit when a new message is created. In blue, we show the tendency of the replication limit over time (for Taxis it is 8, for Info5 it is 2, for Campus it is 2, and for Emergency it is 4). For the scenarios Info5, Campus, and Emergency, a zoomed-in plot with a different scale has been embedded in the main plot.

5.5.2. Buffer Occupancy Evaluation

Figure 7 shows that, for all the scenarios and for all the configurations except for the Control configuration, the buffer fills in a logarithmic manner up to the buffer's total capacity. For the Emergency scenario, we can see an inflexion point that derives to a lower buffer occupancy at the simulation time when the message generation distribution changes from high frequency to low frequency.

Without any replication limit (l), the EP policy fills up the buffer faster than the other policies. The Static policy's static l determines the speed at which the buffer fills up. For the Control configuration, as the control system regulates the replication limit based on congestion information readings from the nodes, the buffer occupancy fluctuates based on the effects of the new replication limit values.

Overall, the buffer occupancy is lower with the Control configuration. For all the scenarios, after a transient period, the buffer utilization by the Control policy tends toward 21% for the Taxis scenario, 87% for the Info5 scenario, 32% for the Campus scenario, and 22% for the Emergency scenario. Nevertheless, this remarkable difference between the buffer utilization by the Control and the No-Control configurations is due to how we measure the buffer occupancy for the Control configuration. For Control, the buffer occupancy measure does not count the messages that are still buffered but have the flag φ set to *false*, so that if buffer space is required, those will be the first messages to be discarded.

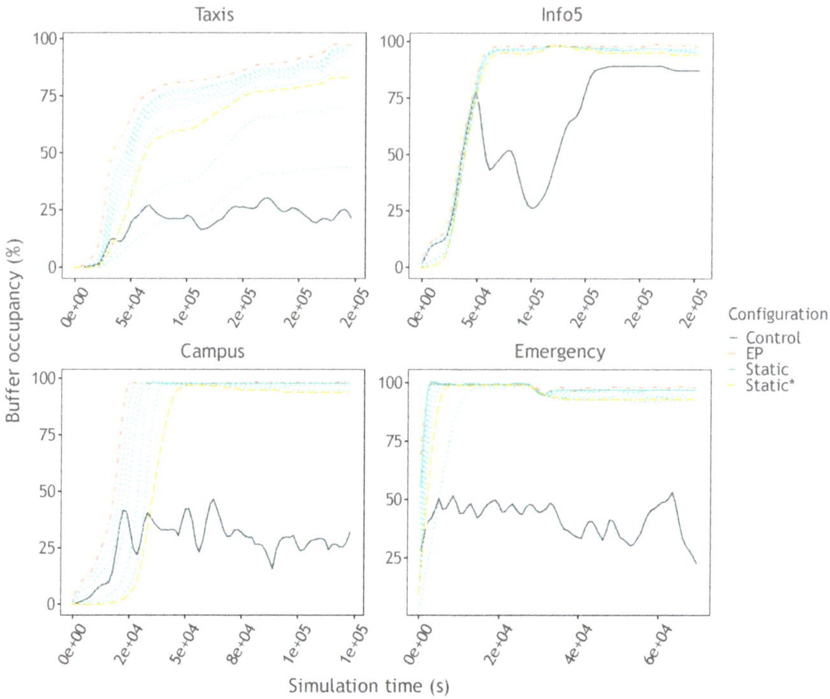

Figure 7. Percentage of the buffer occupancy for the different policies, for each scenario, along the simulation time. In the legend, the configuration named Static* corresponds to the Static configuration with the replication limit that the Control tends toward.

Figure 8 shows the percentage of dropped versus relayed messages. As expected, for the No-Control configurations, the faster the buffer fills (Figure 7), the more messages are dropped. Nevertheless, the Control configuration does not have a lower drop rate despite its lower buffer occupation. Precisely, this is because of the aforementioned detail that the Control configuration measures the occupancy of the buffer such that messages that are still buffered but have the flag φ to *false* will not be counted when calculating the buffer occupancy, but when the buffer requires the space, they will be dropped. Moreover, as expected, for the more congested scenarios (Info5, Campus, and Emergency), the Control drop rate is slightly higher than that of the Static* configuration. This difference is caused by the changes in the replication limit adjusted by the controller.

More precisely, as expected, for all configurations of the less congested scenario (Taxis), the dropped message rate corresponds to the buffer occupancy, as the buffers are not stressed during the simulation. Furthermore, the relation between the buffer occupancy and the dropped message rate remains for the Info5 scenario, a medium–high congestion scenario, where buffers are more stressed. Nevertheless, for the Campus scenario, a highly congested scenario, and for the Emergency scenario, which has a high-congestion phase, for the Control configuration, the high buffer occupancy ends up dropping the buffered messages with the flag φ set to *false*, and, therefore, the dropped message rate is at par with that of the Static* configuration, with slight differences caused by the changes to the replication limit adjusted by the controller.

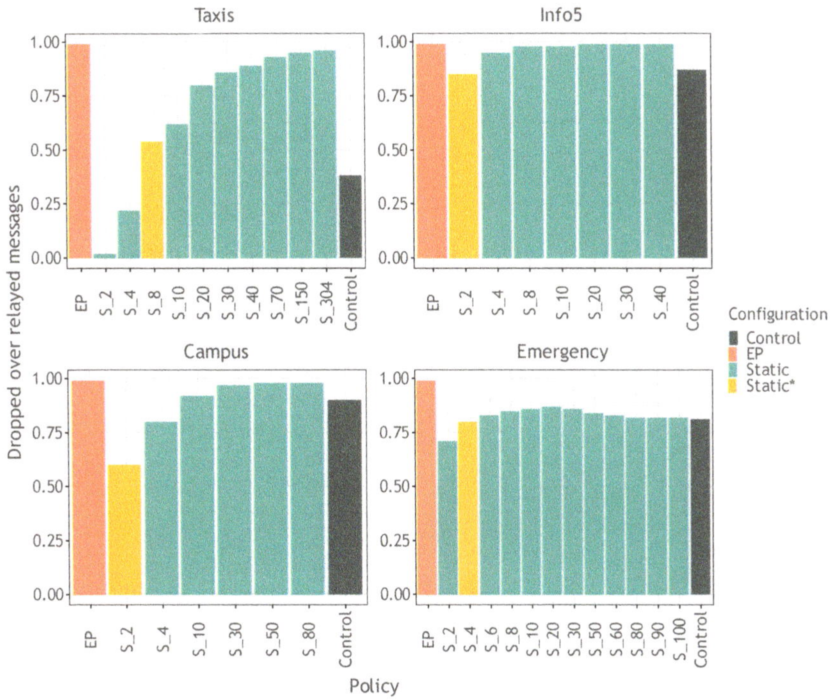

Figure 8. Percentage of dropped versus relayed messages by the different policies. The Static suffix denotes the replication limit. In the legend, the configuration named Static* corresponds to the Static configuration with the replication limit that the Control tends toward.

5.5.3. Performance Evaluation

This section presents and evaluates the results for the performance metrics listed in Section 5.1 for the Control and No-Control configurations in the different scenarios.

Overhead Ratio

Figure 9 shows that the overhead derived from the relay of the message copies depends on the l. EP's overhead surpasses Static and Control by two to three orders of magnitude, while Statics's overhead increases for higher l values. The Control's overhead ratio is similar to that of Static*. The slight difference between the two configurations is due to Control's automatic recalculations of l.

Delivery Ratio

The dropped messages and the overhead directly affect the delivery ratio performance. As we can see in Figure 10, a high replication limit takes its toll on the delivery ratio performance.

With the highest replication limit, the EP policy floods the network, triggering a significant number of drops and, therefore, begets the worst delivery ratio. The behaviour above also applies to the Static policy. The higher the l is, the poorer the delivery ratio we obtain.

Indeed, the Info5 and Campus scenarios (the more congested scenarios), obtain the highest delivery ratio with a Static policy with a low l: 2 in both cases. As l increases, the delivery ratio performance decreases. However, for the Emergency scenario, which combines a high message generation frequency with a low one, and for the Taxis scenario, with a low–medium congestion level, the delivery ratio increases for the values of l up to the inflexion point of the Static's l with the best delivery ratio. This behaviour is coherent with the fact that high values of l prejudice the congested scenarios. In contrast, the low–medium congested scenarios admit higher values of l, favouring a higher delivery ratio.

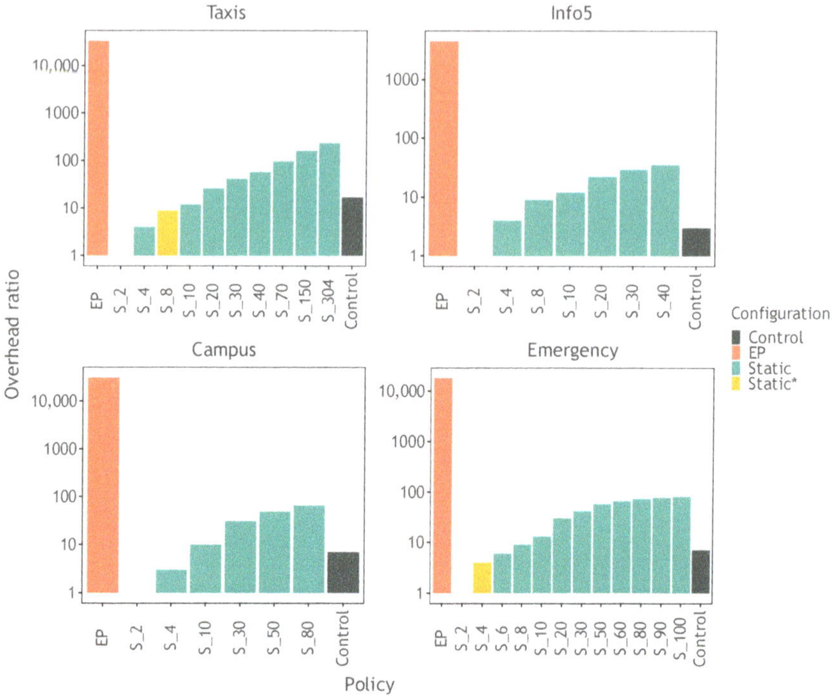

Figure 9. The overhead percentage for the different forwarding policies per scenario in a logarithmic scale.

Overall, the Control policy obtains the best delivery ratio for all the scenarios. We obtain the best increase ratio in the scenarios with low–medium congestion levels. We specifically obtain a 14% and an 11% increment in the delivery ratio for Taxis and Emergency, respectively, over the Static configuration with the best-performing l value. As we have previously seen, these scenarios admit higher l values, bringing on a higher delivery ratio. The ability of the Control policy to dynamically adapt the l provides an optimal l value depending on the current congestion situation at the current time. This flexibility allows the Control policy to stand out in low–medium congestion scenarios over the other configurations. On the other hand, for highly congested scenarios, where the best option is to keep a very low l close to direct delivery, the Control policy provides a low l value. It also benefits from the dynamism and slightly outperforms the Static policy with the best-performing l value by 4% and 9% for the Info5 and Campus scenarios, respectively.

Latency Average

Figure 11 shows that, despite the crushing effects of the message flooding strategy over the buffer occupancy, delivery ratio, and overhead, when it comes to the latency, message flooding benefits the arrival of the messages to their destination and, therefore, it obtains a good performance. Indeed, as pointed out by Krifa et al. [21], flooding-based replication benefits the latency of the messages at the expense of the delivery ratio in case of congestion. This is because dropped messages will not reach the destination, decreasing the delivery ratio. In contrast, with high message dissemination, the more buffered copies, the more chances that a message copy will have to be delivered upon an opportunistic contact, despite plenty of dropped messages. With this premise, we can see that the configurations that fill up the buffer faster (Figure 7), as the EP and Static configurations with the highest replication limit perform worst in terms of delivery ratio (Figure 10) but better in terms of latency (Figure 11).

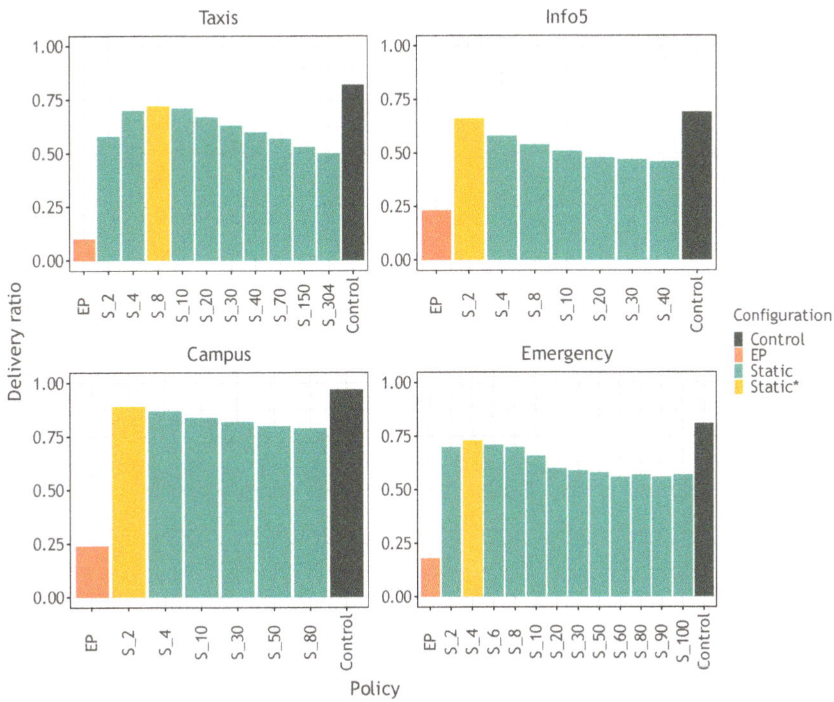

Figure 10. Delivery ratio percentage for the different forwarding policies per scenario.

As for the Control configuration, the premise above applies. More specifically, the Control configuration for the less congested scenario (Taxis) achieves a better performance than EP and Static* due to its low buffer occupancy, backed by a low dropped message rate. Nevertheless, the configurations with a static high replication limit leverage from high replication to obtain lower latency than the Control configuration. For the medium- to high-congestion scenario Info5, where the buffer occupancy and dropped message ratio are close to the No-Control configurations, a high replication benefits a lower latency. Furthermore, for the most congested scenarios (Campus and Emergency), where the Control configuration ends up with a high dropped message rate, the highest replication configurations obtain a better latency performance.

Finally, Figure 11 includes the standard deviation of the latency values, revealing that for all scenarios and configurations there is a high variance between the messages' latency values.

5.5.4. Evaluation of the Control Settings Impact on the Delivery Ratio

The control layer is configurable through the settings listed in Section 5.4.2. We have run simulations over the four selected representative scenarios to analyze the impact of the Control configuration's settings on the delivery ratio over diverse scenarios and to find a general configuration that fits all of them. The following nine sections present our analysis.

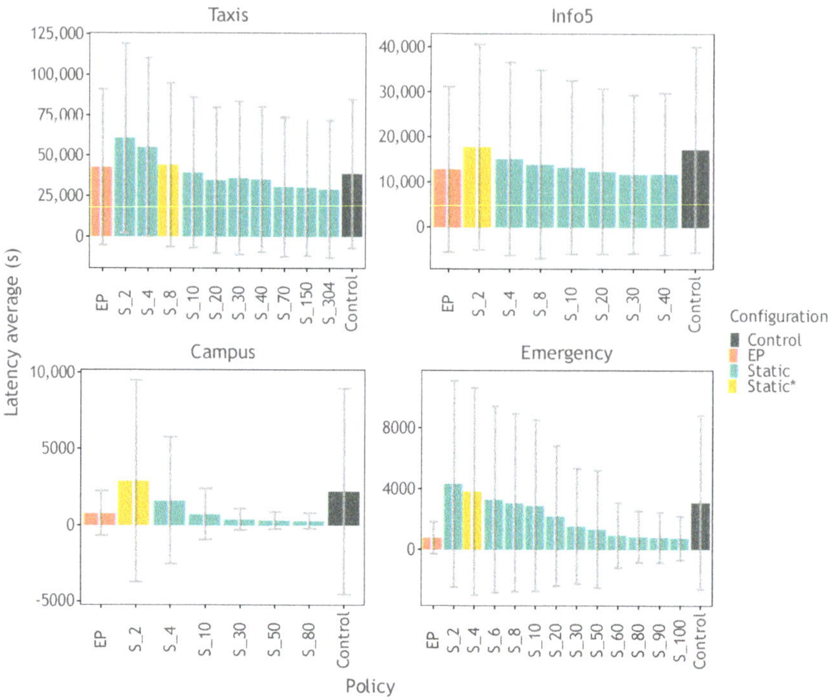

Figure 11. Latency average for the different forwarding policies per scenario.

Number of Controllers

Figure 12 shows the delivery ratio depending on the number of controllers used per scenario. In this plot, the max number of considered controllers is 50. Not all the scenarios have been simulated for all the considered number of controllers as, for example, the Info5 scenario has just 41 nodes. Certainly, as pointed out in Section 3.4, the number of controllers needed to orchestrate the OppNet depends on the nature of the network. In addition, the characteristics of the four simulated scenarios, including the number of nodes, are very different. Therefore, for each scenario, we terminated the simulation as soon as the results showed a clear descending slope corresponding to an increasing number of controllers. Specifically, Figure 12 shows that for all the scenarios, a small number of controllers perform better in terms of delivery ratio. For the most connected scenarios, Campus and Info5, the best performance is achieved with just four and two controllers, respectively.

As the controller receives the congestion readings from the nearby nodes, it obtains an overview of the congestion of a part of the network, its nearest part. We can elaborate on this idea by considering that the network is "segmented" by the number of controllers used. Each network "segment" consists of the number of nodes that can be reached by a controller directly or through short time relays.

Having said that, in a highly connected network, using a high number of controllers results in an overlap of the different network segments, as each controller can reach several of these segments. This overlapping effect results in the nodes receiving directives from different controllers. Of course, a directive emitted from a controller from a segment the node does not belong to has congestion information that is not entirely accurate for the node. This overlapping effect is why using many controllers decreases the network performance.

On the other hand, for the sparser scenario (Taxis) and the scenario with an abrupt change in the communication conditions (Emergency), the implicit segmentation derived by the different controllers obtains disjointed segments. Under these circumstances, having a higher number of controllers (10 for Taxis and 20 for Emergency) helps cover a broader network range, translating to better performance.

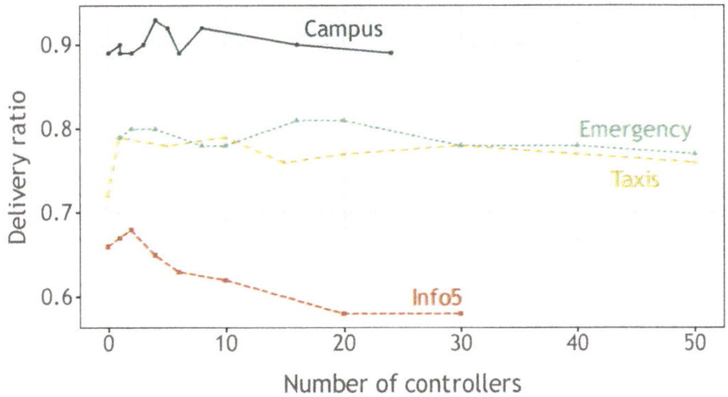

Figure 12. Delivery ratio by number of controllers.

Optimal Congestion Interval ($o_{min} - o_{max}$)

As presented in Section 4.4, the current replication limit is not modified when the congestion calculated by a controller falls in the optimal congestion interval.

Figure 13 shows that for the most congested scenario (Campus), there is a similar delivery ratio for all the optimal congestion intervals. The variance between the performance results for the different intervals is just 0.0002. Nevertheless, we obtain the best result for the interval 0.5–0.8.

These homogeneous results infer that in a congested scenario, if the congestion predictions fit in the configured optimal congestion interval setting for the current replication limit, the best strategy is to keep the current replication limit steady.

The medium- to high-congestion scenario (Info5) shows slightly more variance in the delivery ratio than the previous scenario (0.001). This variance can be appreciated mainly when the interval upper bound (o_{max}) is 0.8 and 0.9. Therefore, when the optimal congestion interval upper bound is set close to the maximum buffer occupation, the replication limit set by the controller is too high. Thus, a more conservative optimal congestion range gives better results which, in this case, is 0.6–0.7.

For the Taxis scenario, the sparsest scenario with fewer contact opportunities, it can be seen that the most conservative interval configuration, 0.3–0.5, performs by far the worst (24% less than the best interval). In contrast, the intervals with a high o_{max} have a good performance. This is because using a high level of message replication promotes a higher message delivery in a sparse scenario.

Finally, the Emergency scenario behaves similarly to the Taxis scenario. The performance of the most conservative interval is the worst (11% less than the best range). Nevertheless, the performance variance of the different intervals is 0.0004, whereas for the Taxis case it is a bit higher: 0.002. For an unpredictable scenario, similarly to the Taxis scenario, the best strategy is to use an interval with a high o_{max} to keep a high replication limit and, therefore, to have more chances for message delivery.

Altogether, we have seen that in a congested scenario, the key is to maintain a steady replication limit if it maintains the congestion within the configured optimal congestion interval. It is better to set a conservative optimal replication limit for a medium- to high-congestion scenario to avoid high replication that could yield in a future congested scenario. On the contrary, for low-congestion scenarios, setting an optimal congestion interval with a high upper bound leads to a higher replication limit favouring message replication, which increases the delivery ratio.

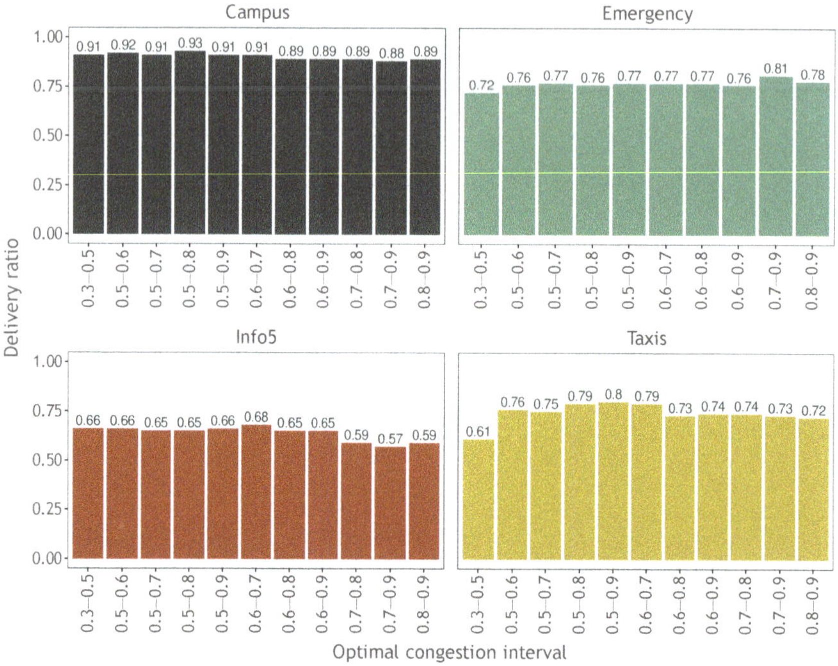

Figure 13. Delivery ratio depending on the optimal congestion interval per scenario.

Additive Increase (k_2); Multiplicative Decrease (k_1)

Figure 14 shows that for the additive increase factor used in (7), the value that gives the best performance in all the scenarios is, undoubtedly, 1. From this result, it can be stated that it is essential that the replication limit grows slowly to mitigate the adverse effects of high replication as much as possible. As for the multiplicative decrease factor (MD), for all the scenarios except for the Campus scenario, the best option is to reduce 75% the replication limit as a drastic measure to decrease the congestion caused by replication.

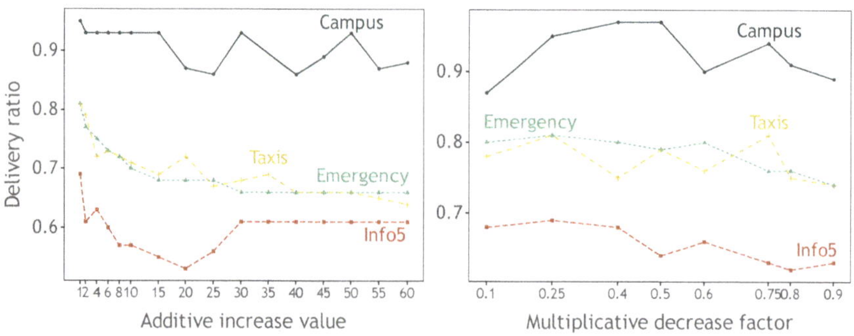

Figure 14. Delivery ratio vs. different values for the AIMD control function.

For all the scenarios except for the Taxis scenario, the delivery ratio continues decreasing for higher MD values (implying less replication reduction). For the particular case of the Taxis scenario, its sparse condition results in more fluctuating delivery ratios depending on the MD factor, but none overcome the ceiling achieved with the 75% of reduction.

Going back to analyzing the results for the Campus scenario, which is highly connected and the most congested, we can see that we obtain the best result by applying a reduction factor of 50%, maintaining a higher replication level. These results are consistent with those in the previous section, where it was stated that once the congestion status fitted in the congestion range thresholds, the best strategy was to keep the replication limit within the range. Precisely, reducing the replication limit to half is a good method for keeping the congestion steady within the optimal congestion interval.

Aggregation Interval ($\hat{t}$); Number of Inputs ($\check{z}$)

Figure 15 shows that gathering congestion readings during a short time interval ($\hat{t}$) results in obtaining the best performance for all the scenarios. Hence, we can say that the best results are obtained when the controller generates directives at a higher rate. There are minor differences between the suitable intervals for each scenario.

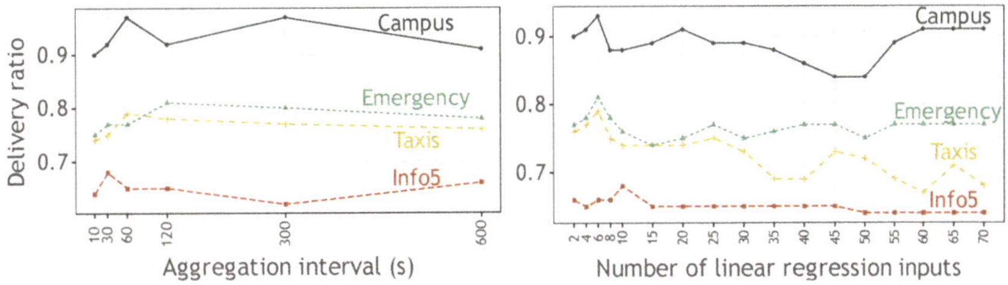

Figure 15. Left: delivery ratio vs. the time interval while the controller is gathering congestion readings. Right: delivery ratio depending on the number of inputs the linear regression is fed with.

The best performance for the Taxis and the Campus scenarios is at a 60 s interval, and for the Info5 scenario (the scenario with the fewest nodes) is 30 s. Thus, it can be said that the fewer nodes there are, the higher rate of directives emitted by the controller is required. For the Info5 scenario, even though in the plot it seems that from the interval of 600 s the function is increasing, simulations up to an interval of 10,800 s with samples every 1800 s have been run, and the delivery ratio remains constant to the value obtained at the 600 s interval.

Finally, the best interval time for the Emergency scenario is 120 s. This scenario drastically changes the message generation in the middle of the simulation. In that variable situation, it is understandable to have a wider time interval range for gathering congestion readings to mitigate the effects of the changes.

On the other hand, regarding the number of entries in $\check{M}$ ($\check{z}$) used to calculate the congestion prediction, Figure 15 clearly shows that, for all the scenarios it is better to have a small $\check{z}$. This is fully understandable, as the fewer inputs we use, the newer the information is.

Prediction Time Factor (ϕ)

As shown in Figure 16, a minor factor, i.e., predicting the future congestion in the short term, gives the best performance for all the scenarios. For all of the scenarios, the factor is 2 except for that of Info5, which is 5. Hence, we can determine that we can predict the near future more accurately than the far future. This assertion implies that we need a small factor combined with a small aggregation interval ($\hat{t}$). This combination is feasible as, in the previous section, we have seen that a small $\hat{t}$ leads to better performance.

Reduction Factor for a Decay (r)

In Figure 17, we consider different decay at different times (300 s, 600 s, 1800 s, and 3600 s). The decay ranges from 1 (no decay) to (0.01) at each considered time. For all the scenarios, we can see that the delivery ratio overwhelmingly drops when the congestion reading is not "penalized" by a decay (decay weight close to 1) after an elapsed time. Hence, we conclude that considering a decay for the congestion readings is crucial.

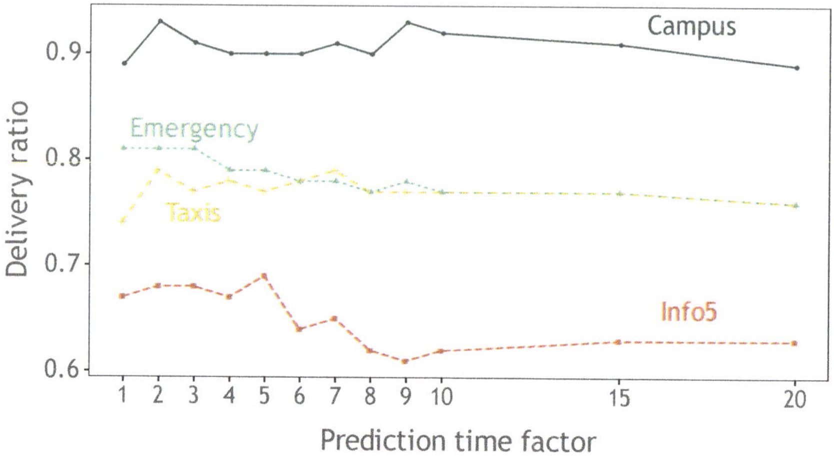

Figure 16. Delivery ratio depending on the prediction time factor.

From the transient of the results, for each scenario, it can be seen that the decay at t heavily affects the performance. Nevertheless, it can be observed that for high decay "penalties" (small decay weight values), better results are obtained at any time than with lower decay (high values).

In conclusion, the best strategy is to apply a high decay (small weight factor value), which implies a considerable reduction in the effect of the congestion measurement after a short elapsed time from its creation up to its reception by a node/controller.

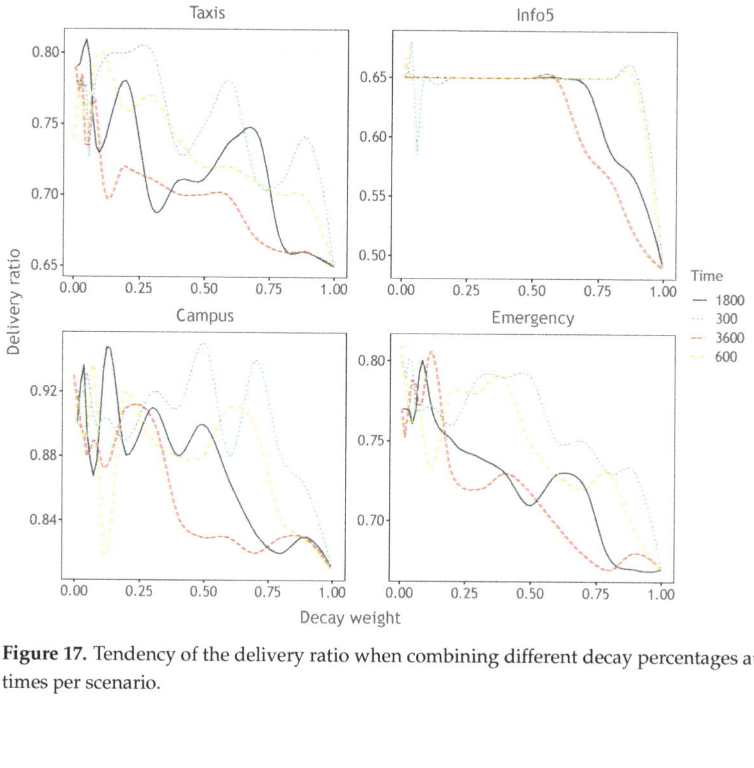

Figure 17. Tendency of the delivery ratio when combining different decay percentages at different times per scenario.

Decay Threshold

Section 3.3 shows how the received measurements are aggregated through (3). This equation double weights the congestion reading by the number of aggregations it is formed by and by its decay, which is calculated with (1). Consequently, a congestion measurement formed by a high number of aggregated congestion measurements would have a high impact, despite its decay in the overall process of the congestion measurements aggregation. Hence, several of these congestion readings in the control's current aggregation process can lead to a long tail effect [39], where almost negligible old congestion readings would highly affect the whole aggregation result. In this case, we would have a congestion reading calculation based on, very likely, expired information. To avoid this undesirable situation, a decay threshold is specified so that the congestion readings with a decay smaller than decay threshold are not considered in the controller's congestion calculation. From Figure 18 it can be stated that, for all the scenarios, the best decay threshold is 10% of the decay.

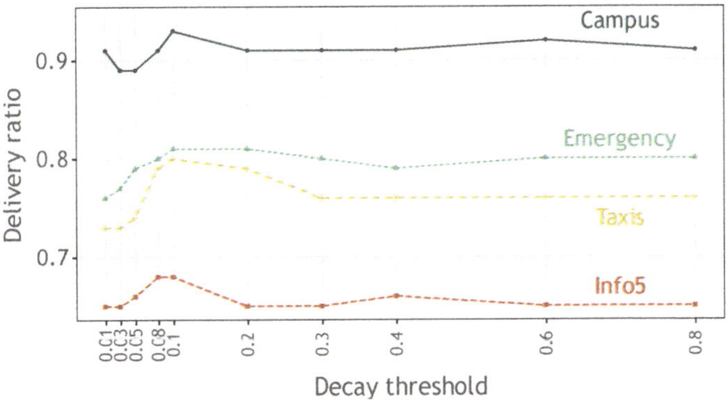

Figure 18. Delivery ratio for different *decay thresholds*.

Figure 18 shows that, as expected, either aggregating old readings (small decay threshold) or discarding new readings (high decay threshold) worsens the performance consistently for all the scenarios. Nevertheless, although small decay thresholds diminish the performance for the Emergency scenario, high decay thresholds do not significantly affect such performance. This behaviour is due to the high variance in the latency of this scenario, so considering old readings does not significantly affect the performance.

Number of Aggregations Weight (α)

The number of aggregations weight setting (α) used in (3) is the weight applied to the *number of aggregations* an aggregated congestion measurement is formed of. As (3) is a two-factor weighted average, the weight related to the *decay* is the *alpha*'s complementary $(1 - \alpha)$.

As we can see in Figure 19, for all the scenarios, we obtain the best performance with an alpha of 0.2. This result concludes that the decay of a congestion measurement is more relevant than the number of aggregations this aggregated measurement is formed of.

Directive Generation Frequency

As we can see in Figure 20, for all the scenarios except for the Taxis scenario, we obtain the best performance by resending the last directive each 900 s, provided no contact has happened before. Nevertheless, for the Taxis scenario, we obtain the optimal performance with a directive frequency of 300 s. We can understand this slight difference, as the Taxis scenario is the sparsest scenario, which implies fewer contacts between nodes. Hence, a more frequent directive beckoning gives the nodes more chances to receive a directive, and consequently, we get better performance.

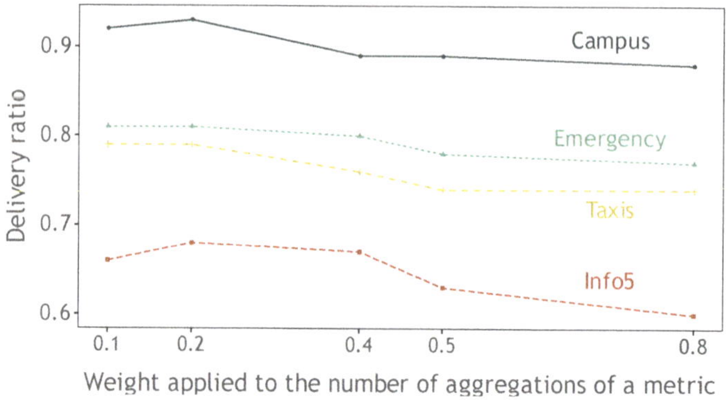

Figure 19. Delivery ratio for different α weights.

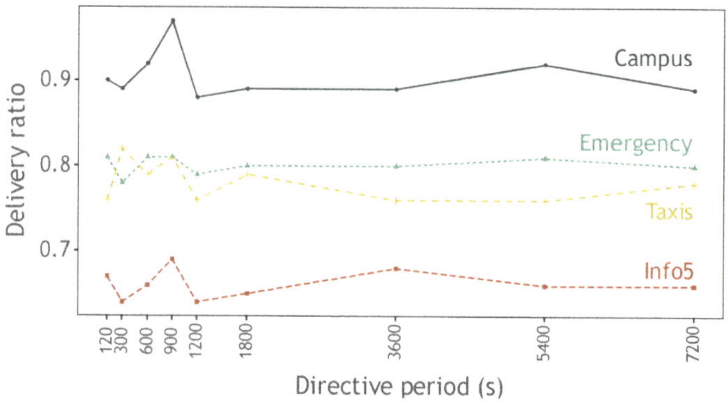

Figure 20. Delivery ratio for different directive generation periods.

6. Conclusions

The motivation for this work was bringing the benefits of the SDN architecture into OppNets by having a controller to retrieve network context information to orchestrate the data plane in terms of tuning the forwarding strategy to achieve a better network performance.

To state the soundness of the proposal, the control layer has been applied to manage the congestion derived from multi-copy-based forwarding algorithms. This controller-driven OppNet has been tested over four scenarios characterized by different mobility patterns and node densities against baseline forwarding strategies based on message replication.

For the scenarios with a message distribution following the ISTH function, the simulations show that for the Control configuration, the replication limit tends toward an asymptote proximal to the replication limit of the best-performing Static policy in terms of delivery ratio (optimal). Therefore, it can be stated that under blind network knowledge, the Control configuration approaches the *optimal* replication limit.

Moreover, the Control configuration adapts to changes in the pattern of message generation distribution. It is precisely under these unpredictable conditions that the Control configuration stands out over the other configurations by leveraging its dynamic adaptability to the network conditions. This adaptability facilitates a significantly lower occupancy of the node's buffer and an important reduction in the overhead intrinsic to replication.

Furthermore, the Control configuration improves the delivery ratio for all the scenarios. Its effectiveness is accentuated for scenarios with medium–low congestion, as a wider replication limit range can be considered. In contrast, a highly congested scenario is stuck to a low replication limit. Undoubtedly, latency benefits from a replication that does not overwhelm the nodes' cache system. The fact that the Control configuration keeps the replication limit at bay to avoid congestion and achieve a better delivery ratio affects the latency. Therefore, the application layer should determine whether to maximize the delivery ratio or minimize the latency, so the Control configuration could apply forwarding strategies to optimize one or the other.

Furthermore, the control layer is highly configurable to provide the best performance depending on the Oppnet's nature. Nevertheless, generic values providing a good performance have been determined from simulations over the aforementioned scenarios. In this regard, simulations show that the controller must use recent congestion readings from contacts. Therefore, applying a decay weight over the measurements used to predict the network conditions is decisive. In this regard, it is more effective to perform a short-term prediction than a long-term prediction.

Moreover, the simulation demonstrates that the sparser the network is (fewer contacts between nodes), the more directives are needed. For the use case of congestion control, the replication limit needs to grow slowly, whereas, in a congestion state, a sharp reduction is required. The optimal congestion interval is highly coupled to the characteristics of the scenario.

Finally, simulations depict that, despite the disconnections, network partitioning, and long delay paths prone to OppNets, a small number of controllers suffices.

Overall, this study asserts that (i) a context-aware system built upon the SDN pillar principles is a good approach for context-management in OppNets and (ii) using this context-aware system to regulate the replication in an OppNet driven by a multi-copy forwarding strategy leads to better network performance.

Author Contributions: Conceptualization, C.B. and S.R.; data curation, M.d.T.; formal analysis, M.d.T.; funding acquisition, S.R.; investigation, M.d.T.; methodology, M.d.T.; project administration, C.B.; resources, C.B.; software, M.d.T.; supervision, C.B.; validation, M.d.T.; visualization, M.d.T.; writing—original draft, M.d.T.; writing—review and editing, C.B. and S.R. All authors have read and agreed to the published version of the manuscript.

Funding: This research was funded by The Catalan AGAUR 2017SGR-463 project and The Spanish Ministry of Science and Innovation PID2021-125962OB-C33 project.

Institutional Review Board Statement: Not applicable.

Informed Consent Statement: Not applicable.

Data Availability Statement: The control layer extension for the ONE simulator is available at https://github.com/MCarmen/the-one/tree/control, (accessed on 3 January 2022). The obtained results, the ONE configuration files for all the scenarios, the script files to run the simulations, and the data traces for the reproducibility of these results are available at https://deic.uab.cat/~mcdetoro/controller-driven_OppNet_results.zip, (accessed on 3 January 2022). The scenarios based on real mobility traces, Taxis and Info5, are also available at https://crawdad.org/epfl/mobility/20090224/ and http://crawdad.org/uoi/haggle/20160828/one, (accessed on 3 January 2022), respectively.

Acknowledgments: We want to thank the guidance and coaching of Ian Blanes Garcia with the writing of this manuscript, Cristina Fernandez Córdoba for showing us the beauty of mathematical formulation, and Joan Borrell Viader for accompanying us in the early stages of this research.

Conflicts of Interest: The authors declare no conflict of interest.

Abbreviations

$\mathfrak{a}$	Application layer message
φ	Message's flag alive
b	Buffer size in bytes
B	Buffer: a list to store messages
$\mathfrak{c}$	Controlled variable
c_{t+n}	Congestion state prediction for time $t+n$
d	Decay
δ	Directive: $\delta = (\mathbb{Id}, \vartheta)$
δ_l	Directive encapsulating a replication limit
ϕ	Factor to calculate t_{t+n}
$\mathbb{Id}_p$	Identifier of the network setting P
$\mathbb{Id}_l$	Message's replication limit Id setting
k_1	Multiplicative decrease factor
k_2	Additive increase factor
g_i	A message
$\#g_d$	Number of delivered messages
$\#g_c$	Number of created messages
$\#g_r$	Number of relayed messages
l	Max. message copies in the network (replication limit)
$\overline{\lambda}$	Latency average
λ_i	Latency of message g_i
L_1	Highest message generation rate for the ISTH dist.
L_2	Lowest message generation rate for the ISTH dist.
m_{l_i}	Node n_i's local network measurement: $m_{l_i} = (v_{l_i}, 1, t_{c_i})$
m_i	Aggregated received network measurements: $m_i = (v_i, \eta_i, t_{c_i})$
m_{t+n}	Predicted network measure value at time $t+n$
M_i	Aggregated network measurements list for node n_i
$\check{m}$	Aggregation of the received measurements during $\hat{t}$s
$\check{M}$	List of $\check{m}$ values
$\check{z}$	Max size of $\check{M}$
μ	Control system's manipulated variable
η	Number of aggregated measurements
n_i	A specific node
o_i	Buffer occupancy in bytes of node n_i
o_{t+n}	Buffer occupancy prediction for time $t+n$
o_{min}	Min. buffer occupancy rate
o_{max}	Max. buffer occupancy rate
θ	Overhead ratio
r	Reduction factor
$\mathfrak{r}$	Control system's reference input
ϱ	Number of times a message has been relayed
ρ	Linear regression function
σ	Delivery ratio
t	Current time
t_{t+n}	Time ahead when calculating the congestion prediction
t_{c_i}	Creation time for either a measurement or an aggregation
$\hat{t}$	Time period for aggregating received measurements
$\check{T}$	List of the timestamps of the aggregations performed
ϑ	Network setting's value
v_{l_i}	Node n_i's local network measurement
v_i	Result value after aggregating the received measurements in M_i

References

1. Union, I. IMT traffic estimates for the years 2020 to 2030. *Rep. ITU* **2015**, *2370*. Available online: https://www.itu.int/dms_pub/itu-r/opb/rep/R-REP-M.2370-2015-PDF-E.pdf (accessed on 3 December 2022).
2. Ansari, R.I.; Chrysostomou, C.; Hassan, S.A.; Guizani, M.; Mumtaz, S.; Rodriguez, J.; Rodrigues, J.J. 5G D2D networks: Techniques, challenges, and future prospects. *IEEE Syst. J.* **2017**, *12*, 3970–3984. [CrossRef]
3. Al-Sultan, S.; Al-Doori, M.M.; Al-Bayatti, A.H.; Zedan, H. A comprehensive survey on vehicular Ad Hoc network. *J. Netw. Comput. Appl.* **2014**, *37*, 380–392. [CrossRef]
4. Rady, A.; El-Rabaie, E.S.M.; Shokair, M.; Abdel-Salam, N. Comprehensive survey of routing protocols for Mobile Wireless Sensor Networks. *Int. J. Commun. Syst.* **2021**, *34*, e4942. [CrossRef]
5. Hui, P.; Chaintreau, A.; Scott, J.; Gass, R.; Crowcroft, J.; Diot, C. Pocket switched networks and human mobility in conference environments. In Proceedings of the 2005 ACM SIGCOMM Workshop on Delay-Tolerant Networking, Philadelphia, PA, USA, 26 August 2005; pp. 244–251.
6. Conti, M.; Boldrini, C.; Kanhere, S.S.; Mingozzi, E.; Pagani, E.; Ruiz, P.M.; Younis, M. From MANET to people-centric networking: Milestones and open research challenges. *Comput. Commun.* **2015**, *71*, 1–21. [CrossRef]
7. Akyildiz, I.F.; Wang, X.; Wang, W. Wireless mesh networks: A survey. *Comput. Netw.* **2005**, *47*, 445–487. [CrossRef]
8. Fall, K. A delay-tolerant network architecture for challenged internets. In Proceedings of the 2003 Conference on Applications, Technologies, Architectures, and Protocols for Computer Communications, Karlsruhe, Germany, 25–29 August 2003; pp. 27–34.
9. Chakchouk, N. A survey on opportunistic routing in wireless communication networks. *IEEE Commun. Surv. Tutor.* **2015**, *17*, 2214–2241. [CrossRef]
10. Kreutz, D.; Ramos, F.; Verissimo, P.; Rothenberg, C.E.; Azodolmolky, S.; Uhlig, S. Software-defined networking: A comprehensive survey. *arXiv* **2014**, arXiv:1406.0440.
11. Trifunovic, S.; Kouyoumdjieva, S.T.; Distl, B.; Pajevic, L.; Karlsson, G.; Plattner, B. A Decade of Research in Opportunistic Networks: Challenges, Relevance, and Future Directions. *IEEE Commun. Mag.* **2017**, *55*, 168–173. 17.1500527CM. [CrossRef]
12. Guidec, F.; Mahéo, Y.; Launay, P.; Touseau, L.; Noûs, C. Bringing Opportunistic Networking to Smartphones: A Pragmatic Approach. In Proceedings of the 2021 IEEE 45th Annual Computers, Software, and Applications Conference (COMPSAC), Madrid, Spain, 12–16 July 2021; pp. 574–579.
13. Touseau, L.; Mahéo, Y.; Noûs, C. A Smartphone-Targeted Opportunistic Computing Environment for Decentralized Web Applications. In Proceedings of the 2021 IEEE 46th Conference on Local Computer Networks (LCN), Edmonton, AB, Canada, 4–7 October 2021; pp. 363–366.
14. Jain, S.; Fall, K.; Patra, R. Routing in a delay tolerant network. In Proceedings of the 2004 Conference on Applications, Technologies, Architectures, and Protocols for Computer Communications, Portland, OR, USA, 30 August–3 September 2004; pp. 145–158.
15. Sobin, C.; Raychoudhury, V.; Marfia, G.; Singla, A. A survey of routing and data dissemination in delay tolerant networks. *J. Netw. Comput. Appl.* **2016**, *67*, 128–146.
16. Vahdat, A.; Becker, D. *Epidemic Routing for Partially Connected Ad Hoc Networks*; Technical Report; Duke University: Durham, NC, USA, 2000.
17. Spyropoulos, T.; Psounis, K.; Raghavendra, C.S. Spray and wait: An efficient routing scheme for intermittently connected mobile networks. In Proceedings of the 2005 ACM SIGCOMM Workshop on Delay-Tolerant Networking, Philadelphia, PA, USA, 26 August 2005; pp. 252–259.
18. Boldrini, C.; Conti, M.; Delmastro, F.; Passarella, A. Context-and social-aware middleware for opportunistic networks. *J. Netw. Comput. Appl.* **2010**, *33*, 525–541. [CrossRef]
19. Li, H.; Ota, K.; Dong, M.; Guo, M. Mobile Crowdsensing in Software Defined Opportunistic Networks. *IEEE Commun. Mag.* **2017**, *55*, 140–145. [CrossRef]
20. Soelistijanto, B.; Howarth, M.P. Transfer Reliability and Congestion Control Strategies in Opportunistic Networks: A Survey. *IEEE Commun. Surv. Tutor.* **2014**, *16*, 538–555. [CrossRef]
21. Krifa, A.; Barakat, C.; Spyropoulos, T. Optimal buffer management policies for delay tolerant networks. In Proceedings of the 2008 5th Annual IEEE Communications Society Conference on Sensor, Mesh and Ad Hoc Communications and Networks, San Francisco, CA, USA, 16–20 June 2008; pp. 260–268.
22. Pan, D.; Ruan, Z.; Zhou, N.; Liu, X.; Song, Z. A comprehensive-integrated buffer management strategy for opportunistic networks. *EURASIP J. Wirel. Commun. Netw.* **2013**, *2013*, 1–10. [CrossRef]
23. Goudar, G.; Batabyal, S. Point of congestion in large buffer mobile opportunistic networks. *IEEE Commun. Lett.* **2020**, *24*, 1586–1590. [CrossRef]
24. Lakkakorpi, J.; Pitkänen, M.; Ott, J. Using buffer space advertisements to avoid congestion in mobile opportunistic DTNs. In Proceedings of the International Conference on Wired/Wireless Internet Communications, Vilanova i la Geltrú, Spain, 15–17 June 2011; pp. 386–397.
25. Thompson, N.; Nelson, S.C.; Bakht, M.; Abdelzaher, T.; Kravets, R. Retiring replicants: Congestion control for intermittently-connected networks. In Proceedings of the 2010 Proceedings IEEE INFOCOM, San Diego, CA, USA, 14–19 March 2010; pp. 1–9.
26. Goudar, G.; Batabyal, S. Estimating Buffer Occupancy sans Message Exchange in Mobile Opportunistic Networks. *IEEE Netw. Lett.* **2022**, *4*, 73–77. [CrossRef]

27. Batabyal, S.; Bhaumik, P.; Chattopadhyay, S.; Misra, S. Steady-state analysis of buffer occupancy for different forwarding strategies in mobile opportunistic network. *IEEE Trans. Veh. Technol.* **2019**, *68*, 6951–6963. [CrossRef]
28. Dorf, R.C.; Bishop, R.H. *Modern Control Systems*; Pearson Studium: London, UK, 2011.
29. Janert, P. *Feedback Control for Computer Systems: Introducing Control Theory to Enterprise Programmers*; O'Reilly Media Inc.: Sebastopol, CA, USA, 2013.
30. Dalal, R.; Khari, M.; Anzola, J.P.; García-Díaz, V. Proliferation of Opportunistic Routing: A Systematic Review. *IEEE Access* **2021**, *10*, 5855–5883. [CrossRef]
31. Dede, J.; Förster, A.; Hernández-Orallo, E.; Herrera-Tapia, J.; Kuladinithi, K.; Kuppusamy, V.; Manzoni, P.; bin Muslim, A.; Udugama, A.; Vatandas, Z. Simulating opportunistic networks: Survey and future directions. *IEEE Commun. Surv. Tutor.* **2017**, *20*, 1547–1573. [CrossRef]
32. Kuppusamy, V.; Thanthrige, U.M.; Udugama, A.; Förster, A. Evaluating Forwarding Protocols in Opportunistic Networks: Trends, Advances, Challenges and Best Practices. *Future Internet* **2019**, *11*, 113. [CrossRef]
33. Mota, V.F.; Cunha, F.D.; Macedo, D.F.; Nogueira, J.M.; Loureiro, A.A. Protocols, mobility models and tools in opportunistic networks: A survey. *Comput. Commun.* **2014**, *48*, 5–19. [CrossRef]
34. Community Resource for Archiving Wireless Data At Dartmouth. Available online: https://crawdad.org/ (accessed on 3 December 2022).
35. Piorkowski, M.; Sarafijanovic-Djukic, M.G.N. Dataset of Mobility Traces of Taxi Cabs in San Francisco, USA (v. 2009-02-24). Available online: https://crawdad.org/epfl/mobility/20090224/cab/index.html (accessed on 3 December 2022).
36. Akestoridis, D.G. CRAWDAD Dataset Uoi/haggle (v. 2016-08-28): Derived from Cambridge/Haggle (v. 2009-05-29). 2016. Available online: http://crawdad.org/uoi/haggle/20160828/one (accessed on 3 December 2022).
37. Boyd, J.P. Asymptotic Fourier coefficients for a C∞ bell (smoothed-"top-hat") & the Fourier extension problem. *J. Sci. Comput.* **2006**, *29*, 1–24.
38. Keränen, A.; Ott, J.; Kärkkäinen, T. The ONE simulator for DTN protocol evaluation. In Proceedings of the 2nd International Conference on Simulation Tools and Techniques, ICST (Institute for Computer Sciences, Social-Informatics and Telecommunications Engineering), Rome, Italy, 2–6 March 2009; p. 55.
39. Brown, G.W.; Tukey, J.W. Some Distributions of Sample Means. *Ann. Math. Stat.* **1946**, *17*, 1–12. [CrossRef]

applied sciences

MDPI

Article

Corpus for Development of Routing Algorithms in Opportunistic Networks

Diego Freire *, **Carlos Borrego** and **Sergi Robles**

Department of Information and Communications Engineering, Universitat Autònoma de Barcelona, 08193 Bellaterra, Spain

* Correspondence: diego.freire@uab.cat

Abstract: We have designed a collection of scenarios, a corpus, for its use in the study and development of routing algorithms for opportunistic networks. To obtain these scenarios, we have followed a methodology based on characterizing the space and choosing the best exemplary items in such a way that the corpus as a whole was representative of all possible scenarios. Until now, research in this area was using some sets of non-standard network traces that made it difficult to evaluate algorithms and perform fair comparisons between them. These developments were hard to assess in an objective way, and were prone to introduce unintentional biases that directly affected the quality of the research. Our contribution is more than a collection of scenarios; our corpus provides a fine collection of network behaviors that suit the development of routing algorithms, specifically in evaluating and comparing them. If the scientific community embraces this corpus, the community will have a global-agreed methodology where the validity of results would not be limited to specific scenarios or network conditions, thus avoiding self-produced evaluation setups, availability problems and selection bias, and saving time. New research in the area will be able to validate the routing algorithms already published. It will also be possible to identify the scenarios better suit specific purposes, and results will be easily verified. The corpus is available free to download and use.

Keywords: opportunistic networks; corpus; routing algorithms; scenarios; new communication paradigms

Citation: Freire, D.; Borrego, C.; Robles, S. Corpus for Development of Routing Algorithms in Opportunistic Networks. *Appl. Sci.* **2022**, *12*, 9240. https://doi.org/10.3390/app12189240

Academic Editors: Yang Yue, Runzhou Zhang, Hao Feng, Zheda Li, Lin Zhang and Dawei Ying

Received: 29 July 2022
Accepted: 13 September 2022
Published: 15 September 2022

1. Introduction

During the last decade, there has been some emerging networking paradigms that were announced that they will become mainstream in the future, such as Delay Tolerant Networking (DTN) or Opportunistic Networking (OppNet). Many of the use cases for them, though, are nowadays better solved by other approaches mainly based on global connectivity. Some examples of these use cases were providing connectivity in sparsely inhabited areas, in underdeveloped regions, and during disasters. However, there are still some scenarios for which flexible ad hoc communications without infrastructure require a different approach, closer to OppNets. Times have changed, and the overuse of the terms DTN and OppNet everywhere for many years has led to a current situation where these paradigms are regarded with suspicion by the research community, and even cause one to be wary of them. And yet, the need for this type of communications is still present. Perhaps it is more convenient to talk about concepts such as disruption-tolerant MANETs, multihop device to device routing in 5G, pervasive IoT, or dynamic source routing, but at the end of the day the concept of devices directly communicating with each other asynchronously using other peer devices as relays is still relevant to this day. Scenarios with these needs are, for instance, proximity-based applications, privacy-preserving communications, limited energy distance communications, or long-distance space communications, among others. For the rest of this study, we use the term OppNet to refer to this paradigm, regardless of the specific technology used to implement it.

Research on routing algorithms for OppNet is very important because it provides the core element that makes this technology work. Due to the asynchronous nature of the forwarding process, selecting the right neighbor to pass messages to, choosing a good number of copies of messages, and defining for how long messages are going to live in the network are crucial aspects of opportunistic networking that are included in the routing algorithm. High-performance algorithms are produced after following a rigorous process based on the scientific method, where evaluation, comparison, and testing allow to determine the best solution for a given scenario.

Unfortunately, most studies on this topic use an experimentation based on self-produced network traces, traces obtained in a particular real scenario by other researchers that have been adapted, or traces that were captured on a real network with specific constrains. Validity of results is often limited to a specific scenario or network conditions, and therefore these outcomes are often not an actual indication of the universal validity of the solution that would allow its utilization by the global community. Obviously, this is not what researchers want, and there is no hint of any bad intention here. The reason of this situation is the lack of common frameworks and objective test environments for facilitating the production of quality routing algorithms enabling the optimal application in the scenarios requiring them.

To solve this problem, methodological approach is required that is based on scientific rigor. Being able to design good algorithms that do not just hold water but that can be objectively evaluated and compared to others under fair conditions is the only way to choose the best option for a particular scenario.

In this study, we propose the cornerstone of this methodology: A corpus of carefully selected scenarios, accessible to the entire community, that can be used for the actual comparison and assessment of routing algorithms. This has not been an easy task. It might seem that just selecting some common, already published scenarios would suffice for this objective. However, that would be incomplete. A valid corpus has to be representative of all possible scenarios; it has to be accessible to the entire community, using a standard format; and all of its scenarios have to be comprehensive, with no missing parts that could be completed in different ways. Similar initiatives are found in other domains, such as image compression. We have studied the different variables, the dimensions, and defining a scenario, finding out a subset that can be considered independent to form a sound vector base for the scenario space. Then, we have selected forty-one of them to constitute the corpus. We have tested this corpus by using a high replication algorithm, and have observed that it performs differently for all of the scenarios.

This corpus is a leg to stand on for new research on the area. At last, fair comparison can be done, and results are easily verifiable. It can also validate the already published routing algorithms and help determine the scenarios they suit better.

The rest of the article is structured as follows: Section 2 describes all of the relevant state-of-the-art information, paying special attention to routing algorithms in opportunistic networks, the current evaluation approach that routing algorithms have, and a review of how other research fields conduct the evaluation of algorithms. Then, in Section 3, the article provides a complete description of a new methodology for evaluating OppNet routing algorithms. The article follows with Section 4, where the appraisement of the contribution is presented with a simulation-based experiment. Next, Section 5 contains a discussion, and finally, Section 6 presents the conclusions drawn from this work.

2. State of the Art

In this section, the state-of-the-art of opportunistic networks is reviewed, emphasizing the performance evaluation of routing algorithms. Then, this article provides an overview of the tools, strategies and metrics used to evaluate the performance of routing algorithms. Additionally, this section describes the challenges when evaluating and comparing opportunistic networks. Finally, we review how other fields have tackled similar problems to assess the performance of algorithms.

2.1. Opportunistic Networks

Opportunistic network(s) (OppNet(s)) are wirelessly connected devices that interchange information, exploiting connection opportunities. In this type of networks, devices with wireless capabilities (such as smartphones, tablets and smartwatches, among others) use direct communications opportunities [1]. OppNets allow information exchange among devices even when an end-to-end path may never exist [2]. Moreover, the variations in the network's topology are considered normal behavior due to the wireless nature of the devices [3].

Additionally, OppNets are challenging networks where disruption and delays in communication are considered normal [4], due to tackling the problem of how to exchange information without a fixed network infrastructure [5]. In OppNets, the information transmitted between devices is also known as messages. These types of networks use the store-carry-and-forward paradigm to transmit information among devices [6,7]. This paradigm allows message routing from source to destination, handling disconnection, delay and disruption in the network. When a device implements the store-carry-and-forward paradigm, the device receives a message. Next, the message will be stored and carried until a transmission opportunity occurs, and finally, the message is forwarded to the other device. In OppNets, a device is also known as a node [8].

The applications of OppNets have been widely studied. In environments where traditional networks do not perform well or, even worse, cannot operate, OppNets may provide a feasible solution for communication. Among the challenging environments, research highlights the following as areas where OppNets may perform well: Cellular network offloading [9], communication in challenged areas [9], censorship circumvention [10], mobile ad hoc social networks [2], offline social networks [2], Internet of Vehicles [11], information-centric networking [12], and proximity-based applications [2], among others.

OppNets are an active research field that is still worth to be studied. For example, one interesting open research topic beyond OppNet applications that inherently implement a store-carry-and-forward delivery paradigm is information-centric networking (ICN) [13]. This communication architecture can effectively suit OppNets. ICN is a non-host-centric communication architecture that, unlike IP, is not tied to a specific network location. It is centered around hierarchical content names used directly at the network layer [14].

A routing algorithm can be described as an implementation of a message routing function whose objective is to deliver the messages to their destination while maximizing the efficiency of resource consumption. Routing algorithms are the intelligence that supports the operation of an OppNet, since they dictate the directives on the behavior of the nodes with the messages. Routing algorithms seek to maximize delivery by optimizing the use of resources [15]. Over the years, researchers have put their efforts into developing routing algorithms. Articles such as [16–18] mention a number of algorithms that have been proposed. These proposed routing algorithms provide routing solutions for particular environments. Some routing forwarding strategies implementations use epidemic [19], probabilistic [20], number of copies [21], or based on neighborhood contact history [22] strategies to deliver messages among nodes.

In the context of forwarding decisions, the routing algorithm must decide upon the best candidate(s) to receive a message among all available nodes. In addition to forwarding, a message has a lifetime in the network, and the routing algorithm will update the message lifetime. Messages can also be stored and deleted; this algorithm does not require any selection of peers.

So far, the definition of OppNets was presented and it was shown that routing algorithms are the basis of communications in challenging environments known as OppNets. The critical role routing algorithms play in an OppNet was also shown. The following section explains how OppNets and routing algorithms can be described.

2.2. Characteristics and Metrics in OppNets

In this section, the article explains the characteristics and metrics used in OppNets and why they are fundamental concepts in OppNets. This section also describes the relationship between the characteristics and the metrics.

OppNets are heterogeneous. A specific OppNet instance can be expressed with a set of characteristics. However, it cannot be said a priori that one instance of an OppNet is necessarily equal to or different from another OppNet. One way to establish the differences between OppNets is to compare the characteristics that describe each OppNet.

The more characteristics are used, the more accurate is the description of an OppNet's behavior. In other words, this feature-based description somewhat simplifies a real-world OppNet. As said in [23], characteristics are deployment facts expressed in numbers for a network.

Metrics return quantitative information about a feature or behavior. Moreover, a metric is a measure function whose output is a numerical value that can be interpreted as the degree to which the routing algorithm has a given attribute [24]. Researchers use metrics to evaluate, compare, or measure the behavior of OppNet routing algorithms [25,26]. Metrics quantify, among others, the performance and the ratio that routing algorithms achieve when the messages are interchanged between source and destination. It can quantify a specific attribute (such as a count of successful processes, time consumption, and messages delivered), providing a quantitative indication.

There are some metrics that most OppNet researchers tend to use to prove performance hypotheses. Among them, three stand out because of their presence in most works related to OppNets this work has found. Those metrics are: Delivery ratio, delivery delay and delivery cost [16]. However, some authors do not use these metrics but rather modified versions of them to fit specific hypotheses. In other cases, some authors even find it necessary to establish entirely new metrics to measure the behavior of their work [27,28].

Characteristics and metrics have a close relationship. This relationship is given because routing algorithms work in OppNets instances. Moreover, since OppNet instances are described by a set of characteristics and those characteristics, in a way, produce a routing algorithm behavior measured by the metrics, the characteristics of an OppNet influence the metrics that a routing algorithm has. For example, there are equivalent network configurations where the same message routing algorithm would be equally efficient. However, the metrics will tell if two scenarios are the same or different from a routing point of view. An adequate characterization of an OppNet withdraws the attention on details that can give a wrong cognitive impression of the difference between networks.

This section has described the characteristics and metrics in OppNets, and it also has described the relationship between them. The following section shows the current performance evaluation techniques among OppNet routing algorithms.

2.3. OppNet Routing Algorithms Evaluation and Comparison

Current evaluation and comparison techniques are worthy of being explained. This section explains the evaluation and the comparison of routing algorithms. Furthermore, it is described the one-way connection between the evaluation and comparison of routing algorithms in OppNets.

Transforming a routing idea into a routing algorithm is a challenge by itself. A complete creation methodology helps in that matter, increasing the quality and speeding up the creation process [29]. Previous work by the authors [29] showed a seven-stage methodology for developing new routing algorithms. This methodology is depicted in Figure 1. The first stage of the seven-stage methodology is the routing idea, where the concept of the routing mechanism is conceived. In the second stage, the idea is modeled; thus, In the third stage, the routing proposal can be analyzed. After the conception, modeling and analysis, the fourth stage simulates the routing algorithm. However, a successful simulation does not guarantee real-world implementation. The fifth stage requires a full-featured code capable of featuring in the real world. In the sixth stage, real-code is executed in controlled conditions—often a proof-of-concept. Finally, the application phase is where the routing algorithm is deployed in a real-world environment with real devices and users. If the creation of a routing algorithm follows a creation methodology, results can be evaluated, compared and repeated. A reliable routing algorithm design methodology enables an objective method of evaluating and comparing routing algorithms.

From the previous paragraph, the reader may note that evaluation and comparison are different terms. Evaluating a routing algorithm can be described as an intra-technique [30] that quantitatively recognizes the routing algorithm behavior for a particular OppNet environment. The evaluation process of one does not require other routing algorithms to assess their performance.

On the other hand, the comparison among routing algorithms can be described as an inter-technique [30] that ranks the performance of a routing algorithm against other routing algorithms.

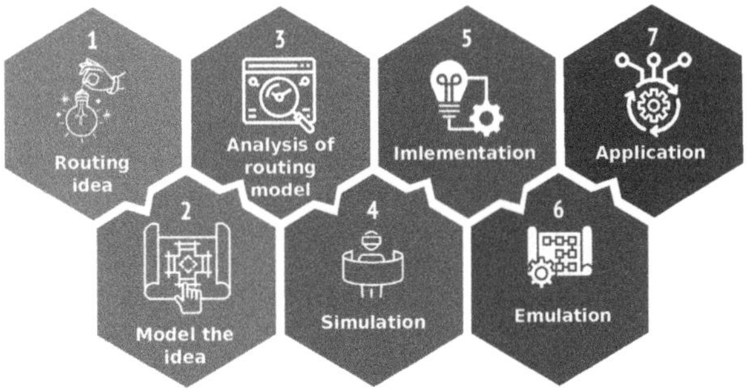

Figure 1. Seven-stage methodology for routing algorithm creation [29].

Although the evaluation and comparison of routing algorithms are closely related, some differences exist. A comparison of routing algorithms requires an evaluation of the performance of any individual routing algorithm, which means that comparison is not possible without the evaluation. However, some authors might be interested only in the evaluation rather than the comparison.

In general terms, metrics can be used to evaluate and compare the performance of routing algorithms. In OppNets, metrics can be obtained from simulation tools. However, comparing the metrics of different algorithms does not address a fair comparison by itself. A fair comparison among routing algorithms could be assessed when OppNets environments, messages and simulation settings are equal or equivalent among them. This section explained the evaluation and the comparison of routing algorithms. Furthermore, it also described the one-way connection between the evaluation and comparison of routing algorithms in OppNets. The following will explain how OppNets have been simulated nowadays.

2.4. OppNet Simulation Deployment Nowadays

This section presents the main existing OppNets simulation tools. It also illustrates the elements and parameters that allow an OppNet simulation deployment and where those tools come from. Furthermore, this section reflects upon how current simulation tools are used.

Figure 2 shows the elements involved when simulating an OppNet. These simulation elements are input, output and software setup. The inputs define the behavior of the network, for example, nodes and message characteristics. Instead, the simulation's output is the information obtained after the simulation, for example, routing algorithm performance metrics and delivery information of messages. Most of the time, performing a post-simulation analysis from the data obtained as the output may be necessary. It is expected that different simulation parameters return different outputs since the simulation is sensible to setup changes [31].

An OppNet has several software simulation alternatives. Among the software tools that allow simulating an OppNet are GloMoSim [32], OMNeT++ [33], DTN2 [34], HaggleSim [35], the ONE Simulator [36], ns-3 [37], Adyton [38] and MobEmu [39]. The simulation tools are mentioned in ordered of creation from 1998 to 2018. For OppNets research, the most used is the ONE simulator, reaching 62% of recent publications [16].

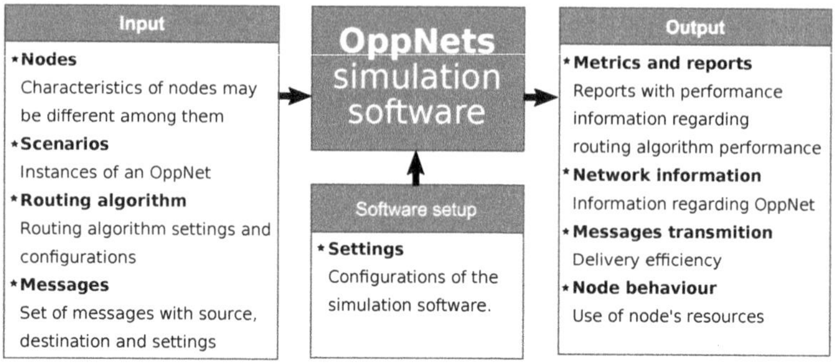

Figure 2. Scenario input, output and configuration elements that enable an OppNet simulation.

Furthermore, as is shown in Figure 2, scenarios are inputs of a simulation. The literature uses the word scenarios interchangeably with the elements they refer to. That is, there is no definition of a scenario in OppNets, and most authors refer to the contact traces (also known as traces or mobility datasets) as scenarios. According to [40], traces are datasets containing registers of nodes, and the information is either positions, contacts or both during a time. Some OppNet simulation softwares, such as the ONE simulator [36] accept trace datasets as input.

Concretely, the source of these traces can be real-world, synthetic or hybrid. The synthetic traces can be produced far faster than real-world traces and may be as valuable as their real-world counterparts for evaluation purposes [41]. A hybrid trace is a mix of real-world and synthetic traces; there is no predefined portion of the real and synthetic traces.

However, the non-standard real-world traces have a cognitive bias. Although they may have some realistic characteristics, their random nature makes generalization difficult and may not be suitable for different environments. For example, if two connectivity traces have been collected from two universities, their characterization will be similar and may not be suitable for simulating a countryside OppNet.

On the other hand, and in addition to the non-standard traces, the synthetic traces are a feasible solution for representativeness because by having control of the characteristics they represent, it is possible to select the traces that, as a whole, are a better representation of a desired OppNet environment. Moreover, since the interest is to represent the real world, it is better to have traces that, due to their characteristics, are representative of the real world instead of several real traces that have similar network behaviors, regardless of the origin. The importance of a trace stands on its network behavior rather than the creation origin.

In the same way, there are several traces suitable for OppNets. Sites such as CRAWDAD [40] gather mobility traces datasets that are shared among the scientific community. Indeed, CRAWDAD has 135 datasets (reviewed on 15 June 2022), but despite this amount, a few datasets are often used rather than others. Some studies even call those "well-known traces" [42] or "well-known scenarios" [43,44]. Datasets such as Asturias [45], Taxis Roma [46], Taxis San Francisco [47] and Cambridge/Haggle [48] are some of those that are usually included in the literature as "well-known traces".

Authors interested in the evaluation and comparison of routing algorithms that might use datasets such as Asturias [45], Taxis Roma [46], Taxis San Francisco [47] and Cambridge/Haggle [48] should have complete knowledge of the representativeness of those datasets. However, evaluation and comparison are not a matter of the number of traces instead of network behaviors. It should be the focus of the authors to test routing performance in representative scenarios. For example, in terms of network behavior, traces do not include information to state a difference between the non-standard traces of Taxis Roma [46] and Taxis San Francisco [47], because both represent the mobility of taxis in a city.

A common practice to assess routing performance is the comparison against peers. Until now, the routing performance has been assessed based on how better an algorithm is compared to other selected routing algorithms within a set of non-standard network traces. However, a routing performance evaluation cannot be extended outside the specific routing algorithms and non-standard network traces. Nowadays, literature does not have a benchmarking scheme for the performance of routing algorithms [16].

In this section, the reader could have seen how the scientific community naturally looks for a group of traces that, in some way, standardize the environments to evaluate the routing algorithms. The following section reviews other research fields with similar problems of evaluating and comparing algorithms, the approach and, above all, the solutions they have found, even though the algorithms described in the following section are not routing-related.

2.5. *Algorithm Performance Evaluation in Other Fields*

The previous sections introduced the features, behaviors and characteristics of OppNets and the importance of routing algorithms. It also showed pitfalls for a fair comparison among routing algorithms. The following section reviews how other fields have proposed solutions for fair comparisons. Specifically, this section reviews how the fields of data compression, linguistics and speech recognition handle the performance comparison problem when developing new algorithms.

This section introduces the term corpus, which refers to a collection of representative data used to analyze the effectiveness of an algorithm's behavior.

2.5.1. Data Compression

Data compression aims to reduce the volume of data while preserving the quality, and it can be classified as either lossy or lossless compression. In lossy and lossless compression, the goal is to maintain quality by using the least amount of data to represent the information. In lossless data compression, the original data can be obtained. However, in lossy compression, some information is lost.

A corpus, in data compression, is a collection of representative files to evaluate the effectiveness of the compression ratio [49]. Calgary [50] and Canterbury [51] are corpuses used in lossless data compression.

Using a corpus to evaluate compression algorithms reduces bias and facilitates the experiments' reproducibility. Furthermore, using a corpus creates compression benchmarks, a standard compression ratio that other algorithms may be compared to. Nowadays, the criteria regarding the corpus are widely accepted in the compression field.

2.5.2. Linguistics Corpuses

As in the field of data compression, linguistics corpuses are sets of text used to study language composition. The use of a corpus allows, in the case of the field of linguistics, to extract complex language structures, which could not be extracted without having a collection that has these complex language structures represented in its files.

Using a corpus can broaden research in other fields. In the case of linguistics, dictionaries and translations have benefited from using a corpus.

2.5.3. Speech Recognition

In speech recognition, the use of corpuses when comparing results is extensive and diverse. The number of corpuses results from one language's heterogeneity compared to another. In languages it is complex to recognize speech, because one language's accents differ from other dialects and phonetics. Nevertheless, despite the variety of results, in this research field, a corpus is a set of selected files seeking representativeness, limiting the number of elements to those necessary, widely available and valuable for developing and evaluating speech recognition techniques.

The techniques of compression, speech recognition or routing will be useless if applied to data that are not relevant or representative. A corpus is helpful within the intended scope of usability.

In this section, our research showed that the creation of routing algorithms voids a fair comparison. Furthermore, this section showed that using self-selected files to perform compression algorithm comparisons seems similar to the well-known traces used to compare routing algorithms. The insights obtained from the review of algorithm evaluation and comparison are that using a corpus improves the performance of algorithms throughout standardization.

3. A Corpus for Routing Evaluation in OppNets

Section 2 has shown that developing routing algorithms in OppNets can be improved using an algorithm creation methodology, particularly when comparing results. Although comparison is essential in research, scientific rigor cannot be assessed now when comparing the performance of routing algorithms. Section 2 also showed that a corpus helps in the algorithm development process, proving to be a crucial part of the methodology. This section defines what an OppNet scenario is and how it can be characterized. Next, this section defines a complete methodology for the development of a corpus. Finally, this section presents a corpus for evaluating and comparing routing algorithms.

3.1. Scenario Definition

As is explained in Section 2, nodes are the principal component of an OppNet scenario. Nowadays, the scenarios are considered a time-ordered list of contacts or positions that nodes have within the same OppNet. It is also mentioned in Section 2 that this information has been called contact traces. However, the contact traces also contain, in a non-explicit way, the corresponding network behavior. Characterizing a trace describes the intrinsic network behavior of the trace with a vector of characteristics. In this article, *an OppNet scenario is denoted as a trace of positions characterized by a vector of seventeen characteristics.*

3.2. Scenario Characterization

Characterizing a scenario basically consists of defining the characteristics that describe their network structure and behavior entirely. This study identifies two types of characteristics, namely direct and indirect. Direct characteristics are the ones that can be identified or defined directly, for example, by counting the number of nodes or measuring the speed of the nodes. On the other hand, indirect characteristics represent characteristics that can not be directly configured but, instead, can be estimated. For example, nodes' centrality refers to betweenness centrality in a trace and cannot be configured with state-of-the-art tools.

A trace could be considered a scenario after the trace has been characterized, that is, when the contact or position trace has a vector describing its network structure and behavior. The characteristics had been identified from the literature review of OppNets routing algorithms. Table 1 shows the list of seventeen characteristics that has been used to characterize the network behavior of a contact trace. Seven characteristics are direct (D) and the rest are indirect (I). The number of nodes, node speed, studied area, movement pattern, node centrality, node contact time, and total encounters are among the direct and indirect characteristics.

The measurement of the characteristics only concerns indirect characteristics. The measurement of those indirect characteristics listed in Table 1 follows the directives depicted in Table 2 and the next paragraph.

Table 1. Set of characteristics for a scenario definition, characteristics are classified as direct (D) and indirect (I).

N°	Characteristic	Type	Description
1	Total number of nodes	D	$[\,nodes\,] \Rightarrow \{nodes \mid 192 < nodes < 960\}$
2	Nodes per group	D	$[\,nodes\_by\_group\,] \Rightarrow \{2^n \in \mathbb{Z} \mid 3 < n < 10\}$
3	Groups of nodes	D	$\{\ groups \in [1,2,3,4]\ \}$
4	Node's movements	D	$[\,movement\,] \Rightarrow \{movement \in [\ m_1,\ m_2,\ \ldots m_m]\}$
5	Node's speed	D	1, 3, 7, 14 and 27 m over second
6	World size	D	$[width\ ,height\,] \Rightarrow \{[width, height] \mid width, height \in [200\ldots 3200]\}m$
7	Area	D	$[\,area\,] \Rightarrow \{[area] \mid area \in [4000\ldots 4,160,000]\}square\ meters$
8	Centrality	I	Measure of how much a given node is in between other nodes
9	Inter-contact time	I	Time a node has no connection
10	Contact time	I	Duration time of the connection between two nodes
11	Contact time per minute	I	Contact time within a minute window
12	Contact node ratio	I	Ratio of nodes contacted by a node
13	Popularity	I	Measure of the ratio of total unique connections
14	Window centrality	I	Mean centrality in a period
15	Encounters	I	Number of encounters
16	Sociability	I	Ratio of contacts
17	Total encounters	I	Total number of encounters within nodes

In Table 2, the number assigned to the characteristic corresponds to the number defined in Table 1. With the exception of characteristic number seventeen (total encounters) the characteristics shown in Table 2 are calculated in a two-step process. The first step is to calculate the characteristic individually in each node. Then, as a second and final step, the mean, variance and standard deviation values of the characteristics are calculated within the values of all or some nodes included in the scenario. For characteristic seventeen, the second step is the sum of the individual values of all nodes. The particular considerations are listed as follows:

- Centrality, inter-contact time, contact time, contact node ratio and encounters: mean of the individual measures of all nodes.
- Popularity and sociability: mean of highest ten percent measurements.
- Contact time per minute, window betweenness centrality: mean of metrics within a period.
- Total encounters: accumulative measurement.

The following section introduces the concept of a corpus in the context of the OppNets. The concept of a corpus will be used across this article as the principle for standardizing the evaluation and comparison of OppNet routing algorithms.

3.3. Corpus Definition

A corpus, in the context of OppNets, is a collection of OppNet scenarios with two main features: First, all scenarios work together to cover all possible network behaviors, and second, the routing algorithms have different performance behaviors when routing messages in each scenario.

Table 2. Indirect scenario characteristics measurement directives with references.

N°	Characteristic	Measurement Directive	Ref.
8	Centrality	Betweenness centrality computed as number of connections held by each node	[2]
9	Inter-contact time	Elapsed time each node has between contacts	[52]
10	Contact time	Elapsed time of the connection between two nodes	[52]
11	Contact time per minute	Contact-time within a period of one minute	[53]
12	Contact node ratio	Node contact ratio	[54]
13	Popularity	Unique peer-connections a node has	[55]
14	Window centrality	Centrality during a period	[56]
15	Encounters	Number of connections a node has	[57]
16	Sociability	Ratio of the number of contacts a node has to the total number of nodes	[58]
17	Total encounters	Summation of the number of connections within nodes	[57]

3.4. Quality Requirements

The corpus aims to be a fair field for evaluating OppNet routing algorithms, providing a set of scenarios that can emulate real-world environments due to their characteristics. This article presented a corpus creation methodology depicted in Figure 3 and explained it in detail in Section 3.5. In addition, the corpus creation methodology presented in this research pursues the following requirements: Coverage, scope, quality and usability.

- Coverage: the coverage of the corpus should have representativeness for real-world environments, considering a significant difference between scenarios.
- Scope: the scope of the corpus should be the performance evaluation of routing algorithms in OppNets.
- Quality: the quality of each scenario of the corpus should be guaranteed by analyzing the representativeness and diversity among other scenarios.
- Usability: the corpus should be easy to use, and the scenarios should be adaptable to simulation software, where the evaluation of the performance of algorithms in OppNets is carried out.

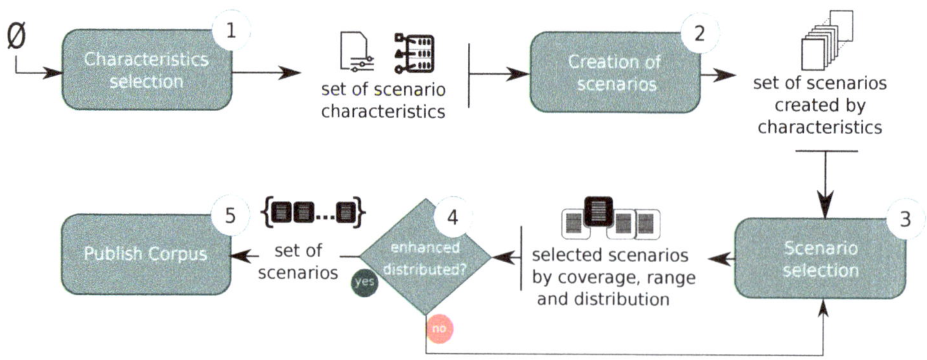

Figure 3. Corpus creation methodology with backtracking stage for scenario selection assuring purpose, coverage, scope, quality and usability requirements.

3.5. Corpus Creation Methodology

This section presents the corpus creation methodology depicted in Figure 3. The corpus creation methodology has five well-delimited stages, each with specific inputs, outputs and tasks. The input information of one stage is the output of the previous one, except for the first stage, which does not have an earlier stage.

The first stage, characteristics selection, decides those characteristics that describe a scenario. The selected seventeen characteristics are displayed in Table 1. A Pearson correlation [59] study of the selected characteristics was performed as shown in Figure 4. Some characteristics have a high correlation because they are based on connections and interaction between nodes. However, despite the redundancy and the high correlation, the characteristics reflect essential connectivity behaviors of the scenarios. This is why these highly correlated characteristics remain within the selected characteristics.

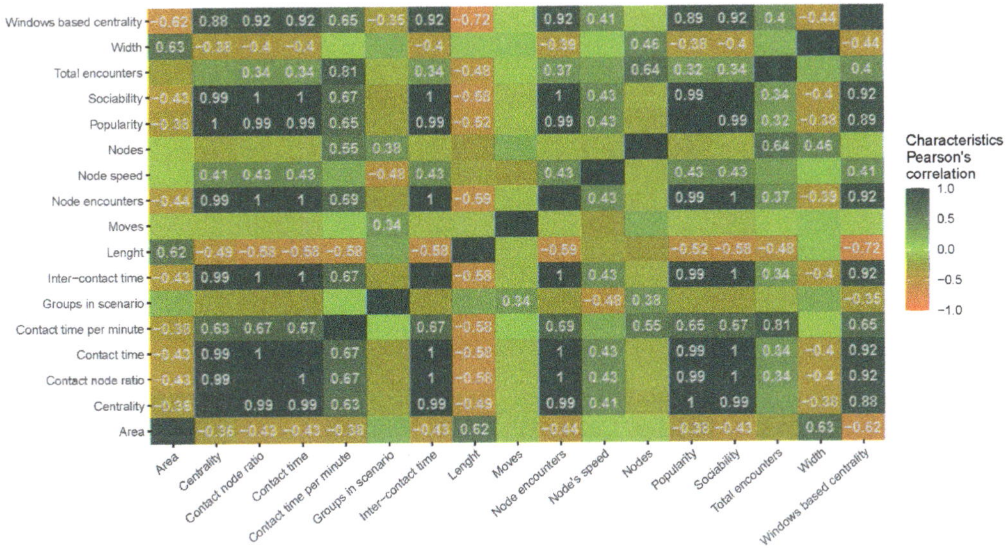

Figure 4. Heatmap of Pearson correlation coefficients between scenarios' characteristics. Only significant Pearson's correlation coefficients are shown.

Figure 4 made it clear that some characteristics were highly correlated. At first glance, one way to deal with highly correlated characteristics is by removing them. However, removing characteristics that describe scenarios was considered a wrong approach because fewer scenario characteristics might hinder the scenario description accuracy. The high correlation helped us understand that selecting scenarios would not be straightforward and that it will require a backtracking process to achieve diversity and representativeness among scenarios in the corpus. The backtracking process uses additional information about the characteristics. Specifically, the variance of the characteristics was used in the case there is a need to achieve representativeness and diversity objectives.

The second stage, creation of scenarios, received the characteristics found in stage one and then created scenarios for the given characteristics. This stage generated over 200,000 OppNet scenarios, many of which had similar behaviors and therefore similar vectors of characteristics. The scenarios with a similar vector of characteristics were considered equivalent.

The third and fourth stages, scenario selection and enhanced distribution, were loop-connected. Each characteristic range was evenly divided into sub-ranges called windows. Then, a subset of scenarios was selected for each window, and this process sequentially looped through the list of characteristics. The number of scenarios was reduced because the scenarios should belong to all windows of the characteristics. If there was no scenario in the window, those empty-scenario windows were re-adjusted until scenarios were found.

When all of the characteristics had been run through, and a representative number of scenarios had been obtained, stage four checked the diversity of the scenario collection. The loop was broken if the diversity of scenarios was fulfilled, which implies not having similar scenarios and that the distribution of characteristics manages to cover the entire range of each characteristic. Each scenario fulfills a part of the range of the characteristic. All scenarios, as a whole, complete the range of the characteristics.

The final stage, publish corpus, made the corpus of OppNet scenarios available for the research community. This assures the usability set as a quality requirement shown in Section 3.4. The following section describes the corpus obtained following the corpus creation methodology presented in this section.

3.6. Corpus Morphology

Section 3.5 describes the creation of the corpus of OppNet scenarios that address the quality requirements mentioned in Section 3.4. Creating the corpus following the methodology returned forty-one scenarios with a balance between representativeness and diversity. The similarities among the scenarios increased with a number higher than forty-one, thus harming the diversity of the corpus. Moreover, some characteristics were not represented when the number was lower than forty-one. Therefore, the corpus is a collection of forty-one OppNet scenarios, and the characteristics and their distribution can be seen in Table 1 and Figure 5, respectively.

Scenarios in the corpus are identified with a number in the range [1–41]. Additionally, the corpus covers the range of each characteristic with the range of each scenario. In Figure 5, the X axis of each sub-figure represents the scenarios, and the Y axis represents the characteristic. Scenarios depicted in Figure 5 are not ordered by their number but by the value of the characteristic.

Furthermore, Figure 5 shows that node centrality, node inter-contact time and node sociability are characteristics with a high Pearson correlation among them. That is the reason why their figures have resemblance among them.

Figure 6 shows a study of the diversity of the corpus scenarios using a heatmap. It shows the relative intensity of characteristics of each of the scenarios in the corpus. Each column in Figure 6 is a scenario of the corpus. As it is mentioned throughout this article, the corpus will be expected to have representative as well as diverse scenarios. Figure 6 shows that (1) there are no equal scenarios and (2) the distribution of the characteristics is uniform since there is no predominance of a single color.

For usability reasons, each scenario of the corpus has two types of traces mentioned in Section 2.4, the contact traces and their homologous position traces. Furthermore, the granularity of the position traces is one second. Additionally, the contact traces can be obtained from their homologous based on the node positions but not the other way around.

The scenarios simulate the speed of pedestrians, cyclists and two types of motorized vehicles. Those speeds are shown in Table 1. For this reason, up to four groups have been organized for each stage. Nodes among the same group share the speed and movement pattern. Movement patterns and node speeds are described in Table 1.

The number of nodes present in a scenario differs from one scenario to another. Still, the total number of nodes is distributed unevenly among the groups present in the scenarios with more than one group.

The morphology of the corpus depicted in Figure 5 is well distributed as a result of the methodologically selected scenarios. The following section assesses corpus behavior when routing messages with routing algorithms.

Section 3 explained the concept and characterization of an OppNet scenario. It also defined and created a corpus to evaluate and compare OppNet routing algorithms. Section 3 also described the creation methodology and the morphology of the corpus obtained. The following section assesses the behavior of the corpus.

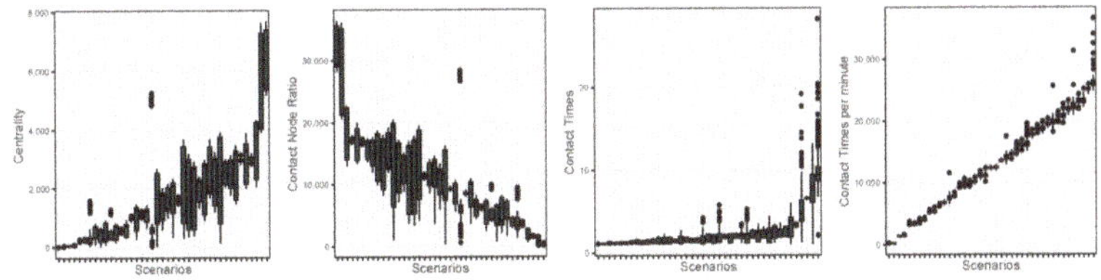

(**a**) Centrality (**b**) Contact node ratio (**c**) Contact time (**d**) Contact time by minute

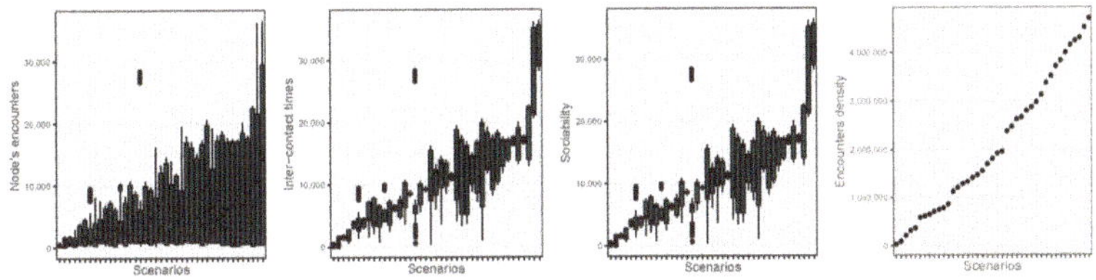

(**e**) Node encounters (**f**) Inter-contact times (**g**) Sociability (**h**) Encounters density

Figure 5. Scenario characteristics range distribution.

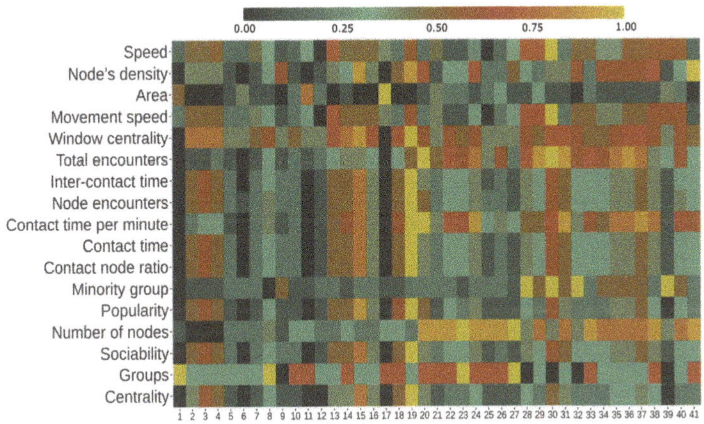

Figure 6. Diversity representation of the forty-one scenario collection, which constitutes the first corpus for the performance evaluation of OppNet routing algorithms. The *X*-axis is the scenario's number, and the *Y*-axis represents the characteristics.

4. Corpus Appraisement

In this section we aim to appraise the corpus behavior when routing messages using a concrete routing algorithm. For this purpose, a series of simulations were conducted to depict the behavior of the corpus scenarios. Therefore, a routing algorithm with high replication of messages was selected. The reason for such a selection is to verify if, under intense replication conditions, the corpus shows a different response within the scenarios.

4.1. Corpus Performance Appraisement

The experiment was conducted over the opportunistic network environment (the ONE simulator) [36] using the corpus of OppNet scenarios presented in Section 3.

Forty-one simulations were performed to assess the network representativeness of the corpus. In those simulations, node and message configurations were equal for all simulation setups. The forty-one simulations mean one simulation for each scenario of the corpus. The routing algorithm was an epidemic algorithm, a routing algorithm replicating messages to every contacted node. The reason behind the selection of an epidemic routing algorithm for the experiment was the ability to flood the network with messages exhaustively. An epidemic routing algorithm will forward a message to every node that it has contact with. Then, each recipient node will store the message until a new connection arises and repeat the forwarding process. An epidemic routing algorithm will delete the message only when the assigned time to live of the message is reached.

As was explained in Section 2, routing algorithms aim to transmit messages from source to destination. For this reason, network behavior could be expressed by how messages are delivered within the scenarios in the corpus. The simulations of the experiments have shown the behavior of the corpus with the metrics related to message delivery. The metrics analyzed were: The number of messages delivered, messages relayed, messages aborted, messages dropped, message hop-count and the message buffer time.

Figure 6 depicts the diversity within the characteristics vectors that define the scenarios in the corpus. In order to establish a difference among scenarios and, therefore, the corpus reliability, the scenario responses should be different between them. The response generated by each simulation was analyzed graphically to find their differences. Figure 7 presents the differences between the behaviors of the scenarios.

Figure 7 shows the differences of the response with eight sub-figures. Each sub-figure is a different metric. The *Scenarios* axis in each sub-figure stands for the forty-one scenarios. Although all sub-figures contain the same scenarios, scenarios are not ordered equally from one sub-figure to another because they are arranged in ascending order according to the metric that sub-figure represents. The *Y* axis in each sub-figure represents the normalized value of each scenario. Furthermore, each sub-figure depicts forty-one values in the [0–1] range since values are normalized.

The results show a different response from one scenario to another, proving a different behavior in each scenario. These results show the diversity among scenarios, which is expressed in Section 3.4 as a corpus design requirement. Some areas are denser than others, but responses are well distributed overall.

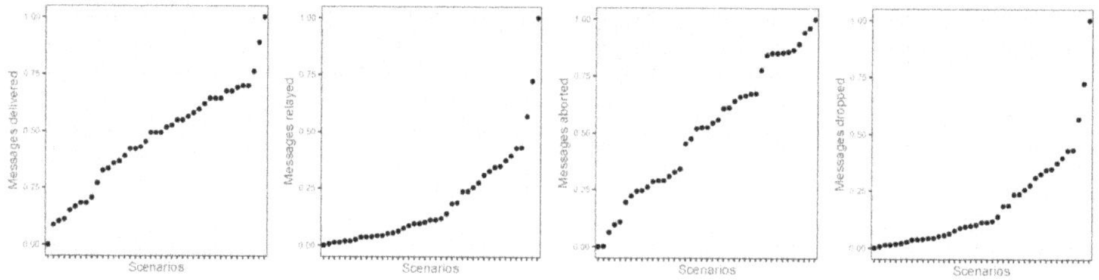

(**a**) Messages delivered (**b**) Messages relayed (**c**) Messages aborted (**d**) Messages dropped

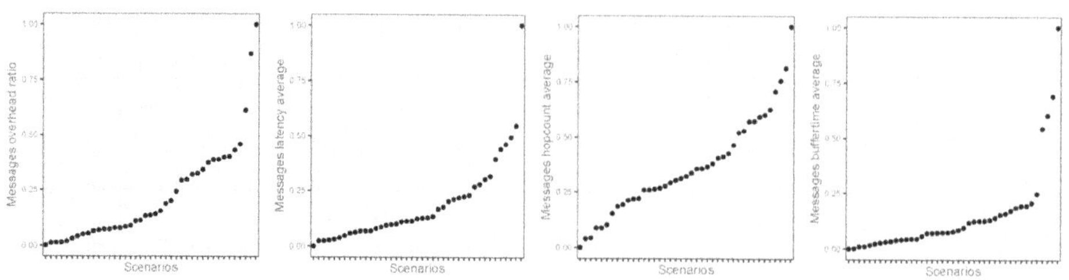

(**e**) Message overhead ratio (**f**) Messages latency average (**g**) Messages hopcount average (**h**) Messages buffertime average

Figure 7. Corpus scaled benchmarks for epidemic routing. Each dot in the figure is a scaled outcome ordered by the the metric; the identification number of scenarios is not shown and the order changes among sub-figures.

4.2. Evaluation and Comparison Using the Corpus

Now that the corpus contribution has been obtained via the methodology shown in Figure 3 and explained in Section 3.5, this section describes how the corpus can be used when a routing algorithm's evaluation and comparison process is needed. For the sake of clarity, some in-depth details are not included in this section, such as software configurations. The reader is asked to keep in mind that this section is intended to outline the usability of this study's main proposal rather than providing a closed recipe for using the corpus of OppNet scenarios.

When the research stage requires an evaluation of a routing algorithm, researchers interested in using a corpus will have to implement a simulation environment such as the one shown in Figure 2. The researchers should start by configuring the OppNet simulation software. After this, to use the corpus, researchers will have to download it. The corpus is available entirely free and without the necessity of login information. It is understood that at this stage, the routing algorithm that is going to be evaluated is already selected. Finally, researchers might configure nodes and messages and establish the metrics that will retrieve the information to evaluate the performance. If the researcher desires to assess a comparison, the process will have to be repeated only by changing the routing algorithm. Then, the researchers should compare the corresponding routing metrics obtained from the respective simulations.

In this section we have evaluated the behavior of the corpus with a high replication algorithm to limit the response of the corpus when transmitting messages, and the results show that the corpus scenarios have different network behaviors between them. This result ratifies the positive assessment of the corpus. From now on, the scientific community has a collection of scenarios where their routing algorithms and features can be tested, thus avoiding scenario selection, reducing time and eliminating unintended bias. The corpus contributes towards establishing a proper benchmarking scheme for OppNet routing algorithms where the routing performance is not relative to other routing algorithms but is examined overall.

5. Discussion

Nowadays, OppNet routing algorithms cannot be objectively evaluated nor compared because there is a lack of a globally accepted evaluation methods. This situation hinders the development of new routing algorithms. The present proposal intends to contribute toward an objective evaluation methodology by providing an analytically selected collection of scenarios, a corpus. This proposal will help ensure that evaluation results can be reliable, reproduced and contrasted in order to improve the objective quality.

Researchers have tried to evaluate their proposals fairly, for example, by evaluating each other's proposals, using scenarios that other researchers have used, or selecting metrics that fit their proposals. However, these evaluation approaches have not overcome problems such as lack of reproducibility or inability to generalize routing algorithms to any scenario.

It is common practice in OppNets to use well-known scenarios with a clear intention of standardizing evaluation methods. The problem, though, is not just a matter of using the same scenarios. If the routing algorithm being evaluated has to be general-purpose, it is also a requirement that the scenarios being used are representative of all possible network situations. Therefore, any collection of scenarios is not the solution, and what is needed is a fine selection of representative scenarios.

Besides the existence of a representative corpus, it is as well important that it be used by the community. The corpus introduced in this study has been proven as representative by means of experimentation, and has been made publicly available.

This work is not intended to create a dilemma of whether or not the corpus should replace the well-known scenarios. Obtaining a simple corpus is not a difficult task. There are different methods of obtaining a collection of scenarios in a straightforward manner, for example by using classical programming techniques such as random selection, trial and error, genetic algorithms, or even machine learning approaches. However, obtaining a representative corpus is complex and challenging. A representative one represents, as a whole, all possible network behaviors. The selection of scenarios for a representative corpus goes beyond a cherry-picking process, and each scenario is carefully analyzed and compared with other scenarios. Still, the selection process may not matter as much as the corpus itself. The differences and representativeness of network behavior that the corpus has are what determines if a corpus is useful or not.

The corpus presented in this work was obtained via a creation methodology based on identifying the variables that characterize OppNet scenarios, methods to create OppNet scenarios and processes to assess differences and diversity among them. The differences and the representativeness of each scenario were carefully assessed. The results measured the representativeness and diversity of the corpus scenarios, showing significant differences. Therefore, it can be said that this is a representative corpus for objective evaluation. Having a representative corpus does not imply necessarily that it is the best. The scenarios of the corpus should be reviewed in the future, especially as new technologies emerge from arising new network behaviors.

The corpus comprises simple scenarios where network behaviors are uniform. There might be environments where it is interesting to have non-uniform behaviors, for example, when defining strategies where the routing algorithm changes depending on network conditions. These complex scenarios can be built, for instance, by concatenating simple scenarios from the corpus without unnecessarily expanding the number of scenarios in the corpus.

When there is a corpus, there is the risk of falling into the trap of developing tailored solutions that only work with the elements of this corpus. The behavior of a routing algorithm should not be finely adjusted to have an outstanding performance in each corpus scenario, since making a fine-tune would reduce the ability of a routing algorithm to extend the solution beyond the scenarios to the real world. Therefore, the routing model would not be able to generalize its routing abilities because the abilities would be too specific for the scenarios.

Another risk while developing routing algorithms for OppNets is to exclusively focus, or pay too much attention, to simulations using the corpus. Simulation is just a part of the developing methodology, which should always be followed by an emulation stage, testing with actual implementations of the algorithms, and real-world experimentation. Researchers should not overlook a complete methodology to convert a routing idea into real-world implementation.

6. Conclusions

From the state-of-the-art, in the review of the methodologies for creating routing algorithms, it was seen that, until now, there was no clear evidence to objectively evaluate and thus compare the performance of these algorithms. Evaluating and comparing routing algorithms is a complex task, and the final quality of the algorithm significantly relies on it.

To right this wrong, this study proposed a potentially global-agreed corpus for a fair evaluation and comparison of routing algorithms—a reference corpus of OppNet scenarios, which is a cornerstone in the design methodology. This corpus is a collection of forty-one methodologically obtained OppNet scenarios. These scenarios can be used to evaluate and thus compare the performance of routing algorithms. These scenarios were obtained using a creation procedure developed in this work that includes a backtracking process to enhance scenario diversity. This means that the corpus has the least number of scenarios, which, as a whole, represents most of the real-world OppNets.

Furthermore, for creating the corpus, it was necessary to characterize OppNets scenarios with a vector of characteristics. Such vectors are the basis for the analysis of similarities that lead to whether a scenario was a corpus member or not. The scenario is a node's contact trace described by a vector of seventeen characteristics. The corpus presented in this work is a step toward creating a benchmarking scheme where the performance of routing algorithms is not relative to a selection of peers.

The corpus presented can be an important tool to help researchers follow the scientific method, especially regarding reproducibility and standardization aspects. These are essential features to improve quality research. The usefulness of the corpus requires that the community embraces it, using it for contrasting and evaluating routing performance results. The corpus is not static and should be revised to adapt to the needs; new technologies may require new scenarios in the future.

We look forward to this contribution simplifying and improving the development of routing algorithms in OppNets.

Author Contributions: Conceptualization, D.F., C.B. and S.R.; methodology, D.F. and S.R.; software, D.F. and C.B.; validation, D.F., C.B. and S.R.; formal analysis, D.F., C.B. and S.R.; data curation, D.F.; writing—original draft preparation, D.F.; writing—review and editing, D.F., C.B. and S.R.; visualization, D.F.; supervision, C.B. and S.R. All authors have read and agreed to the published version of the manuscript.

Funding: This work is partly funded by Secretaria de Educación Superior, Ciencia, Tecnología e Innovación (SENESCYT, ECUADOR), by the Catalan AGAUR 2017SGR-463 project, and by the Spanish Ministry of Science and Innovation TIN2017-87211-R project.

Data Availability Statement: The corpus is available to download at: https://deic.uab.cat/~oppnet-corpus/ accessed on 30 August 2022.

Conflicts of Interest: The authors declare no conflict of interest.

References

1. Helgason, Ó.; Kouyoumdjieva, S.T.; Pajević, L.; Yavuz, E.A.; Karlsson, G. A middleware for opportunistic content distribution. *Comput. Netw.* **2016**, *107*, 178–193. [CrossRef]
2. Borrego, C.; Borrell, J.; Robles, S. Hey, influencer! Message delivery to social central nodes in social opportunistic networks. *Comput. Commun.* **2019**, *137*, 81–91. [CrossRef]
3. Chen, D.; Borrego, C.; Navarro-Arribas, G. A Privacy-Preserving Routing Protocol Using Mix Networks in Opportunistic Networks. *Electronics* **2020**, *9*, 2–15. [CrossRef]
4. Sarros, C.A.; Demiroglou, V.; Tsaoussidis, V. Intermittently-connected IoT devices: Experiments with an NDN-DTN architecture. In Proceedings of the 18th Annual Consumer Communications & Networking Conference (CCNC), Las Vegas, NE, USA, 9 January 2021.
5. Danielis, P.; Karlsson, G. Survey of mobile opportunistic networks for parallel data dissemination and processing. *KuVS-Fachgesp* **2020**, *1*, 1–3.
6. Nayyar, A.; Batth, R.S.; Ha, D.B.; Sussendran, G. Opportunistic networks: Present scenario-a mirror review. *Int. J. Commun. Netw. Inf. Secur.* **2018**, *10*, 223–241. [CrossRef]
7. Borrego, C.; Castillo, S.; Robles, S. Striving for sensing: Taming your mobile code to share a robot sensor network. *Inf. Sci.* **2014**, *277*, 338–357. [CrossRef]
8. Conti, M.; Giordano, S. Mobile ad hoc networking: Milestones, challenges, and new research directions. *IEEE Commun. Mag.* **2014**, *52*, 85–96. [CrossRef]
9. Trifunovic, S.; Kouyoumdjieva, S.T.; Distl, B.; Pajevic, L.; Karlsson, G.; Plattner, B. A decade of research in opportunistic networks: Challenges, relevance, and future directions. *IEEE Commun. Mag.* **2017**, *55*, 168–173. [CrossRef]
10. Mordacchini, M.; Passarella, A.; Conti, M. A social cognitive heuristic for adaptive data dissemination in mobile Opportunistic Networks. *Pervasive Mob. Comput.* **2017**, *42*, 371–392. [CrossRef]
11. Du, Z.; Wu, C.; Chen, X.; Wang, X.; Yoshinaga, T.; Ji, Y. A VDTN scheme with enhanced buffer management. *Wirel. Netw.* **2020**, *26*, 1537–1548. [CrossRef]
12. Borrego, C.; Amadeo, M.; Molinaro, A.; Mendes, P.; Sofia, R.C.; Magaia, N.; Borrell, J. Forwarding in opportunistic information-centric networks: an optimal stopping approach. *IEEE Commun. Mag.* **2020**, *58*, 56–61. [CrossRef]
13. Magaia, N.; Sheng, Z. ReFIoV: A novel reputation framework for information-centric vehicular applications. *IEEE Trans. Veh. Technol.* **2018**, *68*, 1810–1823. [CrossRef]
14. Tsaoussidis, V.; Borrego, C. Network Working Group P. Mendes, Ed. Internet-Draft Airbus Intended Status: Experimental R. Sofia Expires: 19 March 2021 fortiss GmbH 2020. Available online: https://www.ietf.org/archive/id/draft-mendes-icnrg-dabber-05.pdf (accessed on 25 April 2022).
15. Rajeswari, S.R.; Seenivasagam, V. Comparative study on various authentication protocols in wireless sensor networks. *Sci. World J.* **2016**, *2016*, 6854303. [CrossRef] [PubMed]

16. Kuppusamy, V.; Thanthrige, U.M.; Udugama, A.; Förster, A. Evaluating forwarding protocols in opportunistic networks: Trends, advances, challenges and best practices. *Future Internet* **2019**, *11*, 113. [CrossRef]
17. Sachdeva, R.; Dev, A. Routing in Opportunistic Networks: Implementation and Research Challenges. *J. Engg. Res. Icari Spec. Issue* **2021**, *173*, 183. [CrossRef]
18. Alajeely, M.; Doss, R.; Ahmad, A. Routing Protocols in Opportunistic Networks—A Survey. *Iete Tech. Rev.* **2018**, *35*, 369–387. [CrossRef]
19. Vahdat, A.; Becker, D. *Epidemic Routing for Partially Connected Ad Hoc Networks*; Technical Report CS-200006; Duke University: Durham, NC, USA, 2000.
20. Lindgren, A.; Doria, A.; Schelen, O. Probabilistic routing in intermittently connected networks. In Proceedings of the International Workshop on Service Assurance with Partial and Intermittent Resources, Fortaleza, Brazil, 6 August 2004.
21. Spyropoulos, T.; Psounis, K.; Raghavendra, C.S. Spray and wait: an efficient routing scheme for intermittently connected mobile networks. In Proceedings of the SIGCOMM05: ACM SIGCOMM 2005 Conference, Philadelphia, PA, USA, 26 August 2005.
22. De Oliveira, E.C.; De Albuquerque, C.V. NECTAR: A DTN routing protocol based on neighborhood contact history. In Proceedings of the SAC09: The 2009 ACM Symposium on Applied Computing, Honolulu, HA, USA, 8 March 2009.
23. Grasic, S.; Lindgren, A. Revisiting a remote village scenario and its DTN routing objective. *Comput. Commun.* **2014**, *48*, 133–140. [CrossRef]
24. Kaner, C.; Bond, W.P. Software engineering metrics: What do they measure and how do we know. In Proceedings of the 10th International Software Metrics Symposium, Chicago, IL, USA, 11 September 2004.
25. Grasic, S.; Lindgren, A. An Analysis of Evaluation Practices for DTN Routing Protocols. In Proceedings of the Seventh ACM International Workshop on Challenged Networks, Istanbul, Turkey, 22 August 2012.
26. Petz, A.; Enderle, J.; Julien, C. A framework for evaluating dtn mobility models. In Proceedings of the 2nd International Conference on Simulation Tools and Techniques, Rome, Italy, 6 March 2009.
27. Sandulescu, G. Resource-Aware Routing in Delay and Disruption Tolerant Networks. Ph.D. Thesis, University of Luxembourg, Luxembourg, 2011.
28. Angius, F.; Gerla, M.; Pau, G. Bloogo: Bloom filter based gossip algorithm for wireless NDN. In Proceedings of the ACM Workshop on Emerging Name-Oriented Mobile Networking Design-Architecture, Algorithms, and Applications, Hilton Head, CA, USA, 11 June 2012.
29. Freire, D.; Robles, S.; Borrego, C. Towards a Methodology for the Development of Routing Algorithms in Opportunistic Networks. In Proceedings of the The Sixteenth International Conference on Wireless and Mobile Communications ICWMC 2020, Oporto, Portugal, 19 October 2020.
30. Zhang, Y.J. Evaluation and comparison of different segmentation algorithms. *Pattern Recognit. Lett.* **1997**, *18*, 963–974. [CrossRef]
31. Abdelkader, T.; Naik, K.; Nayak, A.; Goel, N.; Srivastava, V. A performance comparison of delay-tolerant network routing protocols. *IEEE Netw.* **2016**, *30*, 46–53. [CrossRef]
32. Bajaj, L.; Takai, M.; Ahuja, R.; Tang, K.; Bagrodia, R.; Gerla, M. *Glomosim: A Scalable Network Simulation Environment*; UCLA Computer Science Department Technical Report; UCLA Computer Science Department: Los Angeles, CA, USA, 1999; 990027, pp. 1–12.
33. Varga, A. OMNeT++. In *Modeling and Tools for Network Simulation*; Frederiksen, N.O., Gulliksen, H., Eds.; Springer: Berlin/Heidelberg, Germany, 2010; pp. 35–59.
34. Fall, K.; Ott, J. Delay-Tolerant Networking Research Group-DTNRG; 2002. Available online: https://www.ietf.org/proceedings/75/DTNRG.html (accessed on 20 May 2020
35. Su, J.; Scott, J.; Hui, P.; Crowcroft, J.; Lara, E.D.; Diot, C.; Goel, A.; Lim, M.H.; Upton, E. Haggle: Seamless networking for mobile applications. In Proceedings of the International Conference on Ubiquitous Computing, Innsbruck, Austria, 16 September 2007.
36. Keränen, A.; Ott, J.; Kärkkäinen, T. The ONE Simulator for DTN Protocol Evaluation. In Proceedings of the 2nd International Conference on Simulation Tools and Techniques, Rome, Italy, 2 March 2009.
37. Riley, G.F.; Henderson, T.R. The ns-3 network simulator. In *Modeling and Tools for Network Simulation*; Frederiksen, N.O., Gulliksen, H., Eds.; Springer: Berlin, Germany, 2010; pp. 15–34.
38. Papanikos, N.; Akestoridis, D.G.; Papapetrou, E. CRAWDAD Toolset Tools/SIMULATE/uoi/Adyton (v. 2016-04-21). Available online: https://crawdad.org/tools/simulate/uoi/adyton/20160421 (accessed on 23 March 2022).
39. Ciobanu, R.I.; Marin, R.C.; Dobre, C. Mobemu: a framework to support decentralized ad-hoc networking. In *Modeling and Simulation in HPC and Cloud Systems*; Joanna, K., Florin Pop, C.D., Ed.; Springer: Berlin/Heidelberg, Germany, 2018; pp. 87–119.
40. Kotz, D.; Henderson, T.; Abyzov, I.; Yeo, J. CRAWDAD Dataset Dartmouth/Campus (v. 2009-09-09). Available online: https://crawdad.org/dartmouth/campus/20090909 (accessed on 1 August 2022).
41. Thiebaut, D.; Wolf, J.L.; Stone, H.S. Synthetic traces for trace-driven simulation of cache memories. *IEEE Trans. Comput.* **1992**, *41*, 388–410. [CrossRef]
42. Manfredi, V.; Crovella, M.; Kurose, J. Understanding stateful vs stateless communication strategies for ad hoc networks. In Proceedings of the 17th annual international conference on Mobile computing and networking, Las Vegas, NE, USA, 19 September 2011.

43. Souza, C.; Mota, E.; Manzoni, P.; Cano, J.C.; Calafate, C.T.; Hernández-Orallo, E.; Tapia, J.H. Friendly-drop: A social-based buffer management algorithm for opportunistic networks. In Proceedings of the 2018 Wireless Days (WD), Dubai, United Arab Emirates, 3 April 2018.
44. Borrego, C.; Borrell, J.; Robles, S. Efficient broadcast in opportunistic networks using optimal stopping theory. *Ad Hoc. Netw.* **2019**, *88*, 5–17. [CrossRef]
45. Cabrero, S.; Garcia, R.; García, X.G.; Melendi, D. CRAWDAD Dataset Oviedo/Asturies-er (v. 2016-08-08). Available online: https://crawdad.org/oviedo/asturies-er/20160808 (accessed on 23 June 2022).
46. Bracciale, L.; Bonola, M.; Loreti, P.; Bianchi, G.; Amici, R.; Rabuffi, A. CRAWDAD Dataset roma/taxi (v. 2014-07-17). Available online: https://crawdad.org/roma/taxi/20140717 (accessed on 3 March 2022).
47. Piorkowski, M.; Sarafijanovic-Djukic, N.; Grossglauser, M. CRAWDAD Dataset Epfl/Mobility (v. 2009-02-24). Available online: https://crawdad.org/epfl/mobility/20090224 (accessed on 22 August 2022).
48. Akestoridis, D.G. CRAWDAD Dataset Uoi/Haggle (v. 2016-08-28): derived from cambridge/haggle (v. 2009-05-29). Available online https://crawdad.org/uoi/haggle/20160828/one (accessed on 23 June 2022).
49. Islam, M.R.; Rajon, S.A. On the design of an effective corpus for evaluation of Bengali Text Compression Schemes. In Proceedings of the 2008 11th International Conference on Computer and Information Technology, Khulna, Bangladesh, 27 December 2008.
50. Usama, M.; Malluhi, Q.M.; Zakaria, N.; Razzak, I.; Iqbal, W. An efficient secure data compression technique based on chaos and adaptive Huffman coding. *Peer -Peer Netw. Appl.* **2021**, *14*, 2651–2664. [CrossRef]
51. Arnold, R.; Bell, T. A corpus for the evaluation of lossless compression algorithms. In Proceedings of the DCC'97 Data Compression Conference, Snowbird, UT, USA, 25 March 1997.
52. Karamshuk, D.; Boldrini, C.; Conti, M.; Passarella, A. Human mobility models for opportunistic networks. *IEEE Commun. Mag.* **2011**, *49*, 157–165. [CrossRef]
53. Sandulescu, G.; Nadjm-Tehrani, S. Opportunistic DTN routing with window-aware adaptive replication. In Proceedings of the 4th Asian Conference on Internet Engineering, Pattaya, Thailand, 18–20 November 2008; pp. 103–112.
54. Yuan, P.; Wang, C. OPPO: An optimal copy allocation scheme in mobile opportunistic networks. *Peer -Peer Netw. Appl.* **2018**, *11*, 102–109. [CrossRef]
55. Schurgot, M.R.; Comaniciu, C.; Jaffres-Runser, K. Beyond traditional DTN routing: Social networks for opportunistic communication. *IEEE Commun. Mag.* **2012**, *50*, 155–162. [CrossRef]
56. Settawatcharawanit, T.; Yamada, S.; Haque, M.E.; Rojviboonchai, K. Message dropping policy in congested social delay tolerant networks. In Proceedings of the 2013 10th International Joint Conference on Computer Science and Software Engineering (JCSSE), Khon Kaen, Thailand, 29–31 May 2013; pp. 116–120.-
57. Bhattacharjee, S.; Roy, S.; Ghosh, S.; DasBit, S. Exploring the impact of connectivity on dissemination of post disaster situational data over DTN. In Proceedings of the 18th International Conference on Distributed Computing and Networking, Delhi, India, 3–5 July 2017; pp. 1–4.
58. Boldrini, C.; Conti, M.; Passarella, A. Social-based autonomic routing in opportunistic networks. In *Autonomic Communication*; Springer: Berlin/Heidelberg, Germany, 2009; pp. 31–67.
59. Freedman, D.; Pisani, R.; Purves, R. *Statistics (International Student Edition)*, 4th ed.; WW Norton & Company: New York, NY, USA, 2007.

applied sciences

MDPI

Article

User-BS Selection Strategy Optimization with RSSI-Based Reliability in 5G Wireless Networks

Jie Shen [1,2], Yijun Hao [1,2], Yuqian Yang [2,3] and Cong Zhao [1,2,*]

1 School of Mathematics and Statistics, Xi'an Jiaotong University, Xi'an 710049, China; sj19998@stu.xjtu.edu.cn (J.S.); yijunhao@stu.xjtu.edu.cn (Y.H.)
2 National Engineering Laboratory for Big Data Analytics, Xi'an Jiaotong University, Xi'an 710049, China; yuqian.yang@stu.xjtu.edu.cn
3 School of Computer Science and Technology, Xi'an Jiaotong University, Xi'an 710049, China
* Correspondence: congzhao@xjtu.edu.cn

Abstract: Although fifth-generation (5G) wireless communication can] support well a high data rate of transmission, issues such as base station (BS) failure and poor BS signals may cause serious interruption problems. This paper studies the user-BS selection strategy with received signal strength indication (RSSI)-based reliability in 5G wireless networks. First, reliability is defined on the basis of the RSSI and failure probability of the BS. The problem is modeled as a selection strategy optimization problem with BS capacity and receiving sensitivity as constraints. Second, the original problem can be transformed into a resource allocation problem with probabilistic constraints. For the situation where user distribution is known, we used dynamic programming to obtain the optimal BS selection strategy. For the situation where user distribution is unknown, starting from user trajectory data, wc used the space–time density estimation method based on the Epanechnikov kernel to estimate user density and bring it into dynamic programming to obtain the optimal selection strategy. Simulation results show that our density estimation algorithm is more accurate than the commonly used density estimation algorithm. Compared with the distance-based optimization method, our RSSI-based optimization method also improved the communication signal quality under different scenarios.

Keywords: wireless network; selection-strategy optimization; RSSI; dynamic programming; space–time density estimation

Citation: Shen, J.; Hao, Y.; Yang, Y.; Zhao, C. User-BS Selection Strategy Optimization with RSSI-Based Reliability in 5G Wireless Networks. *Appl. Sci.* **2022**, *12*, 6082. https://doi.org/10.3390/app12126082

Academic Editors: Yang Yue, Runzhou Zhang, Hao Feng, Zheda Li, Lin Zhang and Dawei Ying

Received: 20 May 2022
Accepted: 14 June 2022
Published: 15 June 2022

1. Introduction

With the continuous increase in the number of network users, people's requirements for network operation speed and stability are also increasing, and the emergence of 5G wireless communication technology can effectively meet current users' comprehensive needs for network and device communication. It can also greatly enhance the user's service experience [1]. In order to meet the ever-increasing demand for data services, wireless communication networks using the 30–300 GHz millimeter-wave frequency band have become an indispensable part of the 5G communication system [2,3]. However, millimeter wave transmission is still affected by severe signal attenuation and congestion, which requires a sophisticated BS deployment plan for heterogeneous cellular networks [4]. In addition, in order to combat high path loss, 5G wireless BSs are usually densely deployed. Different from the fixed signal transmission paths in wired networks, signals have different transmission paths in wireless networks due to different user-BS connection strategies. Especially when users move, even the connectivity between users and 5G wireless BSs rapidly changes. Therefore, how to associate users with BSs to ensure the reliability of communication as much as possible is another key issue for 5G wireless networks [5], and this is the key object of this paper.

The user-BS connection strategy is an important issue in wireless communication systems and has been extensively studied in the past few decades. In traditional cellular systems, BSs are usually deployed to achieve seamless network coverage. Whether a user can be covered by a BS depends on the distance between them [4]. However, such a user-BS connection strategy based solely on distance is difficult to meet the high standards of 5G wireless networks [6], since distance is only from a geometric point of view and can only reflect limited information. The standard requirements of the 5G networks are quite different from 4G in terms of low latency, bandwidth, reliability, and availability [7]. In order to deal with the problems encountered in actual situations, the user-BS connection strategy mostly starts from some utility indicators such as spectral and energy efficiency, and quality of service (QoS) [8]. However, instead of being inherent properties, these utility indicators are defined by people. In addition to difficulties in obtaining, there may also exist problems such as inconsistent standards. In order to address the above problems, we need to consider a common and real indicator that can easily be obtained in practical scenarios. In addition, the above studies rarely take reliability into account, while 5G wireless networks have high requirements for reliability. Reliability is one of the most important factors of wireless network communication quality; it can not only improve user experience, but also help operators in operation and maintenance. On the basis of the above motivations, we considered modeling the user-BS connection strategy problem as an optimization problem with reliability as the optimization object.

The received signal strength indicator (RSSI) is commonly used in the communication localization field [9]. RSSI can easily be obtained from most WiFi receivers such as mobile phones, tablets, and laptops [10,11]. RSSI meets the above requirements, so we chose to optimize reliability with RSSI.

This paper studies the optimization of 5G wireless communication networks based on RSSI by rationally designing a BS-user strategy selection scheme. For the selection strategy, our goal was to maximize the communication signal quality of the overall wireless network when BS capacity and receiving sensitivity constraints are met. Since the constraint function exists in the form of an expectation function that cannot be handled by the traditional random approximation simulated annealing (SA) algorithm, we began from user distribution and considered the two cases where user distribution is known and unknown.

In the case of a given user distribution, the distribution table is used to directly derive the confidence interval at a given significance level. Random constraints are converted into general constraints, and the dynamic programming method is then used to solve the optimization problem. For the case where the user distribution is unknown, we estimated the density of users on the basis of trajectory data, and derived the dynamic programming recursion formula on the basis of the estimated density. Our main contributions are listed below:

(1) In order to meet the high requirements for reliability in 5G wireless networks, we propose to model the user-BS selection strategy problem as an optimization problem with reliability as the object.
(2) In view of the drawbacks in commonly used indicators, we chose to define reliability by the RSSI for the first time, which could easily be obtained in practical scenarios. Considering the time-varying nature of the user's position, we transformed the original problem into a resource allocation problem under probability constraints, and solved it with dynamic programming and the time–space density estimation method.
(3) We conducted two comparative simulations to verify the superiority of our algorithm in terms of the density estimation effect and the reliability approach GeoLife GPS Trajectories dataset [12]. Simulation 1 showed that, under different sampling frequencies, our time–space density estimation method improved accuracy by 2.84% compared to the two-dimensional kernel density estimation method. Simulation 2 showed that, compared with the distance-based selection strategies, our RSSI-based selection strategies improved communication reliability by an average of 3.57% under the three scenarios above.

The rest of this paper is organized as follows. Section 3 presents the system model and formulates the BS-user selection strategy problem. In Section 4, the wireless BS selection-strategy optimization problem based on identified distribution is solved by dynamic programming. The wireless BS selection-strategy optimization problem based on user trajectory is addressed in Section 5, where a time–space density estimation method is proposed. Simulation results are presented in Section 6 to illustrate the performance of the proposed time-space density estimation method under different RSSI scenarios. Lastly, Section 7 concludes this paper.

2. Related Work

2.1. User-BS Selection Strategy

User-BS selection, aiming to associate a user with a particular serving BS, is a critical procedure in wireless networks that substantially affects network performance [13]. In traditional LTE systems, the radio admission control entity is located in the radio resource control layer of the protocol stack, which decides whether a new radio-bearer admission request is admitted or rejected [14]. The distance-based user-BS selection strategy where a user chooses to associate with the nearest BS is the most prevalent. Five metrics are commonly used in user-BS selection, namely, outage/coverage probability, spectral efficiency, energy efficiency, QoS, and fairness [15]. In actual situations, one or a combination of several indicators are used. The new technologies and standards of 5G networks inevitably render ineffective the above rudimentary user-BS selection strategy and metrics, and more effective user-BS selection algorithms are needed for addressing the unique features of emerging 5G wireless networks.

Utility is widely used in the modeling of user association problems. In order to make decisions, utility quantifies the satisfaction that a particular service provides to decision makers [16]. According to the used metrics, utility considered in user association may consist of, for example, spectral efficiency [17], energy efficiency [18,19], QoS [20]. Logarithms, exponentials, and sigmoidal utility functions are used to model these properties. For studies that do not specifically discuss the selection of utility functions, it can be assumed that they use linear utility functions, that is, utility is spectral efficiency, energy efficiency, or QoS itself [21]. Game theory [22], combinatorial optimization [23], and random geometry [24] are commonly used models to solve user-BS selection strategy problems. Details are shown in Table 1.

Table 1. Common metrics and models for user-BS selection strategy problems.

Metrics	Spectral efficiency	[17]
	Energy efficiency	[18,19]
	QoS	[20]
Models	Game theory	[22]
	Combinatorial optimization	[23]
	Random geometry	[24]

2.2. RSSI-Based Optimization Model

In wireless communication networks, signal quality is an important indicator that affects communication reliability, and received signal strength (RSS) is its most important part. The majority of existing work for RSS focused on large-scale cooperative sensor network localization subject to communication constraints. To the best of our knowledge, RSS was first considered for cooperative localization in [25], where a single RSS was optimized so as to limit the number of neighboring sensors. In a recent work [26], RSS was considered for noncooperative infrastructure-based indoor positioning. However, the focus of the above study lay on the overall positioning performance in a given service area and thorough treatment on the measurement campaign, RSS modeling, model fitting and parameter calibration, signaling, and performance evaluation using real data measured from a live network.

In order to measure RSS, we have RSSI, an optional part of the transmission layer, which is one of the most important indicators used to determine the link quality and whether to increase the broadcast transmission strength [27]. PredominatingRSSI models can be divided into three types: the spatial-propagation, ShadowWing, and distance-loss models. Considering the influence of complex factors such as reflection, blocking, and diffraction, the distance-loss model is more capable of reflecting the actual application environment and the closest to the true value of the distance [28].

RSSI indicates the strength of the received wireless signal, which is easily affected by the environment and has unstable characteristics. Its value gradually decreases as the distance between the terminal and the access point increases. The larger the RSSI is, the higher the signal reception strength and the better the data transmission channel are. On the other hand, the lower the RSSI is, the weaker the received signal and the worse the quality of the physical channel for data transmission are, and the probability of packet loss and bit errors during data transmission obviously increases.

RSSI calculates propagation loss by measuring the transmission power and the received power, and then uses the signal attenuation model to convert propagation loss into transmission node distance [29]. This is a low-power, low-cost ranging technology with the characteristics of low cost, less equipment, long distance, and easy access.

Due to the above advantages, RSSI is often used to solve optimization problems in wireless communication. In [30] RSSI was used to resolve the optimization problem of Bluetooth low-energy (BLE) beacon density, and the authors in [31] used RSSI to help in person tracking and monitoring in industrial environments. However, most of the existing RSSI-based optimization models mainly focused on indoor localization and rarely applied it to other scenarios.

In this paper, we define the overall reliability of the wireless network by the RSSI, which can easily be obtained in the real-world scenario, together with the failure probability of the BS, and model the problem as a selection strategy optimization problem with reliability as the optimization goal.

3. System Model and Problem Formulation

We considered a wireless network with N BSs serving a group of users in a 2-dimensional geometry. We used R_{BS} to denote the invalid probability of each BS, and N_{BS} to denote the number of connection restrictions for each BS. Let $\mathcal{S}$ denote the set of all BS selection strategies, from which users choose the appropriate strategy $s \in \mathcal{S}$. We used x to denote the position coordinates of UEs, and then $s(x)$ can be expressed in detail as $(s_1(x), s_2(x), \ldots, s_N(x))$. Such a selection strategy means the probability that the user chooses to connect to BS i at location x is $s_i(x)$, and we have $s_i(x) \in [0,1]$ together with $\sum_{i=1}^{N} s_i(x) = 1$. Our optimization goal was to maximize the quality of the communication signal, so we define the reliability of the communication signal from the perspective of reliability modeling.

Throughout this paper, $\mathbb{P}[A]$ denotes the probability of event A, $d_i(x)$ denotes the distance between BS i and user location x, $R_{BS}(i)$ denotes the probability that BS i is invalid, s denotes the selection strategy, $\mathbb{E}[\cdot]$ denotes the expectation operator, α denotes the given significance level, and λ denotes the confidence interval.

3.1. Reliability Modeling

The main objective of reliability modeling is to express the reliability of a given system in terms of the reliability measures of its constituent components. Consider a system Y that consists of n components. Each component can only have two distinct states: it can either be functional or be off. Let binary variable π_i be the state indicator of component i as follows:

$$\pi_i = \begin{cases} 1, & \text{if component } i \text{ is on,} \\ 0, & \text{if component } i \text{ is off.} \end{cases} \tag{1}$$

A state of system Y is a description of the states of all its components; hence, $\pi = \{\pi_i\}$ for $i = 1, \ldots, n$. Let Π be the set of all possible states of Y. The structural function of Y, denoted by $f(\pi)$, is a binary function that indicates whether the system is working under a given state according to the following equation:

$$f(\pi) = \begin{cases} 1, & Y \text{ is functional,} \\ 0, & Y \text{ has failed.} \end{cases} \tag{2}$$

On the basis of the above definitions, the reliability of Y, denoted by $R(Y)$, can be calculated using the following equation:

$$R(Y) = \mathbb{P}[f(\pi) = 1] = \sum_{\pi \in \Pi} f(\pi)\mathbb{P}[\pi]. \tag{3}$$

3.2. RSSI

RSSI is a method of receiving signal strength indicating ranging. In an actual application environment, since a wireless signal is affected by various obstacles, reflections, multipath propagation, temperature, and propagation mode, electromagnetic wave transmission loss conforms to the lognormal shadow model, which can be described by the modified path-loss model [29]:

$$PL(d) = PL(d_0) + 10n \log_{10}(\frac{d}{d_0}) + X_\sigma, \tag{4}$$

where $PL(d)$ is the loss after signal propagation distance d, $PL(d_0)$ is the loss after signal propagation distance d_0, n is the propagation factor (usually 2~5), and X_σ is the shielding factor that is a Gaussian random noise variable with mean 0 and variance α.

$PL(d_0)$ in (4) can be calculated by the outdoor radio free space propagation model. The free space propagation model [32] is:

$$PL(d_0) = 32.44 + 10n \log_{10}(d) + 10n \log_{10}(f_c), \tag{5}$$

where f_c is the frequency of the propagated signal, and d is the distance between the sending and receiving nodes; usually, $d_0 = 1$ m. Our approach is not restricted by this specific formula and it can be straightforwardly extended to any path-loss forms.

The signal strength of the anchor node received by the unknown node is:

$$RSSI(d) = P_s + P_a - PL(d), \tag{6}$$

where P_s is the transmitting power of the node signal, P_a is the antenna gain, and $PL(d)$ is the loss after signal propagation distance d. According to Formulas (4)–(6), the distance can be calculated.

In the WINNERII C2 model [33] that simulates the wireless channel of the cellular connection in tahe 5G network environment, the path-loss value at distance d from the urban macrocell base station can be expressed as:

$$PL(d) = 27 + 22.7 \log_{10}(d) + 20 \log_{10}(f_c) + X_\sigma. \tag{7}$$

3.3. Optimization Model with RSSI-Based Reliability

The higher the received signal strength is, the more reliable the connection is. Combined with failure probability $R_{BS}(i)$ and $\mathbb{P}[\pi]$ in (3), the reliability of the user at location x, $R(s, x, R_{BS})$, can be expressed by selecting strategies s, x, and R_{BS} :

$$R(s, x, R_{BS}) = \sum_{i=1}^{N} s_i(x) RSSI(d_i(x))(1 - R_{BS}(i)). \tag{8}$$

Combining Equation (8) with user distribution, the overall reliability of wireless side R_w, can be calculated as:

$$\begin{aligned} R_w &= \iint R(s,x,R_{BS})f(x)dx \\ &= \iint \sum_{i=1}^{N} s_i(x)(1-R_{BS}(i))RSSI(d_i(x))f(x)dx \\ &= \sum_{i=1}^{N} \iint s_i(x)(1-R_{BS}(i))RSSI(d_i(x))f(x)dx. \end{aligned} \quad (9)$$

The overall optimization problem is:

$$\max_{s\in\mathcal{S}} \sum_{i=1}^{N} \iint s_i(x)(1-R_{BS}(i))RSSI(d_i(x))f(x)dx. \quad (10)$$

BS capacity refers to the number of channels that should be configured for a base station or a cell. In large cities and megacities, due to the rapid growth of users, each *BS* should be equipped with as many available channels as possible. Therefore, *BS* capacity becomes user capacity calculated by the number of channels. When there are too many users connected to the same *BS*, this leads to a decrease in communication quality and reduced reliability. Therefore, the expected number of users connected to each BS i cannot exceed the limit of the number of connections of *BS* capacity $N_{BS}(i)$. For each *BS* i, the probability of a user connecting to it can be expressed as $\iint s_i(x)f(x)dx$, so the capacity condition for M users can be expressed as:

$$M \iint s_i(x)f(x)dx \leqslant N_{BS}(i), \quad (11)$$

for $i = 1, 2, \ldots, N$.

Receiving sensitivity refers to the minimal received signal strength with which the receiver can correctly take out the useful signal, which means that the RSSI must be greater than receiving sensitivity SEN:

$$RSSI > SEN. \quad (12)$$

SEN [34] can be expressed as:

$$SEN = 10\log_{10}(KT0) + 10\log_{10}(BW) + NF + SNR_{min}. \quad (13)$$

where, $10\log_{10}(KT0)$ represents that the noise floor at a room temperature of 25 °C is -174 dBm, BW refers to bandwidth, NF is the noise figure of the system that generally refers to the noise figure of the first low noise amplifier, and SNR_{min} is the minimal signal-to-noise ratio (SNR) requirement of the receiver. Similar to RSSI, the above approach is not restricted by this specific formula and can be straightforwardly extended to any sensitivity forms.

Taking a gNodeB BS of 5G NR with 20 MHz bandwidth as an example, the SEN of gNodeB is:

$$\begin{aligned} SEN &= -174 + 10\log_{10}(19.08 \times 10^6) + 6 + (-1) \\ &= -96.2, \end{aligned} \quad (14)$$

where $BW = 20$ MHz, and the actual bandwidth occupied by the business is 19 MHz; $NF = 6$ dB for the system. NF here is the insertion loss before the first-level LNA and the noise coefficient NF of the LNA itself, $SNR_{min} = -1$ dB.

According to the RSSI formula [35]:

$$d = d_0 10^{\frac{P_a + P_s - PL(d_0)}{10n}} 10^{-\frac{1}{10n} RSSI(d)}, \tag{15}$$

so $RSSI(d_i(x)) >$ Sensitivity can be transformed into

$$d_i(x) < D = d_0 10^{\frac{P_t - PL(d_0)}{10n}} 10^{-\frac{1}{10n} SEN}, \tag{16}$$

which means that signal effective range C_i is given to any $BS\ i$, which satisfies that $\forall x \in C_i$, $d_i(x) < D$, and D represents the maximal signal connection distance. Therefore, the optimization problem under constraints can be expressed as:

$$s_i(x) = 0, \quad \text{for } x \notin C_i. \tag{17}$$

On the basis of Equation (10), considering BS capacity (11) and SEN (17) constraints at the same time, the optimization problem can be expressed as follows:

$$\max_{s \in \mathcal{S}} R_w, \tag{18}$$

$$s.t. \begin{cases} M \iint s_i(x) f(x) dx \leq N_{BS}(i), \\ s_i(x) = 0, \end{cases} \tag{19}$$

for $i = 1, 2, \ldots, N$ and $x \notin C_i$.

4. Wireless BS Selection-Strategy Optimization Based on Identified Distribution

First, the optimization problem is simplified. According to the calculation formula of $RSSI(d)$, $RSSI(d_i(x))$ is negatively correlated with $\log_{10} d_i(x)$, so original optimization Problem (18) can be simplified to

$$\begin{aligned} &\min_{s \in \mathcal{S}} \iint \sum_{i=1}^{N} s_i(x) R_{BS}(i) \log_{10} d_i(x) f(x) dx \\ &= \min_{s \in \mathcal{S}} \sum_{i=1}^{N} \iint s_i(x) R_{BS}(i) \log_{10} d_i(x) f(x) dx. \end{aligned} \tag{20}$$

The actual optimization objective of the above optimization problem is functional s instead of variable x. The general idea of finding the extreme value of a functional is to construct a Lagrangian functional according to the Lagrangian multiplier theorem and then set the Frechet derivative of the Lagrangian functional to zero.

Assuming a uniform distribution and that $s_i(x)$ is fixed for all x locations, the optimization function is set to be:

$$f(s) = \sum_{i=1}^{N} \iint_C s_i R_{BS}(i) \log_{10} d_i(x) \frac{1}{A_C} dx. \tag{21}$$

with constraints:

$$\begin{aligned} g_i(s) = M \iint s_i(x) f(x) dx - N_{BS}(i) \leq 0, \\ 1 - \sum_{i=1}^{N} s_i(x) = 0, \end{aligned} \tag{22}$$

for $i = 1, 2, \ldots, N$ and $x \notin C_i$, where $\mathcal{C} = \cup_{i=1}^{N} C_i$, $A_{\mathcal{C}}$ represents the area of the region $\mathcal{C}$. According to the Lagrangian multiplier theorem, the Lagrangian functional is:

$$L(s, \alpha, \beta) = f(s) + \sum_{i=1}^{N} \alpha_i g_i(s) + \beta(1 - \sum_{i=1}^{N} s_i(x)). \tag{23}$$

The Frechet derivative operator L of Formula (23) is calculated, and KKT conditions (24)–(27) are solved:

$$\nabla_{s_i} L(s, \alpha, \beta) = 0, \tag{24}$$

$$\alpha_i g_i(s) = 0, \tag{25}$$

$$\alpha_i > 0, \tag{26}$$

$$1 - \sum_{i=1}^{N} s_i(x) = 0, \tag{27}$$

Then, we can obtain $s_i = N_{BS}(i)/N$.

The above method can only be applied to simple functionals. For more complex functionals, the commonly used functional optimization methods are variational methods and optimal control.

Variational method is a branch of mathematics developed at the end of the 17th century. It is a field of mathematics dealing with functions as opposed to ordinary calculus dealing with functions of numbers. The Euler–Lagrange (E–L) equation is the key theorem of the variational method that corresponds to the critical point of the functional. The E–L equation is only a necessary condition for the functional to have extreme values, but not sufficient. That is, when the functional has extreme values, the E–L equation holds.

Classical variational theory can only solve the problem of unconstrained control, but most of the problems in engineering practice are control-constrained. Therefore, modern variational theory with optimal control as the research object appeared. Optimal control refers to seeking a control under given constraint conditions to allow for the given system performance index reach the maximal (or minimal) value. The main methods to solve the optimal control problem are the classical variational method, the maximal-value principle, and dynamic programming.

The simplest form of variational method for a functional is:

$$J[y(x)] = \int_{x_1}^{x_2} F(x, y(x), y'(x))dx. \tag{28}$$

In our optimization problem, F only depends on y and with no y'; at this time, $F_y \equiv 0$, so Euler equation $F_y(x, y) = 0$ or $F_y(y) = 0$. This is a functional equation whose solution does not contain any constants. The solution of this function usually does not meet the boundary conditions, and the variational problem has no solution. It is difficult to directly calculate it through mathematical methods, so we attempted to use the most advanced method, optimal control algorithms, which combines modern theoretical ideas to help in solving problems such as dynamic programming, which is an efficient mathematical method for the study and optimization of multistage sequential decision-making problems such as resource allocation.

4.1. Unconstrained Optimization

The original problem can be regarded to be an optimal allocation problem, N processes correspond to the connection with N BSs, and the allocation object corresponds to the probability of connecting to each BS in selection strategies.

We can define

$$F_k(I) = \min_{s \in \mathcal{S}} \sum_{i=1}^{k} \iint s_i(x) R_{BS}(i) \log_{10} d_i(x) f(x) dx,$$

with $\sum_{i=1}^{k} s_i(x) = I$. Next, we need to assign the connection probability of $\sum_{i=1}^{k} s_i(x) = I$ to $k+1$ BSs, and the total probability assigned to the first k BSs is $I - s_{k+1}(x)$. According to the optimization principle, we have:

$$\begin{aligned} F_{k+1}(I) &= \min_{s \in \mathcal{S}} \sum_{i=1}^{k+1} \iint s_i(x) R_{BS}(i) \log_{10} d_i(x) f(x) dx \\ &= \min_{s_{k+1}} [F_k(I - s_{k+1}(x)) \\ &+ \iint s_{k+1}(x) R_{BS}(k+1) \log_{10} d_{k+1}(x) f(x) dx]. \end{aligned} \tag{29}$$

Since the selection strategies were noncontinuous and unguided, we only needed to ensure that the selection strategy was optimal at each point to ensure that the overall selection strategy was optimal. Therefore, we could transform the original overall optimization (29) into optimizations on input node.

Given user location x_I, when the SEN and BS capacity constraints were not considered, we could first calculate distance d_i from the user to each BS i. According to dynamic programming, $\sum_{i=1}^{k} s_i(x_I) = I$ was divided, and the following recurrence formula was obtained:

$$\begin{aligned} F_{k+1}(I) &= \min_{s \in \mathcal{S}} \sum_{i=1}^{k+1} s_i(x_I) R_{BS}(i) \log_{10} d_i(x_I) \\ &= \min_{s_{k+1}} [F_k(I - s_{k+1}(x_I)) \\ &+ s_{k+1}(x_I) R_{BS}(k+1) \log_{10} d_{k+1}(x_I)]. \end{aligned} \tag{30}$$

4.2. Constrained Optimization

The receiving sensitivity constraint condition is relatively simple with Formula (17). For input location x_I and calculated distance d_i to each BS i, it can be judged whether x_input is within the signal effective range C_i given by BS i or outside. Let those s_i corresponding to outside BSs be 0, which is not considered; only those BSs within the effective range of SEN distance are calculated step by step according to the above DP method.

According to Formula (11), the BS capacity constraint actually considers that the number of users connected to a certain one BS cannot exceed the limit, which can be transformed into the following form:

$$\iint s_i(x_I) M f(x) dx = \mathbb{E}[m s_i(x_I)], \tag{31}$$

where m represents the user density at position x, whose density function is $Mf(x)$.

Corresponding to input position x, the BS capacity constraint condition needs to ensure $\mathbb{P}[m s_i(x) < N_{BS}(i)] > 1 - \alpha$. We have:

$$\begin{aligned} \mathbb{P}[m s_i(x_I) < N_{BS}(i)] &> 1 - \alpha \\ \mathbb{P}[m < \frac{N_{BS}(i)}{s_i(x_I)}] &> 1 - \alpha \\ \frac{N_{BS}(i)}{s_i(x_I)} &> \lambda_{1-\alpha} \\ s_i(x_I) &< \frac{N_{BS}(i)}{\lambda_{1-\alpha}}, \end{aligned} \tag{32}$$

λ represents the confidence region of m's distribution under α, which can be obtained from the distribution table when the distribution is identified.

Adding the BS capacity constraint to the DP method, we can obtain:

$$\begin{aligned} F_{k+1}(I) &= \min_{s \in \mathcal{S}} \sum_{i=1}^{k+1} s_i(x_I) \cdot R_{BS}(i) \log_{10} d_i(x_I), \\ & s.t.\ s_{C_i} < \frac{N_{BS}(C_i)}{\lambda_{1-\alpha}}\ for\ i = 1, \ldots, k+1 \\ &= \min_{s_{k+1}} [F_k(I - s_{k+1}(x_I)) \\ & + s_{k+1}(x_I) R_{BS}(k+1) \log_{10} d_{k+1}(x_I)], \\ & s.t.\ s_{C_{k+1}} < \frac{N_{BS}(C_{k+1})}{\lambda_{1-\alpha}}. \end{aligned} \tag{33}$$

According to the above analysis, the process of solving wireless BS selection-strategy optimization on the basis of identified distribution can be obtained as shown in Algorithm 1.

Algorithm 1: Wireless BS selection-strategy optimization based on identified distribution.

Require: x: user's location; BS: BS information, including locations, reliability, and connection limit numbers; D: maximal signal connection distance;
Ensure: s: BS-user connection strategy;
1: **initialize**: Set $s = 0$;
2: **Step 1**: Calculate distance d_i from x to each BS i, and find BS set K_C within the signal effective range of the user SEN on the basis of (17); let $k = N(K_C)$.
3: **Step 2**: Calculate the BS capacity constraint at position x according to (32).
4: **Step 3**: Use the DP method (16) to calculate the optimal user selection strategy s under the condition of meeting the connection restriction.
5: **return** s

First, the distance from x to each BS is calculated, and the nodes within the effective range of the signal are selected according to Formula (16). Then, the DP recursive formula in (33) is combined to optimize the user-BS selection strategy.

5. Wireless BS Selection-Strategy Optimization Based on User Trajectory

In view of unknown user distribution, we estimate the user density of any given location from the trajectory.

5.1. Trajectory Data

Effective data in the trajectory dataset mainly include space latitude and longitude coordinate data, and time data.

Most of the trajectory data density estimation methods only start from spatial data, which are commonly analyzed and visualized using the methods for home range or utilization distribution estimation [36]. However, these two concepts often only focus on the spatial distribution of the measured positions in 2D space and ignore the time series of the measurements. So, we used a time–space density estimation method to process the trajectory data.

5.2. Density Estimation

Kernel density estimation is a nonparametric estimation method that does not use prior knowledge about the data distribution and does not attach any assumptions to the data distribution. It is a method to study the characteristics of the data distribution from the data sample itself, which is consistent with our unknown user distribution situation. The so-called kernel density estimation is to use a smooth peak function, which is called a 'kernel', to fit the observed data points, thereby simulating the true probability distribution curve. Assuming that $x_1, x_2, \ldots, x_n$ are the n sample points of independent and identically distributed F, whose probability density function is f, the kernel density can be estimated as follows:

$$f_h(x) = \frac{1}{n}\sum_{i=1}^{n} K_h(x - x_i) = \frac{1}{nh}\sum_{i=1}^{n} K(\frac{x - x_i}{h}), \tag{34}$$

where $K(\cdot)$ is the kernel function (non-negative, integral is 1, conforms to the nature of probability density, and the mean is 0). There are many kinds of kernel functions, for example, uniform, triangular, biweight, triweight, Epanechnikov, and normal.

Commonly used kernel functions include the Gaussian and Epanechnikov kernels [37]:

$$K(x) = \frac{1}{\sqrt{2\pi}} \exp{(-\frac{x^2}{2})} \text{ (Gaussian Kernel)}, \tag{35}$$

$$K(x) = \frac{3}{4}(1 - x^2)I(|x| \leq 1) \text{ (Epanechnikov Kernel)}. \tag{36}$$

We used a 3-dimensional kernel density estimation method to estimate the user density at the input location coordinate point, and selected the optimal Epanechnikov kernel in the sense of mean square error to estimate.

At the same time, we also added the automatic calculation of bandwidth h in Formula (34) according to the Scott principle of the d-dimensional space:

$$h = \sqrt{5} \cdot n^{\frac{1}{d+4}}, \tag{37}$$

where N represents the number of data.

We consider both space data and time data; first, the space–time kernel density function is estimated and then integrated into the time dimension. However, because it is impossible to accurately calculate the space–time nuclear density function, we could only use numerical methods to calculate the space–time nuclear density of the input position coordinates at these time nodes according to a certain time interval, which represents density in this time interval multiplied by the time interval and accumulated as the integration process. The specific algorithm pseudocode is shown as Algorithm 2.

Algorithm 2: Time–space density estimation.

Require: *points*: target location; $data_{space}$: spatial data; $data_{ti}$: time data; d: data dimension;
Ensure: *Den*: estimated density;

```
initialize: Set Den = 0;
Convert latitudinal and longitudinal data (points and data_space) into coordinate axis data;
if d = 2 then
    Use data_space to estimate space density Den at points with kernel density estimation Formula (34) and the Epanechnikov kernel (36).
end if
if d = 3 then
    Normalize time data data_ti.
    for t = 0,2,...,19 do
        time = (i − 5)/10;
        Use data_space and data_ti to estimate time–space density den at (points,time) with (34) and (36).
        Den = Den + den/10.
    end for
end if
return Den
```

In time–space density estimation, there are two options for 2- and 3-dimensional density estimation. Two-dimensional density estimation only considers spatial data to estimate the user's spatial kernel density. Three-dimensional density estimation considers both space and time data, first estimating the spatiotemporal kernel density function and then integrating it into the time dimension.

The overall flow of the algorithm is shown in Figure 1.

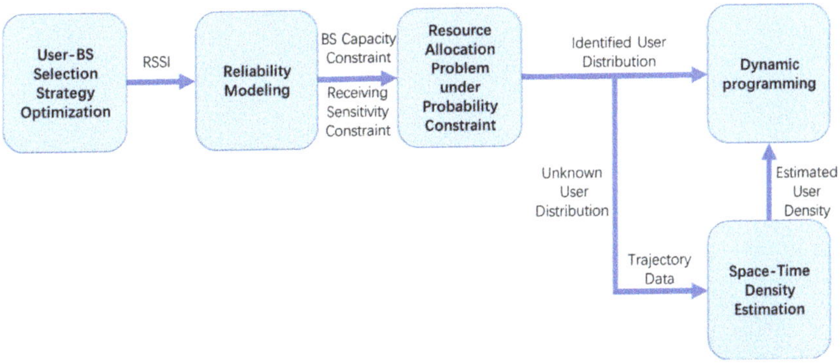

Figure 1. Overall flow of Algorithm 3.

On the basis of the above analysis and the time–space density estimation algorithm, the process of solving wireless BS selection-strategy optimization based on trajectory data can be obtained as Algorithm 3.

Algorithm 3: Wireless BS selection-strategy optimization based on trajectory data.

Require: x: user's location; BS: BS information, including locations, reliability, and connection limit numbers; D: maximal signal connection distance;

Ensure: s:BS-user connection strategy.

1: **initialize**: Set $s = 0$;

2: **Step 1**: Calculate distance d_i from x to each BS i and find BS set K_C within the signal effective range of the user SEN on the basis of (17), let $k = N(K_C)$.

3: **Step 2**: Calculate the density at the position of x according to the given user's trajectory data using the time–space density estimation algorithm.

4: **Step 3**: Substitute density into the the DP Formula (30) to calculate the optimal user selection strategy s under the condition of meeting the connection restriction.

5: **return** s

First, the distance from x to each BS is calculated, and nodes within the effective range of the signal are selected according to Formula (16). Then, the user density obtained in Algorithm 2 is brought into the DP recursive formula of (30), and the user-BS selection strategy is optimized.

6. Performance Evaluation

We evaluated the performance of our proposed approach on the basis of the GeoLife GPS Trajectories dataset. The GeoLife GPS Trajectories dataset was assembled from the Microsoft Research Asia Geolife project by 182 users during a period of over three years (from April 2007 to August 2012). A GPS trajectory of this dataset is represented by a sequence of time-stamped points, each of which containing the information of latitude, longitude, and altitude. Of the trajectories, 91% are logged in a dense representation, e.g., every 1 to 5 s or every 5 to 10 m per point. This dataset recorded a broad range of users' outdoor movements, including not only life routines such as going home and to work, but also some entertainment and sports activities, such as shopping, sightseeing, dining, hiking, and cycling. Most of the data were created in Beijing, China. Figure 2 plots the trajectories of user [004] from 23 October 2008 to 5 July 2009.

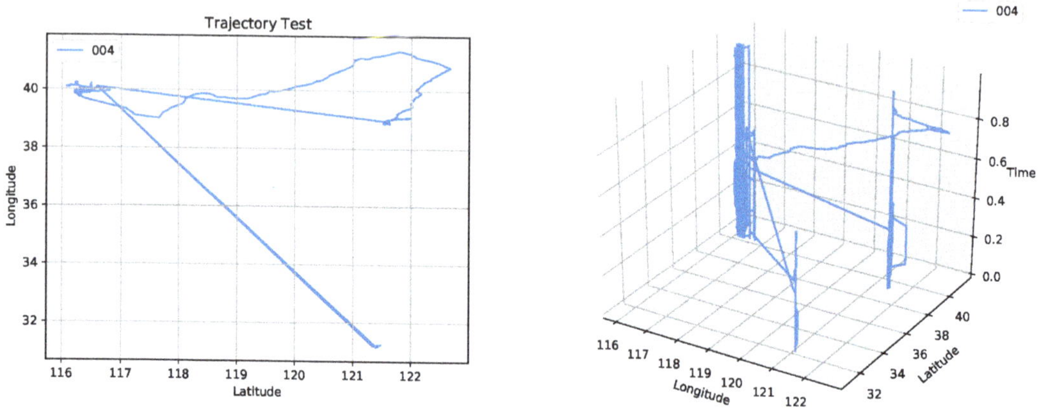

(a) (b)

Figure 2. Trajectories of user [004] from 23 October 2008 to 5 July 2009. (**a**) Two-dimensional trajectory; (**b**) three-dimensional trajectory.

Figure 2a shows that the user's track coverage was relatively large, but in fact, according to the analysis of the track data and Figure 2b, the user spent most of the time within a small area in the center and had rarely traveled far away.

The reason for this difference is that Figure 2a is a two-dimensional space map that only considers the spatial scale, and ignores information in the time scale. Our algorithm considers both space and time information to estimate the density.

We conducted comparative simulations from two aspects. Simulation 1 compares the difference of the density estimation effect for 2-dimensional kernel density estimation and time–space density algorithms under different sampling frequency conditions, while Simulation 2 compares the reliabilities of RSSI- and distance-based selection strategies. Both simulations were conducted in a Python environment.

6.1. Comparative Simulations between Two-Dimensional Kernel Density Estimation and Time–Space Density Estimation

In order to verify the superiority of our time–space density estimation method, we conducted two comparative simulations. First, we compared our time–space density estimation method with the two-dimensional kernel density estimation method. Comparative simulations were carried out on a trajectory dataset with the same sampling time interval, together with passing the same point, and on a trajectory dataset with different sampling time intervals, together with passing the same point.

The variable between the above comparative simulations was the sampling frequency condition. We set up two simulation scenarios, A and B, which corresponded to the same sampling frequency and different sampling frequencies. Under both scenarios, we assumed that RSSI followed path-loss model $PL(d) = 27 + 22.7\log_{10}(d) + 20\log_{10}(f_c)$ in the WINNERII C2 model [33]. The setups of this comparative simulation are shown in Table 2.

Table 2. Simulation setups for simulation under different sampling scenarios.

Path Loss	$PL(d) = 27 + 22.7\log_{10}(d) + 20\log_{10}(f_c)$		
Scenarios	**Point**	**Users**	**Sampling Interval**
Scenario A	(40.01, 116.31)	[000, 001, 002, 003, 004]	Same
Scenario B	(39.96, 116.40)	[132, 135, 163, 167, 168]	Different

In Scenario A with the same sampling time interval, we selected five user trajectories [000, 001, 002, 003, 004]. Their trajectory data were all sampled at 0.05 s, and they all passed through point (40.01, 116.31). The specific trajectory is shown in Figure 3.

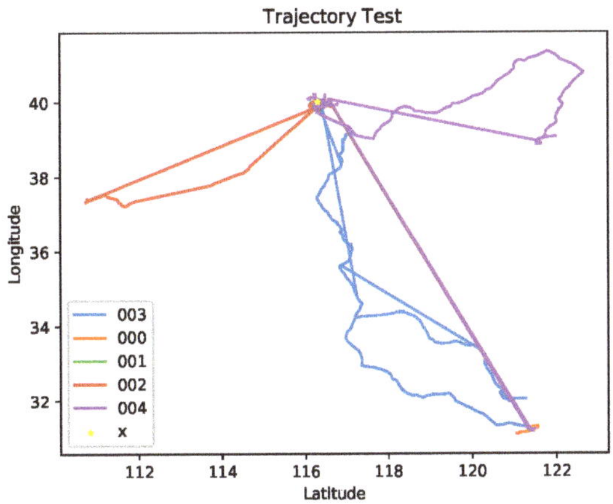

Figure 3. Trajectories of users [000, 001, 002, 003, 004] and point (40.01, 116.31).

The calculated 2-dimensional kernel density and time-space density at (40.01, 116.31) were 2.75 and 2.77 respectively, which are very similar.

In Scenario B with different sampling time intervals, we selected the trajectory data of five users [132, 135, 163, 167, 168]. Their data collection was rather chaotic, and the sampling time interval was not fixed. Users 135, 163, and 167 all passed through point (39.96, 116.40). The specific trajectory is shown in Figure 4.

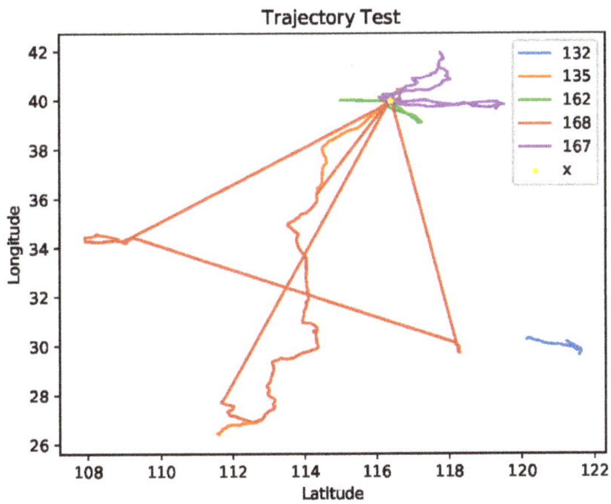

Figure 4. Trajectories of users [132, 135, 163, 167, 168] and point (39.96, 116.40).

The calculated two-dimensional and time-space densities at (39.96, 116.40) were 2.39 and 2.54. Simulation results at position [41, 116.33] are shown in Table 3.

Table 3. Simulation results for density estimation under different sampling scenarios.

Scenarios	Two-Dimensional Kernel Density Estimation	Time-Space Density Estimation	Improvements in Density Estimation Accuracy
Scenario A	2.75	2.77	0.67%
Scenario B	2.39	2.54	5.00%

Table 3 shows that, since the sampling time interval in Scenario A was the same, the influence of the time factor on the estimation of trajectory density could be ignored. In Scenario B, time–space density was significantly closer to the expected density 3 than the two-dimensional kernel density was because 2D kernel density estimation ignores the time factor. For different sampling frequencies, the density calculated at a high sampling frequency may be higher, and the density calculated at the low sampling frequency may be lower. Our time–space method first aligned the data in the time dimension, which means that it could overcome the problem of different sampling frequencies and take the effects of both time and space on density estimation into account.

6.2. Comparative Simulations between RSSI-Based and Distance-Based Selection Strategy

In order to verify the improvement effect of our algorithm, we compared it with optimization on the basis of only distance rather than RSSI. In order to minimize the overall connection distance between users and BSs, the optimization object of the user-BS selection strategy optimization based on distance was set to be as follows:

$$\min_{s \in \mathcal{S}} \sum_{i=1}^{N} s_i(x) \cdot d_i(x), \tag{38}$$

and the constraint conditions were the same as those of the user-BS selection strategy optimization based on reliability. Comparative simulations were carried out under the conditions of known and unknown user distribution.

In order to verify the versatility of our algorithm, we also verified RSSI on the basis of different path-loss models. Path loss is presented in decibels as a function of distance, and was calculated by summing the taps in SEN domain and averaging over the measurement snapshots along the measurement run. Path loss and shadow fading are given for the urban microcell scenario (UMi), suburban macrocell scenario (SMa), and urban macrocell scenario (UMa) [33] for the frequency range of 0.45–6.0 GHz in Table 4.

Table 4. Path loss under different scenarios.

Scenario	Path Loss
UMi	$PL(d) = 27 + 22.7\log_{10}(d) + 20\log_{10}(f_c)$
SMa	$PL(d) = 27.2 + 23.8\log_{10}(d) + 20\log_{10}(f_c)$
UMa	$PL(d) = 25 + 26\log_{10}(d) + 20\log_{10}(f_c)$

In the case of the given user distribution, we considered a wireless network with 5 BSs and 10 users. The attributes of the 5 BSs were set as shown in Table 5.

Table 5. Simulation setup for BSs with given user distribution.

BS Serial Number	Position	R_{BS}	N_{BS}
1	[0, 0]	0.8	3
2	[5, 1]	0.8	5
3	[1, 7]	0.7	4
4	[2, 3]	0.8	5
5	[6, 2]	0.9	4

We assumed that all users followed normal distribution and the maximal signal connection distance was 8, and the user-BS selection strategies at position [5, 5] are shown in Table 6.

Table 6. Simulation results for reliability under given user distribution.

BS	Selection Strategy Based on RSSI	Selection Strategy Based on Distance
1	0.362	0.362
2	0	0.155
3	0	0.483
4	0.603	0
5	0.035	0
Reliability	0.565	0.540

The communication signal quality for the selection strategies based on distance and on RSSI was 0.540 and 0.565, respectively, and our RSSI-based optimization result was 4.6% higher in communication signal quality than the non-RSSI optimization result. This demonstrates that the RSSI-based method could better ensure communication reliability when user distribution is known.

In the case of unknown user distribution, we considered a wireless network with 3 BSs and 10 users. The attributes of the 3 BSs were set as shown in Table 7.

Table 7. Simulation setup for BSs with unknown user distribution.

BS Serial Number	Position	R_{BS}	N_{BS}
1	[39, 116.5]	0.9	3
2	[40, 116.3]	0.95	5
3	[42, 116.4]	0.85	4
4	[39, 116.3]	0.7	5
5	[40, 116.5]	0.9	4

We selected ten user trajectories [000, 001, 002, 003, 004, 005, 006, 007, 008, 009]. Simulation results at position [41, 116.33] are shown in Table 8 and Figure 5.

Table 8. Simulation results for reliability under different scenarios.

Scenarios	Reliability Based on RSSI	Reliability Based on Distance	Improvements in Reliability
UMi(B1)	1.990	1.943	2.4%
SMa(C1)	1.897	1.804	5.2%
UMa(C2)	1.788	1.734	3.1%

Figure 5. Comparison chart for simulation results under different scenarios.

Simulation results show that the reliability of the selection strategy based on RSSI was 3.6% higher on average than that of the selection strategy based on distance in the UMi, SMa, and UMa scenarios. This demonstrates that RSSI could reflect more information with the user-BS selection strategy under both user-known and user-unknown cases, which means that our algorithm is more effective than traditional distance-based algorithms in common scenarios due to the better fulfilment of the high requirements for reliability in 5G wireless networks.

7. Conclusions

In this paper, we presented a user-BS connection strategy optimization method based on RSSI to maximize the overall communication signal quality of 5G wireless networks. The original problem is a functional optimization problem that is difficult to solve under nonideal conditions. Therefore, we transformed it into a resource allocation problem with random constraints based on RSSI, and solved it with the DP method and space–time density estimation. Simulation results show that, compared with the estimation of the 2-dimensional kernel density method that only considers spatial data, our time–space density could simultaneously imply the information in the space and time dimensions, and solve the problems caused by random sampling frequency with an improvement of 2.84% in density estimation accuracy. At the same time, compared with the distance-based method, our RSSI-based optimization method improved the communication signal quality by an average of 3.57% under different RSSI path-loss models.

In fact, there are other factors in the reliability modeling of complex systems, such as remaining life, series, parallel connections, and the bridging model that may fit with the reliability model of 5G wireless networks. In our future work, we are focusing on the impact of the above factors on the reliability of 5G wireless networks.

Author Contributions: Conceptualization, J.S. and Y.Y.; data curation, J.S.; formal analysis, J.S. and Y.H.; investigation, J.S.; methodology, J.S. and Y.Y.; supervision, Y.Y. and C.Z.; validation, Y.H. and C.Z.; visualization, J.S.; writing—original draft preparation, J.S.; Writing review editing, Y.H. and C.Z. All authors have read and agreed to the published version of the manuscript.

Funding: This research was supported in part by the National Key Research and Development Program of China under Grant 2020YFA0713900; the National Natural Science Foundation of China under Grants 61772410, 61802298, 62172329, U1811461, U21A6005, 11690011, and the China Postdoctoral Science Foundation under Grants 2020T130513, 2019M663726.

Institutional Review Board Statement: Not applicable.

Informed Consent Statement: Not applicable.

Data Availability Statement: Not applicable.

Conflicts of Interest: The authors declare no conflict of interest.

References

1. Tsumachi, N.; Ohseki, T.; Yamazaki, K. Base Station Selection Method for RAT-Dependent TDOA Positioning in Mobile Network. In Proceedings of the 2021 IEEE Radio and Wireless Symposium (RWS), San Diego, CA, USA, 17–22 January 2021; pp. 119–122. [CrossRef]
2. Boccardi, F.; Heath, R.W.; Lozano, A.; Marzetta, T.L.; Popovski, P. Five disruptive technology directions for 5G. *IEEE Commun. Mag.* **2014**, *52*, 74–80. [CrossRef]
3. Xiao, M.; Mumtaz, S.; Huang, Y.; Dai, L.; Li, Y.; Matthaiou, M.; Karagiannidis, G.; Björnson, E.; Yang, K.; Chih-Lin, I.; et al. Millimeter wave communications for future mobile networks. *IEEE J. Sel. Areas Commun.* **2017**, *35*, 1909–1935. [CrossRef]
4. Zhao, N.; Liang, Y.C.; Niyato, D.; Pei, Y.; Wu, M.; Jiang, Y. Deep reinforcement learning for user association and resource allocation in heterogeneous cellular networks. *IEEE Trans. Wirel. Commun.* **2019**, *18*, 5141–5152. [CrossRef]
5. Andrews, J.G.; Buzzi, S.; Choi, W.; Hanly, S.V.; Lozano, A.; Soong, A.C.; Zhang, J.C. What will 5G be? *IEEE J. Sel. Areas Commun.* **2014**, *32*, 1065–1082. [CrossRef]
6. Dai, Y.; Xu, D.; Maharjan, S.; Zhang, Y. Joint computation offloading and user association in multi-task mobile edge computing. *IEEE Trans. Veh. Technol.* **2018**, *67*, 12313–12325. [CrossRef]
7. Esmaeily, A.; Kralevska, K. Small-scale 5g testbeds for network slicing deployment: A systematic review. *Wirel. Commun. Mob. Comput.* **2021**, *2021*, 6655216. [CrossRef]
8. Mei, W.; Zhang, R. Performance analysis and user association optimization for wireless network aided by multiple intelligent reflecting surfaces. *IEEE Trans. Commun.* **2021**, *69*, 6296–6312. [CrossRef]
9. Hoang, M.T.; Yuen, B.; Dong, X.; Lu, T.; Westendorp, R.; Reddy, K. Recurrent neural networks for accurate RSSI indoor localization. *IEEE Internet Things J.* **2019**, *6*, 10639–10651. [CrossRef]
10. Liu, C.; Fang, D.; Yang, Z.; Jiang, H.; Chen, X.; Wang, W.; Xing, T.; Cai, L. RSS distribution-based passive localization and its application in sensor networks. *IEEE Trans. Wirel. Commun.* **2016**, *15*, 2883–2895. [CrossRef]
11. Hao, C.; Zhang, Y.; Wei, L.; Tao, X.; Ping, Z. ConFi: Convolutional Neural Networks Based Indoor Wi-Fi Localization Using Channel State Information. *IEEE Access* **2017**, *5*, 18066–18074. [CrossRef]
12. Li, Q.; Zheng, Y.; Xie, X.; Chen, Y.; Liu, W.; Ma, W.Y. Mining user similarity based on location history. In Proceedings of the 16th ACM SIGSPATIAL International Conference on Advances in Geographic Information Systems, Irvine, CA, USA, 5–7 November 2008; pp. 1–10. [CrossRef]
13. Gatzianas, M.; Mesodiakaki, A.; Kalfas, G.; Pleros, N.; Moscatelli, F.; Landi, G.; Ciulli, N.; Lossi, L. Offline Joint Network and Computational Resource Allocation for Energy-Efficient 5G and beyond Networks. *Appl. Sci.* **2021**, *11*, 10547. [CrossRef]
14. Wannstrom, J. *LTE-Advanced*; Third Generation Partnership Project (3GPP): Valbonne, France, 2013.
15. Liu, D.; Wang, L.; Chen, Y.; Elkashlan, M.; Wong, K.K.; Schober, R.; Hanzo, L. User association in 5G networks: A survey and an outlook. *IEEE Commun. Surv. Tutor.* **2016**, *18*, 1018–1044. [CrossRef]
16. Alizadeh, A.; Vu, M. Load balancing user association in millimeter wave MIMO networks. *IEEE Trans. Wirel. Commun.* **2019**, *18*, 2932–2945. [CrossRef]
17. Feng, M.; Mao, S.; Jiang, T. Joint Frame Design, Resource Allocation and User Association for Massive MIMO Heterogeneous Networks with Wireless Backhaul. *IEEE Trans. Wirel. Commun.* **2017**, 17, 1937–1950. [CrossRef]
18. Zhou, T.; Nan, J.; Dong, Q.; Liu, Z.; Li, C. Joint Cell Selection and Activation for Green Communications in Ultra-Dense Heterogeneous Networks. In Proceedings of the 2017 IEEE International Conference on Computational Science and Engineering (CSE) and IEEE International Conference on Embedded and Ubiquitous Computing (EUC), Guangzhou, China, 21–24 July 2017. [CrossRef]
19. Zola, E.; Kassler, A.J.; Kim, W. Joint User Association and Energy Aware Routing for Green Small Cell mmWave Backhaul Networks. In Proceedings of the 2017 IEEE Wireless Communications and Networking Conference (WCNC), San Francisco, CA, USA, 19–22 March 2017; pp. 1–6. [CrossRef]
20. Tan, J.; Xiao, S.; Han, S.; Liang, Y.C.; Leung, V. QoS-Aware User Association and Resource Allocation in LAA-LTE/WiFi Coexistence Systems. *IEEE Trans. Wirel. Commun.* **2019**, *18*, 2415–2430. [CrossRef]
21. Zhang, H.; Huang, S.; Jiang, C.; Long, K.; Leung, V.C.; Poor, H.V. Energy efficient user association and power allocation in millimeter-wave-based ultra dense networks with energy harvesting base stations. *IEEE J. Sel. Areas Commun.* **2017**, *35*, 1936–1947. [CrossRef]
22. Luo, Q.; Su, G.; Lin, X.; Chen, B.; Dai, M.; Wang, H. A Stable Matching Game for User Association in Heterogeneous Cellular Networks. In Proceedings of the 2019 IEEE 5th International Conference on Computer and Communications (ICCC), Chengdu, China, 6–9 December 2019; pp. 1098–1102. [CrossRef]
23. Soleymani, B.; Zamani, A.; Rastegar, S.H.; Shah-Mansouri, V. RAT selection based on association probability in 5G heterogeneous networks. In Proceedings of the 2017 IEEE Symposium on Communications and Vehicular Technology (SCVT), Leuven, Belgium, 14 November 2017; pp. 1–6. [CrossRef]
24. Afshang, M.; Dhillon, H.S. Poisson cluster process based analysis of HetNets with correlated user and base station locations. *IEEE Trans. Wirel. Commun.* **2018**, *17*, 2417–2431. [CrossRef]

25. Giorgetti, G.; Gupta, S.; Manes, G. Optimal RSS threshold selection in connectivity-based localization schemes. In Proceedings of the 11th International Symposium on Modeling Analysis and Simulation of Wireless and Mobile Systems, MSWiM 2008, Vancouver, BC, Canada, 27–31 October 2008. [CrossRef]
26. Feng, Y.; Zhao, Y.; Gunnarsson, F. Proximity report triggering threshold optimization for network-based indoor positioning. In Proceedings of the 2015 18th International Conference on Information Fusion (Fusion), Washington, DC, USA, 6–9 July 2015.
27. Ghatak, G. On the Placement of Intelligent Surfaces for RSSI-Based Ranging in Mm-Wave Networks. *IEEE Commun. Lett.* **2021**, *25*, 2043–2047. [CrossRef]
28. Sadowski, S.; Spachos, P. Rssi-based indoor localization with the internet of things. *IEEE Access* **2018**, *6*, 30149–30161. [CrossRef]
29. Marks, M.; Niewiadomska-Szynkiewicz, E. Localization based on stochastic optimization and RSSI measurements. In Proceedings of the ACM/IEEE International Conference on Information Processing in Sensor Networks, Stockholm, Sweden, 12–16 April 2010; p. 402. [CrossRef]
30. Sadowski, S.; Spachos, P. Optimization of BLE beacon density for RSSI-based indoor localization. In Proceedings of the 2019 IEEE International Conference on Communications Workshops (ICC Workshops), Shanghai, China, 20–24 May 2019; pp. 1–6. [CrossRef]
31. Aravinda, P.; Sooriyaarachchi, S.; Gamage, C.; Kottege, N. Optimization of RSSI based indoor localization and tracking to monitor workers in a hazardous working zone using Machine Learning techniques. In Proceedings of the 2021 International Conference on Information Networking (ICOIN), Jeju Island, Korea, 13–16 January 2021; pp. 305–310. [CrossRef]
32. Rappaport, T.S. *Wireless Communications: Principles and Practice*; Horwood Publishing Limited: Chichester, UK, 2002.
33. Heino, P.; Meinil, J.; Kysti, P.; Hentil, L.; Narandzic, M. CP5-026 WINNER+ D5.3 v1.0 WINNER+ Final Channel Models. *Waseda Commercial Review*, 30 January 2010. Available online: https://docplayer.net/34577643-D5-3-winner-final-channel-models.html (accessed on 13 June 2022).
34. Yuan, M.; Xu, W.; Zhi, Q.; Poor, H.V. Data-Driven Measurement of Receiver Sensitivity in Wireless Communication Systems. *IEEE Trans. Commun.* **2019**, *67*, 3665–3676. [CrossRef]
35. Dieng, N.A.; Chaudet, C.; Charbit, M.; Toutain, L.; Meriem, T.B. Experiments on the RSSI as a Range Estimator for Indoor Localization. In Proceedings of the 2012 5th International Conference on New Technologies, Mobility and Security (NTMS), Istanbul, Turkey, 7–10 May 2012. [CrossRef]
36. Zou, Y.; Chen, Y.; He, J.; Pang, G.; Zhang, K. 4D time density of trajectories: Discovering spatiotemporal patterns in movement data. *ISPRS Int. J. Geo-Inform.* **2018**, *7*, 212. [CrossRef]
37. Wglarczyk, S. Kernel density estimation and its application. *ITM Web Conf.* **2018**, *23*, 00037. [CrossRef]

applied sciences

MDPI

Review

Optical Beamforming Networks for Millimeter-Wave Wireless Communications

Fei Duan *, Yuhao Guo, Zenghui Gu, Yanlong Yin, Yixin Wu and Teyan Chen

Huawei Technologies Co., Ltd., Shenzhen 518000, China; guoyuhao2@huawei.com (Y.G.); guzenghui@huawei.com (Z.G.); wuyixin@huawei.com (Y.W.); chenteyan@huawei.com (T.C.)
* Correspondence: f.duan@huawei.com or arkstephen@sina.com

Abstract: With the rapid data growth driven by smart phone, high-definition television and virtual reality/augmented reality devices and so on, the launched 5G and upcoming 6G wireless communications tend to utilize millimeter wave (mmWave) to achieve broad bandwidth. In order to compensate for the high propagation loss in mmWave wireless communications and track the moving users, beamforming and beamsteering are indispensable enabling technologies. These have promising potential to be realized through the use of optical beamforming networks (OBFNs) that have a wider bandwidth and smaller size, lower power consumption, and lower loss compared to those of their electric counterparts. In this paper, we systematically review various OBFN architectures using true time delays and optical phase shifters, as well as discuss performances of different architectures, scalable technologies that promote the advancement of OBFNs, and the application potentials of OBFNs. Two-dimensional OBFNs with discrete components or integrated optical devices have been elaborated, in addition to one-dimensional architectures. Moreover, the state-of-the-art technologies relative to reducing the size, loss and nonlinearity of OBFNs have also been discussed here.

Keywords: optical beamforming; beamsteering; wireless communication; millimeter wave; phase array antenna; true time delay; optical phase shifter; 5G; 6G

Citation: Duan, F.; Guo, Y.; Gu, Z.; Yin, Y.; Wu, Y.; Chen, T. Optical Beamforming Networks for Millimeter-Wave Wireless Communications. *Appl. Sci.* **2023**, *13*, 8346. https://doi.org/10.3390/app13148346

Academic Editor: Christos Bouras

Received: 15 May 2023
Revised: 20 June 2023
Accepted: 28 June 2023
Published: 19 July 2023

1. Introduction

In the era of 5G and beyond, wireless communications will witness an explosive growth in data traffic with the technology advancements of 4 k/8 k high-definition video, virtual reality/augmented reality (VR/AR), mixed reality (MR), industrial internet of things, remote healthcare and so on [1–4]. These corresponding application scenarios will drive the evolution of mobile broadband networks toward wide bandwidth and high frequency. To this end, mmWave and terahertz (THz), which have a much larger bandwidth and energy efficiency compared to those of sub-6 GHz, are proposed as the signal carriers of next-generation wireless communication networks [4–8]. Among them, mmWave has gained more popularity in terms of research and application in the past two decades [1,7,8]. However, it is challenging to exploit mmWave, which suffers severe propagation loss in wireless communications, due to the air absorption, rain attenuation and blockages of buildings, foliage and vehicles, etc. [5,7,9]. One key enabling technology for mmWave communications is the phase array antenna (PAA) that adopts a large number of antenna elements to provide sufficient gain in a certain direction through beamforming, and also has the beamsteering ability to track moving users such as pedestrians and passengers in vehicles [8,10,11].

Conventionally, the beamforming and beamsteering of PAAs are implemented using electric beamforming networks that have analog and digital architectures. Analog architectures of electric beamforming generally utilize a phase shifter array that has a bulky size and narrow bandwidth, inducing a high cost and the beam squint problem [12,13]. Meanwhile, digital beamforming networks require that the numbers of high-speed digital–analog converters (ADCs), analog–digital converters (DACs) and mixers are identical to those of antenna elements, which are too expensive to implement in PAAs with massive antenna elements [9,14]. These issues hinder the utilization of electric beamforming networks in mmWave communications with broad bandwidth and compact antenna arrays. One promising solution is to use OBFNs that have inherent advantages in enabling broadband wireless communications, thanks to the wide bandwidth of optical devices and multi-dimensional multiplexing capability of optical signals. With photonic integration technology, OBFNs also possess the potential advantages of small size, low weight, low power consumption and low loss [1,14,15].

OBFNs are used to phase tune or induce a time delay in the radio frequency (RF) signal at each antenna element with optical phase shifters or true time delay (TTD) components, and further control the beam pattern of PAAs. The implementations of OBFNs in PAAs can be dated back to the 1970s [16,17]. Early demonstrations of OBFNs are mainly based on TTD architectures which adopt discrete devices to build fiber-optic or free-space beamforming systems [18–21]. These architectures are easy to be realized by using commercially mature components; however, their bulkiness inhibits their integration with antennas especially in mmWave wireless communications. OBFNs with optical phase shifters predominantly produce RF signals via coherent beating at photodetectors (PDs), which have a relatively small tuning range and bandwidth [16,22–25]. Moreover, OBFN architectures combining TTDs and phase shifters have also been demonstrated to have superiorities in reducing the cost and complexity of TTD architectures designed for PAAs with limited bandwidth [20,21]. In recent years, OBFNs with integrated TTDs or integrated phase shifters have received much concern and exhibited advantages in compactness, high scalability and low power consumption [26–31]. Several application scenarios of OBFNs such as indoor coverage and mobile fronthaul have been proposed [29,32–38]. Up to now, reviews about OBFNs have paid more attention on TTD architectures [8,12,15,20,21,39], but the roles and advancements of phase shifter architectures have not been discussed. Furthermore, relative advancements in materials, optical devices and electro-photonic integrations, which may improve the scalability of OBFNs, have not been discussed either.

In this paper, we present a systematic review of OBFN architectures using various TTDs and phase shifters, introduce typical architectures, discuss the scalability of different architectures, as well as propose several scalable techniques and application scenarios for OBFNs. The following sections are organized as follows: Section 2 shows the principles of PAAs, OBFNs with TTDs and phase shifter arrays, Section 3 introduces the representative architectures of five subclasses of OBFNs with TTDs and the corresponding two-dimensional (2D) schemes, Section 4 presents four classic categories of OBFNs with phase shifters and a promising combination of these with TTD architectures, Section 5 discusses the scalability, scalable techniques and application potential of OBFNs in next-generation wireless communication networks, and finally, Section 6 gives the conclusion and outlook.

2. Principles

To sweep the beam in free space for wireless communications, PAAs tune the phase or induce a time delay at each antenna element. For PAAs using OBFNs, the working principles relate to the beam pattern of one-dimensional (1D) and two-dimensional (2D) PAAs, as well as the RF signal processing of OBFNs. The beam pattern is formed by the integral of the electromagnetic field of each antenna in free space. Categorized by the delay, OBFNs have two types: TTD architectures and phase shifter architectures [20]. The former produce different time delays for RF signals transmitted or received by different antenna elements, while the latter produce various phase shifts.

2.1. Beam Pattern of PAAs

PAAs using OBFNs consist of a 1D antenna array or a 2D antenna array and corresponding OBFNs, as shown in Figure 1a,b. For a 1D PAA with equal amplitude and equal spacing (d) between adjacent antennas, the beam pattern (array factor) can be expressed as follows: [40]

$$F(\theta) = \sum_{n=1}^{N} e^{i(n-1)(kd\sin\theta - \Delta\phi)} \quad (1)$$

where $\Delta\phi$ is the progressive phase of the linear array antenna, k is the wave number of the RF signal exited from the antennas, and θ is the beam angle. The beam angle (θ) can be given as Equation (2) which is derived from $F(\theta)$ obtaining the maximum value.

$$\theta = \arcsin\left(\frac{\Delta\phi}{kd}\right) \quad (2)$$

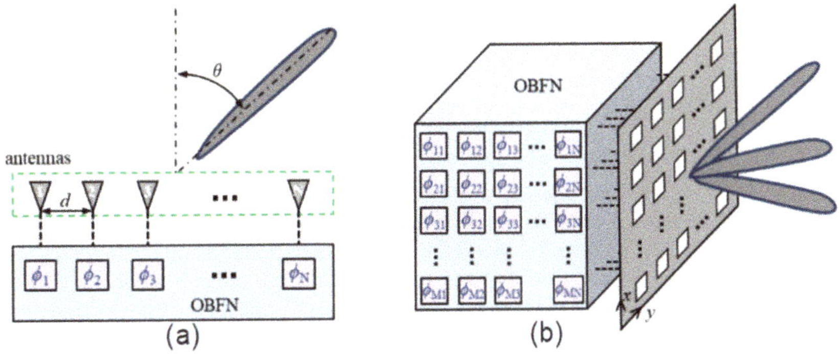

Figure 1. Schematic diagrams of PAAs using OBFNs: (**a**) a 1D PAA and (**b**) a 2D PAA.

Owing to $k = 2\pi/\lambda_{RF}$, where λ_{RF} is the wavelength of the RF signal, Equation (2) can also be expressed as follows:

$$\theta = \arcsin\left(\frac{\Delta\phi}{2\pi d}\lambda_{RF}\right) \quad (3)$$

Meanwhile, for 2D M × N PAA with equal amplitude and equal spacing (d) between adjacent antenna elements, the beam pattern can be expressed as follows: [40]

$$F(\theta', \varphi) = \sum_{m=1}^{M} \sum_{n=1}^{N} I_{mn} e^{i((m-1)(kd\sin\theta'\cos\varphi)+(n-1)(kd\sin\theta'\sin\varphi))} \quad (4)$$

where θ' and φ are the azimuthal angle and polar angle in the spherical coordinate system, respectively, and I_{mn} is the excitation of antenna element at m-th row and n-th column.

2.2. Principle of TTD Architectures

As presented in Figure 2, TTD architectures utilize the direct detection of modulated optical carriers, in which the signal processing of a channel is as follows. The optical carrier from a laser is modulated by a RF signal, which is then delayed by a TTD with a time delay of Δt, then the optical carrier is fed to PD and mixed back to RF signal. The other channels have a similar operation with the optical carrier and generate a group of RF signals with a fixed phase difference, in order to form a beam with a certain angle.

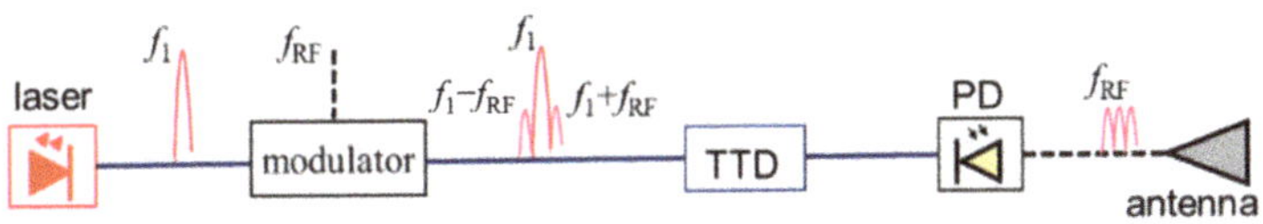

Figure 2. Operation of the RF signal in one channel of a TTD architecture with non-coherent optical carriers.

Ignoring the high-order harmonics, the optical signal before the PD can be expressed as follows: [30]

$$S(t)_1 = A_1 \exp^{i(2\pi f_1(t+\Delta t))} + A_2 \exp^{i(2\pi(f_1+f_{RF})(t+\Delta t)+\phi_0)} - A_2 \exp^{i(2\pi(f_1-f_{RF})(t+\Delta t)+\phi_0)} \quad (5)$$

where A_1 and A_2 are the amplitude of the optical carrier and sidebands, Δt is the time delay, and ϕ_0 is the initial phase of the RF signal. Detected by the PD, an electric current is produced as follows:

$$I(t)_1 = R|S(t)_1 \times S^*(t)_1| \quad (6)$$

where R is the responsivity of PD. Neglecting components of the direct current and beating term, the electric current can be written as follows:

$$I(t)_1 = R[4A_1A_2\cos(2\pi f_{RF}(t+\Delta t)+\phi_0)] \quad (7)$$

This equation states that the phase of the RF signal is shifted by $2\pi f_{RF}\Delta t$.

2.3. Principle of Phase Shifter Architectures

Typically, the signal processing of phase shifter architectures is characterized by the heterodyne detection of coherent optical carriers. For example, two coherent optical carriers can be produced by two phase-locked lasers, respectively, as plotted in Figure 3. The first optical carrier from one laser is modulated by an electrical signal (RF1), and then the phases of the optical carrier and its modulation sidebands are shifted by the phase shifter or TTD; next, the optical carrier and one sideband are filtered out. The second optical carrier goes directly to the PD and combines with one sideband of the first optical carrier, generating a new electrical signal (RF2) via coherent beating. If the lower sideband of the first optical carrier is left behind, the optical signals before the PD can be expressed as follows:

$$S(t)_2 = A_3 \exp^{i(2\pi(f_1-f_{RF1})t+\phi)} + A_4 \exp^{i2\pi f_2 t} \quad (8)$$

where A_3 and A_4 are the amplitude of the sideband signal and the second optical carrier. Meanwhile, ignoring the components of direct current and high-order frequency, the electric current can be simplified as follows:

$$I(t)_2 = R \times [2A_3A_4\cos(2\pi(f_2 - f_1 + f_{RF1})t - \phi)] \quad (9)$$

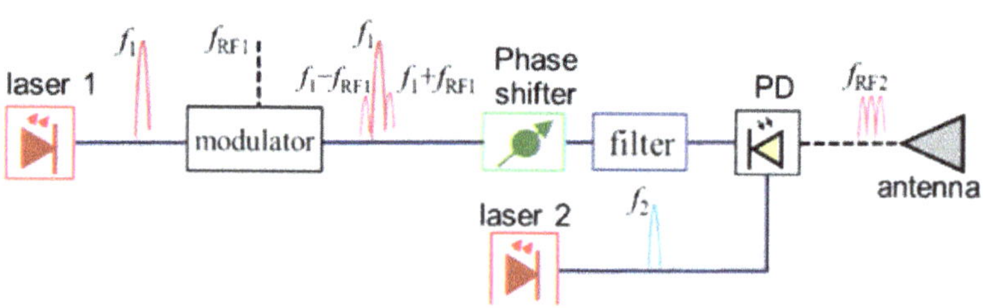

Figure 3. Operation of the RF signal in one channel of a coherent OBFN.

This equation demonstrates that the phase shift in the optical domain can be transferred to one in the RF domain via heterodyne detection which can also change the frequency of the RF signal into a new frequency of $f_{RF2} = f_2 - f_1 + f_{RF1}$. Derived from Equation (3), the relationship of the phase difference ($\Delta\phi$) between adjacent antennas with the beam angle (θ_0) is expressed by the following equation.

$$\Delta\phi = 2\pi d \sin(\theta_0)/\lambda_{RF} \quad (10)$$

3. OBFNs with TTD Architectures

TTD architectures have the advantages of no beam squint and a large delay tuning range [8,20,41]. As addressed above, OBFNs with TTD architectures can be based on fiber-optic or free-space beamforming systems. Meanwhile, TTD architectures with integrated photonic circuits have also been proposed and demonstrated with the advancement of photonic integration technology, as shown in Figure 4 [17,18,23,24,42–50]. It is also shown in this figure that the 1D and 2D OBFN architectures have been proposed since the 1990s. Up to now, TTD architectures mainly include fiber dispersion delay architectures, microring resonator (MRR) group delay architectures, Mach–Zehnder interferometer (MZI) delay architectures, photonic crystal (PC) delay architectures and time delay selection architectures, which will be elaborated in subsequent sections.

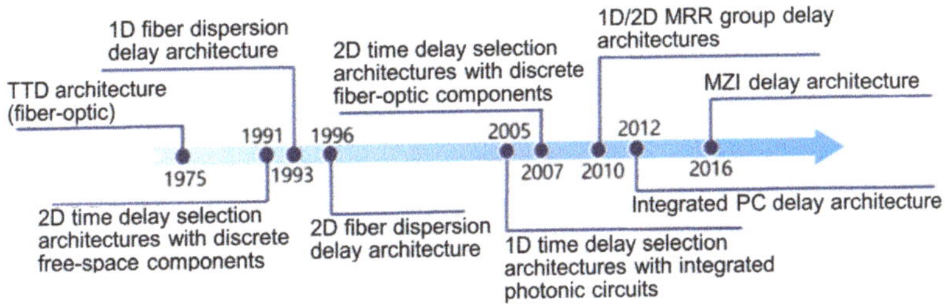

Figure 4. Timeline of TTD architectures proposed in the past [17,18,23,24,42–50].

3.1. Fiber Dispersion Delay Architectures

The fiber dispersion delay architectures are the most mature ones in TTD architectures, owing to adopting commercially mature fibers and devices. The time delay difference between adjacent beam angles or adjacent channels is produced by the chromatic dispersion of different wavelengths and can be given by the following [43,51]:

$$\Delta\tau = LD(\lambda_2 - \lambda_1) \tag{11}$$

where D is the chromatic dispersion coefficient (ps/nm/km), L is the length of the fiber (km), and λ_1 and λ_2 represent the adjacent wavelengths (nm).

As shown in Figure 5a,b, the fiber dispersion delay architecture with N channels (equal to the number of antennas) can be established by one tunable laser, one electro-optic modulator, $1 \times N$ splitter, N different fiber delay lines and N PDs, namely the $1\lambda \times N$ framework [42], while it can also be built by N tunable lasers, one multiplexer, one electro-optic modulator, one fiber, one de-multiplexer and N PDs, namely the $N\lambda \times N$ framework [52]. The $1\lambda \times N$ framework uses one tunable laser and N fiber channels with different lengths of dispersive fibers and non-dispersive fibers, causing progressive time delays for adjacent antenna elements, whereas the $N\lambda \times N$ framework use only one dispersive fiber, such as a common single-mode fiber (SMF), to afford various delays for N channels. These frameworks realize 1D beam sweeping by changing the laser wavelength, which has the advantage of wide bandwidth that is only limited by the bandwidth (~100 GHz) of optical devices such as modulators and PDs [53,54]. Moreover, the $N\lambda \times N$ framework can achieve very low loss, due to the low transmission loss of commercial SMFs. This advantage can also be obtained in the $1\lambda \times N$ framework if SMFs with different lengths are utilized in this architecture, instead of using relatively high-loss dispersion fibers. Although the $N\lambda \times N$ framework can largely reduce the cost of the fiber, the increase in number of the tunable laser may be counter-productive in terms of the cost of whole system. Nevertheless, the wavelength multiplexing applied in the $N\lambda \times N$ framework is one of paramount advantages in OBFNs compared to electrical beamforming networks. Note that, a PAA with a large number of antenna elements needs high-resolution scanning with a wide coverage, which further demands a tunable laser with very high tuning precision. For example, as calculated by Equation (12) [52], where n is the bit number corresponding to the resolution of the PAA, a PAA with a six-bit scanning resolution and a RF of 30 GHz requires a tunable laser with a wavelength resolution of less than 10 p.m.

$$\Delta\lambda = 1/(2^{n+1} LD f_{RF}) \quad (12)$$

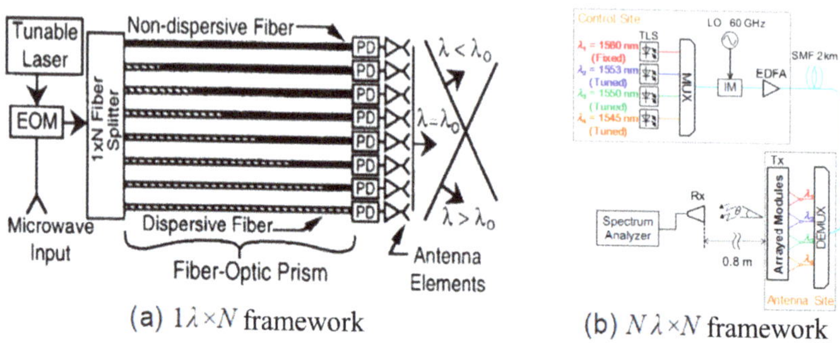

Figure 5. Schematic diagrams of OBFNs based on fiber dispersion with (**a**) one wavelength and N antenna elements [42] (Copyright 1993, IEEE, #5487530392418), and (**b**) N wavelength and N antenna elements [52] (Copyright ©2017, IEICE, #22RB0082).

For practical implementation, 2D beamforming is a basic requirement for 2D planar PAAs, which can exhibit the superiority of OBFNs thanks to their multi-dimensional multiplexing abilities such as wavelength multiplexing and spatial multiplexing. As illustrated in Figure 6a, an OBFN architecture for 2D PAAs has been proposed [51], using an optical frequency comb, fiber dispersion unit, programmable optical filter and microwave photonic filter (MPF) to control the delay of each channel. The optical frequency comb can simultaneously produce multiple optical carriers with various wavelengths and replace multiple lasers presented in the $N\lambda \times N$ framework, reducing the size of OBFNs. In this architecture, multiple optical carriers are first modulated with N RF frequencies at the polarization modulator (PolM), then delayed by the fiber dispersion unit and selected by the programmable filter which induces N optical carriers to N different MPF paths; finally, RF signals with N center frequencies are generated at MPF paths and sent to an antenna element. This indicates that a 2D PAA with this OBFN architecture can achieve multi-beam beamforming, which is a promising technique for a mmWave massive multiple-input multiple-output (MIMO) system [9,55]. However, there is an issue for this concept in that multi-frequency beamforming requires wideband PAAs which are not commercially available presently. Two-dimensional beamforming with one center frequency is easier to be implemented with a fiber dispersion delay architecture, as shown in Figure 6b [56]. Similarly, an electro-optic frequency comb is utilized to provide multiple wavelengths, whereas optical carriers are modulated by Mach–Zehnder modulators (MZM), are then delayed by two-stage TTDs including dispersion compensation fibers (DCFs) and tunable delay lines (TDLs), and are split into various cores of the multi-core fiber (MCF). Next, they are filtered by de-multiplexers and detected by PDs. Finally, the RF signals produced are distributed to different columns and rows of the PAA. The delays of optical carriers with different wavelengths are separately tuned by DCFs, while relative signal delays in various cores of fiber are controlled by TDLs. This architecture features a time delay variation of less than 1 ps with a drift in room temperature, which is much more stable than that (~15 ps) with a single-mode fiber. Similarly, the 2D OBFN with fiber dispersion delay architectures has a large size and high sensitivity to changes in the environment, since discrete devices and fibers are applied in this architecture.

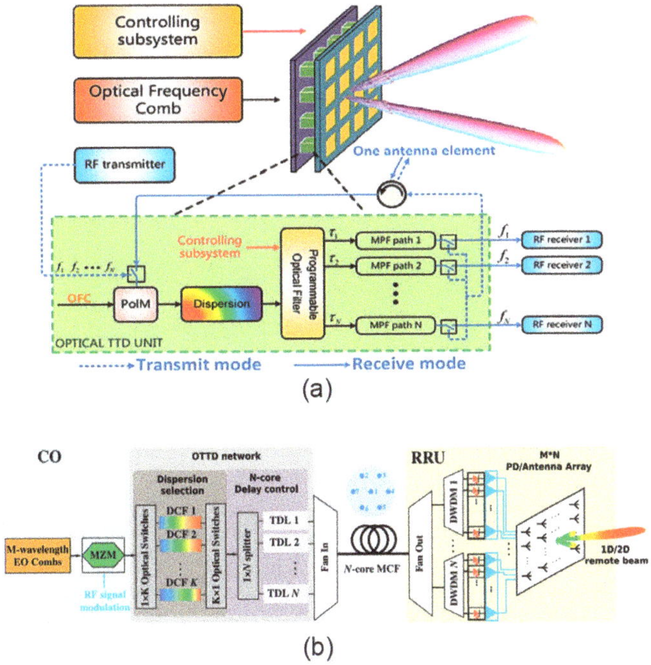

Figure 6. Schematic diagrams of 2D OBFNs based on fiber dispersion delays; (**a**) an OBFN transceiver using an optical frequency comb and fiber dispersion units (reprinted with permission from [51] © The Optical Society), and (**b**) an OBFN transmitter using electro-optic frequency combs and MCF (reprinted with permission from [56] © The Optical Society).

3.2. MRR Group Delay Architectures

With photonic integration technology, a large number of TTDs can be integrated on one substrate, which greatly reduces the size of OBFNs. The MRR group delay architecture, employing multiple MRRs as TTDs, is one approach of such OBFNs within a relatively wide bandwidth, which is typically several gigahertz [26,47,57]. Figure 7a plots the structure of the MRR, which comprises one microring and one bus waveguide. The time delays of a MRR with a round-trip time of 20 ps (free spectral range = 50 GHz) are shown in Figure 7b, as calculated using the following equation [31] in the case that the optical loss of the MRR is ignored:

$$\tau_g(\omega) = \Big(\frac{\gamma^2 - \gamma\sqrt{1-K}\cos(\omega\tau_r)}{\gamma^2 + 1 - K - 2\gamma\sqrt{1-K}\cos(\omega\tau_r)} + \frac{\gamma\sqrt{1-K}\cos(\omega\tau_r) - \gamma^2(1-K)}{1 + \gamma^2(1-K) - 2\gamma\sqrt{1-K}\cos(\omega\tau_r)}\Big)\tau_r \quad (13)$$

where K is the coupling coefficient, γ is the optical loss factor, τ_r is the round-trip time (s), and ω is the angular frequency of light (rad/s). When ignoring the optical loss of the MRR, $\gamma = 1$. By changing the K, the variation of the group delay induces different time delays of light. The time delay has a larger variation at the on-resonance wavelength than that at the off-resonance/anti-resonance wavelength. The MRR group delay architectures can adopt the time delays at the on-resonance wavelength and off-resonance wavelength. Figure 8a,b shows two examples of these cases, respectively, with non-coherent optical carriers and a binary tree structure fabricated on the silicon nitride platform [26,31,58]. To obtain a large time delay (~one hundred picoseconds), the MRR group delay architecture at the on-resonance wavelength uses one MRR as a basic delay unit, while the MRR group delay architecture at the off-resonance wavelength requires several MRRs, such as three MRRs. The former has a smaller size and much fewer power supplies to control the K of the MRR than those of the latter. However, the MRR group delay architectures at the off-resonance wavelength possess a high resolution of time delay, leading to finer beam-sweeping than that of the architecture at the on-resonance wavelength. It is worth noting that these MRR group delay architectures with a binary tree structure have an increasing delay ripple (around several picoseconds) with the number increase of MRRs at the first stage [58]. Moreover, thermal crosstalk is another issue in these architectures to be carefully dealt with during the design stage or the post-processing stage.

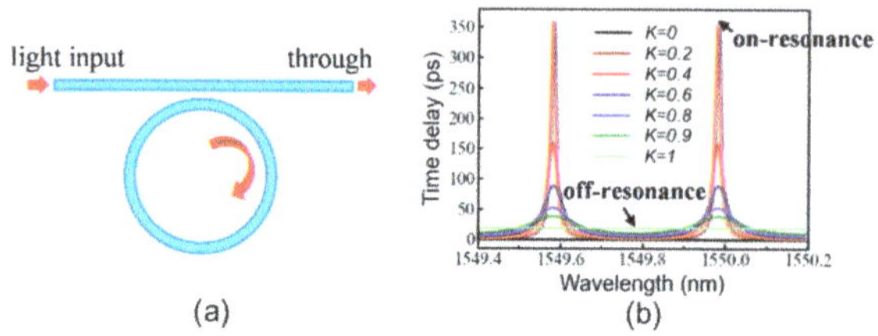

Figure 7. (**a**) Schematic diagram of MRR, and (**b**) time delay of MRR.

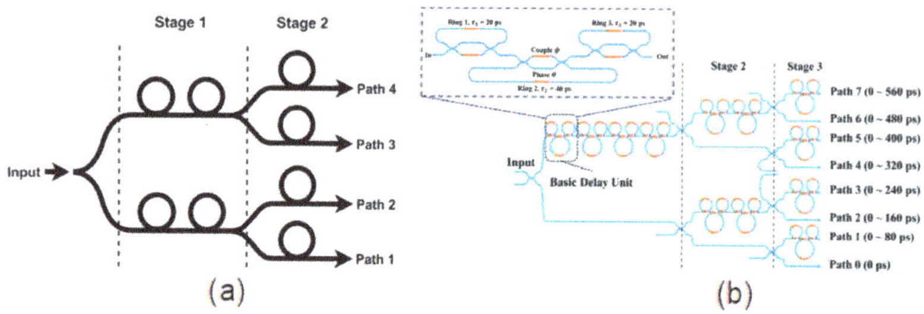

Figure 8. Schematic diagrams of MRR group delays at the (**a**) on-resonant wavelength [58] (Copyright 2017, IEEE, #1298491-1), and (**b**) off-resonant wavelength [31] (Copyright 2022, IEEE, #5491090143023).

Furthermore, MRR group delay architectures can also be applied in OBFNs with coherent optical carriers. The main advantage of this kind of OBFN is that the intensity of the noise in the optical signals can be nearly eliminated through the balance PDs (BPD) [59,60], which reduces the noise of these systems. One example of such an architecture for an OBFN receiver with N antennas has been proposed [24]; the RF signals received are modulated to the optical carrier by MZMs and phase shifted by the MRR delays, and then upper sidebands of N signal channels are selected by the optical sideband filter (OSBF) and are combined with optical carrier. Finally, the phase-aligned RF signals are produced by the BPD. The MRR delays have a binary tree structure made up of eight MRRs, seven phase shifters and seven couplers. Note that, the number of MRR is decreased from 12 to 8 compared to that in the architecture proposed in [61,62], which can reduce the size of the architecture and electric control units, especially for application in massive PAAs. In addition, this architecture can better relax the complexity and cost of modulators compared to the architecture using single-sideband (SSB) suppressed carrier modulation [23].

To realize 2D optical beamforming, one MRR group delay architecture has been proposed, as shown in Figure 9 [47]. This architecture is used as a receiver for 4 × 4 PAAs, which shows the capability of horizontal and vertical beamforming via the use of a fixed-wavelength laser, 16 MZMs, 16 × 1 MRR group delays and a PD. Among them, 16 × 1 MRR group delays have a binary tree structure that consists of 20 MRRs and corresponding power supplies. Although this 2D OBFN architecture has a small-scale integration of delays, the numbers of MRRs and power supplies are huge when adopted in massive PAAs. In addition, MRR group delay architectures require an electric monitoring and control circuit for each MRR, thanks to the MRR being highly sensitive to temperature fluctuation. Therefore, MRR group delay architectures need a large amount of electric circuits to control the time delays and operation points of MRRs. One approach to reduce the complexity of the MRR group delay architecture for 2D beamforming is to use the wavelength multiplexing technique which utilizes the wavelength multiplexing and frequency-periodic response of an optical ring resonator [63,64]. Horizontal and vertical beamforming are realized using two cascaded 4 × 1 MRR group delay architectures for the 4 × 4 2D PAA. Note that, compared with the MRR group delay architecture using one wavelength, more than half of the total number of MRRs is reduced by that adopting multiple wavelengths.

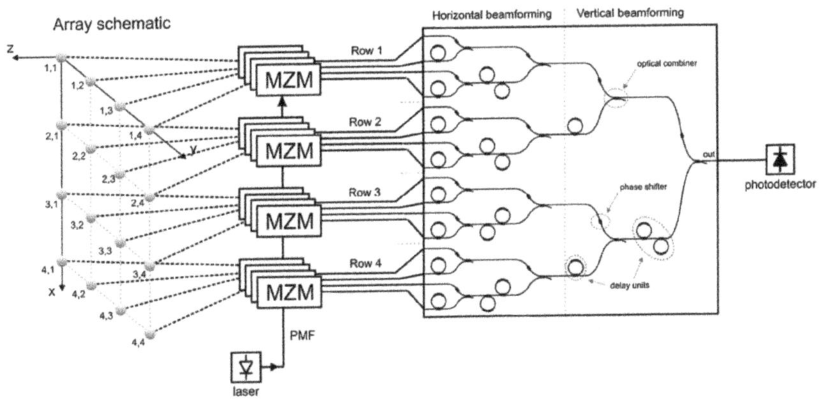

Figure 9. Schematic diagrams of 2D OBFN receivers based on MRRs with a 16 × 1 binary tree structure (reprinted with permission from [47] © The Optical Society).

3.3. MZI Delay Architectures

In addition to the MRR group delay architectures, TTDs can also be composed of a Mach–Zehnder delay interferometer (MZDI) and a MZI [45,65], and these can be called MZI delay architectures. In this architecture, MZDI is used to tune the phase or time delay of the optical carrier, in which the power coupling ratio of the upper to lower arms is varied by the MZI. With MZI delays, a TTD architecture for an OBFN receiver of N antennas has been proposed, as shown in Figure 10a [45]. The architecture includes N signal channels and one reference channel using the same laser light. N signal channels are encoded with RF signals received by antennas, while a reference channel is modulated with a local oscillator (LO) signal. After tuning the time delays, N signal channels and one reference channel are combined and injected into a BPD, producing an intermediate frequency (IF) signal with a frequency which is equal to the frequency difference between the RF and LO. The responses of amplitude and time delay of a MZDI are plotted in Figure 10b, it is clearly illustrated that these responses resemble to those of a MRR. This MZI delay architecture has a squint-free operation bandwidth of at least five percent of the RF frequency [66]. Based on this architecture, an experimental implementation has been established, as shown in Figure 11, demonstrating the receiving capabilities of I/Q signals and two beams [27]. Owing to the fact that a laser is used for the $N + 1$ channels, the architecture has a high sensitivity and low phase noise. However, the limited power of a laser will induce a weak optical signal in each channel, requiring low-loss optical devices such as low-loss phase shifters, optical splitters and couplers. One approach to relax the requirement of optical loss for optical devices is to use a power amplifier in each channel, which, however, will increase the cost and fabrication complexity.

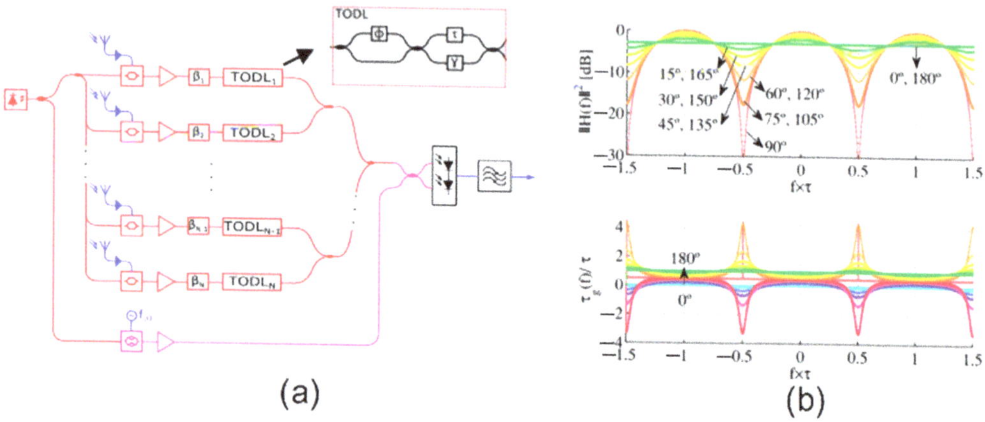

Figure 10. (**a**) Schematic diagram of OBFN receiver based on MZI delays, (**b**) amplitude and group delay responses of the MZDI in MZI delay architecture [45]. Copyright 2016, IEEE, #5491091197507.

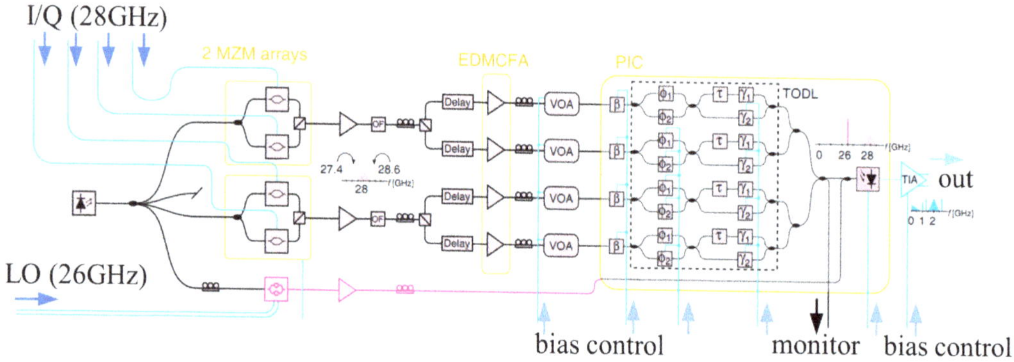

Figure 11. Schematic diagram of OBFN receiver setup based on MZI delays [27]. Figure licensed under a Creative Commons Attribution 4.0 License.

3.4. PC Delay Architectures

OBFNs using PC structures, founded on the high dispersion of the PC in the vicinity of the band edge [67], are also a promising architecture for small-scale integration. An early report of this architecture was based on PC fibers [68]. With several PC waveguides of various lengths, the architecture provides different delays for PAAs originally. Meanwhile, the beam angle can be changed by tuning the wavelength of the laser. In order to reduce the size of the PC delay architecture, integrated PC waveguides have been adopted, which can scan the beam by changing the wavelength and by thermo-optic effect simultaneously [46,69–72]. One example is based on a silicon platform and four PC waveguides of varied lengths which are integrated on one substrate, as presented in Figure 12 [46]. The 1×4 delay lines of this architecture occupy less than a 1 mm^2 area with a length of few millimeters, largely reducing the size compared to that of the counterpart employing PC fibers. However, there is an issue that the integrated PC waveguides have a mode mismatch between general waveguides on the same substrate, which induces a larger optical loss than that of the PC fibers. Similarly, PC delay architectures are sensitive to fabrication imperfection, owing to the fact that the PC waveguide has small units (~few hundred nanometers in the optical domain) and a periodic structure. Moreover, OBFNs with PC delay architectures generally have a limited delay bandwidth product for a certain length of PC waveguides [67].

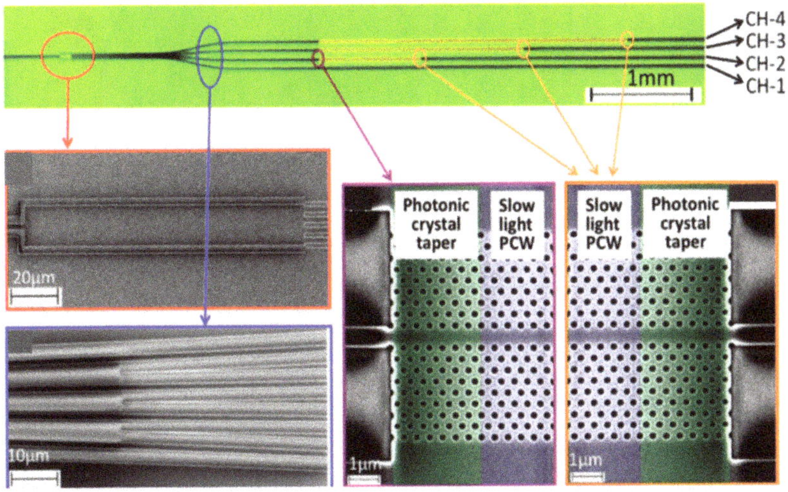

Figure 12. Schematic diagrams of OBFNs based on integrated photonic crystal waveguides [46] (Copyright 2012, AIP Publishing, #5443530971561).

3.5. Time Delay Selection Architectures

The time delay selection architecture is a significant type of OBFNs employing TTD and has squint-free behavior. According to the optical devices exploited for time delay selection, these architectures predominantly include three subclasses as plotted in Figure 13a–c: (1) the architecture with optical switches; (2) the architecture with a wavelength demultiplexer; (3) the architecture with space light modulators (SLMs). The first architecture uses an optical switch to select the delay lines of various lengths, which is widely studied and several prototypes of small-scale integration have been manufactured [50,73–77]. The optical switch can take advantage of the thermo-optic effect and electro-optic effect, which have a tuning speed of a microsecond scale and nanosecond scale, respectively. Thus, beam sweeping speed can be engineered according to the various requirements of wireless systems. The second architecture (Figure 13b) adopts the wavelength demultiplexer (DE-MUX) to select different delay lines according to the laser wavelengths; namely, the time delay of each channel in the architecture is determined by the laser wavelength [32,33,78,79]. Once the time delay is selected, the optical signals under different wavelengths are connected to the same PD by a multiplexer (MUX). This architecture can realize nanosecond-scale beam sweeping, resulting from the fast tunable laser which has a switching time of ~1 nanosecond [80]. The third architecture (Figure 13c) selects the time delays of SLMs and polarization beam splitters (PBSs) [18,43,44,81]. Owing to the fact that free-space devices are extensively applied, this architecture has a large size and weight, causing high difficulty for integration. Moreover, the number of antenna elements and structure of 2D planar PAAs should be identical to the pixel number and shape of SLM, which limits the upgrade of the architecture. Nevertheless, the SLMs are commercially available and have small optical loss [82,83], making it easy for this architecture to meet the demand of 2D PAAs with massive antenna elements. One solution to match the pixel number and shape of a SLM with 2D planar PAAs is to utilize a flexible SLM module which can change the pixel number and shape of the SLM [81].

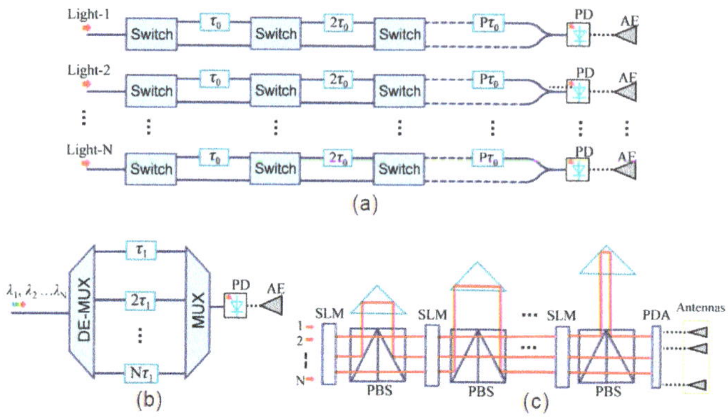

Figure 13. Schematic diagrams of OBFNs based on time delay selection by (**a**) optical switches, (**b**) wavelength de-multiplexer, and (**c**) SLMs.

As mentioned earlier, the time delay selection architecture with the wavelength de-multiplexer can tune the time delay by changing the laser wavelength. This architecture adopts multiple optical delay lines for high-resolution beam sweeping. However, these delay lines are used for one antenna, leading to the fact that a massive PAA will utilize a large number of delay lines and increasing the complexity of the system. One approach to simplify the architecture is to allocate one laser wavelength for an antenna, changing the time delay of each antenna using other delay-tuning methods [84,85]. Figure 14 shows one implementation of such an approach with dispersion components, namely linearly chirped fiber Bragg grating (LCFBG) [84]. It is similar to the fiber dispersion architecture in that the time delay difference between adjacent optical channels of the architecture is given by Equation (14) [85], where β is the dispersion coefficient of the LCFBG. This implementation architecture tunes the time delay differences of antennas by changing the wavelength spacing between the adjacent optical channels, thanks to the strain-induced period variation of the fiber Bragg grating with different periods. As a result, the time delay difference, $\Delta\tau_{g1}$, between adjacent channels changes with the variation in $\lambda_2 - \lambda_1$. The architecture is based on fibers and fiber components, which have a relatively large size and weight. To realize small-scale integration, integrated de-multiplexers have been applied to simplify the time delay selection architectures with a wavelength multiplexer [32,33]. Furthermore, when using one AWG as a de-multiplexer and the multiplexer simultaneously, the architecture has the potential for more compact integration than does its counterpart with two AWGs [49,78,79].

$$\Delta\tau_{g1} = \beta(\lambda_2 - \lambda_1) \quad (14)$$

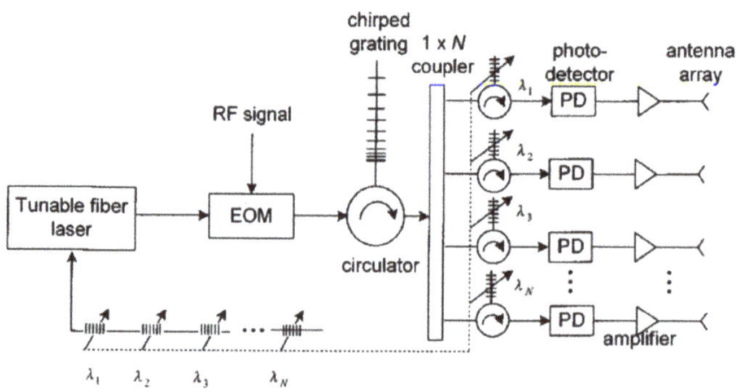

Figure 14. Schematic diagrams of OBFNs based on time delay selection with wavelength de-multiplexer and dispersion components [84]. Copyright 2002, IEEE, #5491100280556.

As addressed earlier, 2D optical beamforming is required for 2D planar PAAs. Among the OBFNs with time delay selection architectures, 2D optical beamforming has been achieved by the architectures combined wavelength multiplexers and optical switches or dispersion components. As plotted in Figure 15a, a 2D time delay selection architecture implements wavelength-dependent (WD) TTD and wavelength-independent (WI) TTD to realize horizontal beamforming and vertical beamforming, respectively [48,86]. The WD-TTD adopts fiber Bragg gratings (FBGs) to control time delays for different wavelengths, which is induced by the reflection of the FBGs with various periods, while the WI-TTD uses optical switches to set progressive delays for antenna elements in different rows. It is obvious that a large number of optical switches is demanded in this architecture. For an l-bit $\times$ n-bit beamforming system to support a 2D $p \times q$ PAA, a total number of $l + n \times q$ optical switches is needed. Alternatively, a 2D OBFN architecture with wavelength multiplexers and chirped FBGs (CFBGs), as shown in Figure 15b [87], can be used to reduce the components needed. Owing to the multi-wavelength operation capability of tunable CFBGs, this architecture can tune time delays for antenna elements in one row of PAAs by changing the dispersion of the corresponding CFBG. Meanwhile, the time delay difference between adjacent rows of PAAs is controlled by the center wavelength of the CFBG [88]. Note that, for a similar l-bit $\times$ n-bit beamforming system to support a 2D $p \times q$ PAA, a total number of p CFBGs is needed in this 2D OBFN architecture in case the tunable CFBG has an l-bit tuning ability. For the small-scale integration of this architecture, the CFBG can be substituted by integrated chirped Bragg grating or chirped sub-wavelength grating [89,90], while the de-multiplexer can be replaced by integrated AWG.

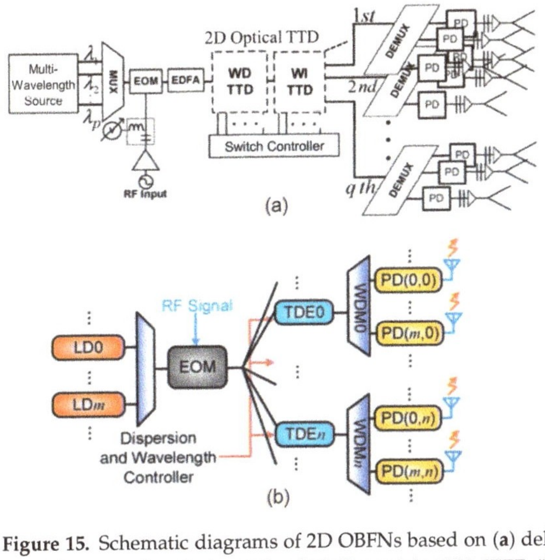

Figure 15. Schematic diagrams of 2D OBFNs based on (**a**) delays combined with wavelength de-multiplexers and optical switches [86] (Copyright 2009, IEEE, #5491100487672), and (**b**) wavelength de-multiplexers and tunable CFBGs (reprinted with permission from [87] © The Optical Society).

4. OBFNs with Phase Shifter Architectures

The phase shifter architectures are built by an array of optical phase shifters which tune the phases of RF signals. Commonly, the main advantage of this kind of architecture is that the phases of RF signals are equal to phase differences between two optical carriers owing to coherent beating at PDs, as expressed by Equation (9). In other words, the phase shifts of RF signals produced by phase shifter architectures are same as the phase shifts in the optical domain. Phase shifter architectures predominantly include four subclasses: polarization-modulated phase shifter architectures, modulator-induced phase shifter architectures, integrated phase shifter array architectures, matrix architectures (Butler matrix, Blass matrix and Nolen matrix), as shown in Figure 16 [16,22,91–96].

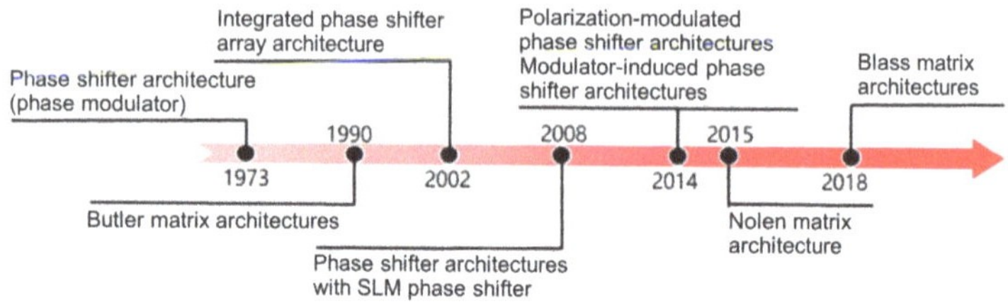

Figure 16. Timeline of phase shifter architectures proposed in the past [16,22,91–96].

4.1. Polarization-Modulated Phase Shifter Architectures

One phase shifter architecture for the OBFN receiver is presented in Figure 17a, which is a polarization-modulated phase shifter architecture that is composed of a laser diode (LD), a polarization division-multiplexing MZM, a PBS, an optical band-pass filter (OBPF), a $1 \times N$ splitter, N polarization controllers (PCs), N polarizers, and N PDs [97]. Via carrier-suppressed double-sideband modulation at the MZM, sidebands corresponding to the RF signal and local oscillator (LO) signal with orthogonal polarization directions are produced, separately. Then, the upper sidebands of the RF signal and LO signal, extracted by the OBPF, beats at the PD and down-converts into an intermediate frequency (IF) signal in this OBFN receiver architecture. The phase of each IF signal is controlled by the corresponding PC which adjusts the light polarization direction and further tunes the phase difference produced at the polarizer [98]. A similar architecture is presented in Figure 17b, in which the laser light is oriented at an angle of 45° originally and is modulated by a polarization modulator [93]. The functions of the tunable PC, polarizer and OBPF are same as those applied in the architecture before. The phase of the RF signal at one antenna can be given by Equation (15) [25], where ϕ_i is the phase of one antenna in the PAAs, and α_i is the polarization angle between one principal axis of the PolM and the polarization direction aligned by a PC. The phase induced by the phase shifter can vary in a range between 0 and 2π, if α_i changes from 0 to π. These architectures both have a key advantage in that the amplitude of the RF signal in each antenna remains unchanged when tuning the phase, since the PC, PBS and polarizer will not influence the magnitude of a circularly polarized optical signal [93]. However, the tuning speed of the PC may be too low to meet the demand of beam sweeping for PAAs. One solution to improve the tuning speed is to use the other PolM to change the phase through the electric control of its DC voltage [99].

$$\phi_i = \frac{\pi}{2} + 2\alpha_i \tag{15}$$

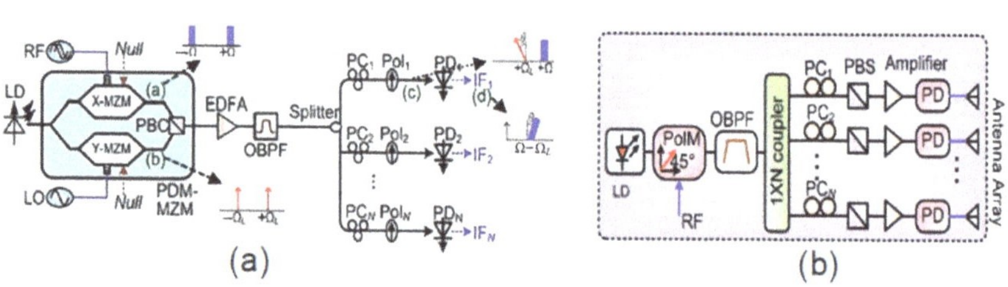

Figure 17. Schematic diagrams of OBFNs based on polarization-induced phase shifter using (**a**) polarization division-multiplexing MZM (reprinted with permission from [97] © The Optical Society), and (**b**) PolM (reprinted with permission from [93] © The Optical Society).

4.2. Modulator-Induced Phase Shifter Architectures

For optical beamforming, realizing the functions of modulation and the phase shift by the same modulator may largely reduce system complexity, especially for OBFN receivers. OBFNs with this kind of modulator can be called modulator-induced phase shifter architectures. A phase modulator is one kind of such modulators that can be adopted to tune the phase difference of two coherent optical carriers [94]. The phase modulator, manufactured by $LiNbO_3$, has two different electro-optic coefficients for the TE mode and TM mode; namely, the former is one-third of the latter, inducing a phase difference ($\Delta\phi_i$) between the TM mode and TE mode in one channel as $(2\pi V_i)/(3V_\pi)$, where V_i is the applied voltage in this channel, and V_π is the half-wave voltage of the phase modulator [94]. Thus, the phase of the RF signal at each antenna is induced by a phase difference of two coherent optical carriers in the orthogonal polarization direction. The phase difference of adjacent channels in an antenna array can be given by Equation (16), where ΔV is the difference of the applied voltages on the adjacent phase modulators. Additionally, modulation and the phase shift can be realized by a dual-drive MZM, as plotted in Figure 18 [100]. The phase shift in a channel can be induced by the bias voltage of the modulator [101]. The phase difference between adjacent channels can be expressed by Equation (17), where ΔV_{DC} is the difference of bias voltages added on the dual-drive MZMs in adjacent channels. In addition, DMZM is also modulated by the LO signal which has a small frequency difference from the RF signal and down-converts the RF signals into IF signals at the PD. With these phase shifter array architectures, a simplified OBFN without or with less discrete phase shifters can be built for PAAs operating with a given bandwidth. Although fibers are used to connect discrete components, this OBFN architecture has the advantage of immunity to the influence of the environment, due to the fact that the two coherent optical signals employed pass through the same path.

$$\Delta\phi = (2\pi\Delta V)/(3V_\pi) \quad (16)$$

$$\Delta\phi = (\pi\Delta V_{DC})/(3V_\pi) \quad (17)$$

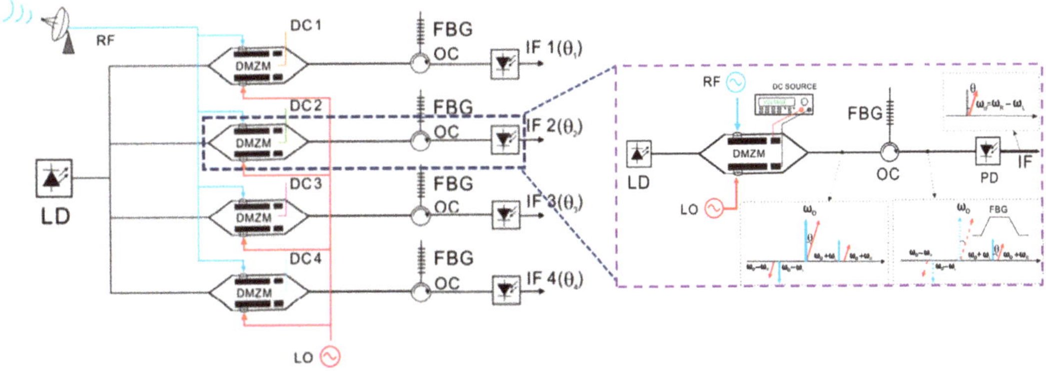

Figure 18. Schematic diagrams of OBFNs based on modulation-induced phase shifter consisting of dual-drive MZM (reprinted with permission from [100] © The Optical Society).

4.3. Integrated Phase Shifter Array Architectures

To miniaturize the OBFNs, multiple phase shifter arrays integrated on one substrate can be employed for phase shifter architectures, owing to the fact that the optical phase shifter has a small footprint for a phase shift in the optical domain. As depicted in Figure 19a, four phase shifters on a silica platform are manufactured for 1 × 4 OBFN [22]. This architecture mainly consists of two lasers, an integrated phase shifter array and four optic/millimeter-wave converters (OMC), namely optic-electric converters. The integrated phase shifter array occupies an area of 2 × 30 mm^2. Two lasers are coupled to high modulation sidebands of a master laser (LDM) and have a frequency spacing of 60.8 GHz (19 × 3.2 GHz). Therefore, except for the phase shift, the frequency up-conversion is also realized in the optical domain, which may enhance the value of the OBFNs applied in PAAs. One disadvantage of this architecture is the slow tuning speed of phase shifters using a thermo-optic effect, which is typically tens of or hundreds of microseconds. This issue can be solved by integrating electro-optic phase shifters with a much higher tuning speed. Figure 19b shows the implementation of a phase shifter array architecture using the electro-optic effect, composed of a continuous work (CW) laser, an electro-optic modulator (EOM) with the ability of single-sideband (SSB) modulation, and four beamforming network elements (BFN-E) fabricated on a silicon platform [28,102]. The BFN-E includes an optical filter, an electro-optic phase shifter, and a BPD, which occupies an area of ~4.5 mm^2. The optical filter consists of a MRR and MZI structure and separates the optical carrier and the sideband. The electro-optic phase shifter has a tuning speed of 5 ns, which is at least three orders of magnitude faster than that of the thermo-optic phase shifter. Therefore, the integrated phase shifter array architectures is promising for realizing small-scale OBFNs with a beam sweeping speed as fast as hundreds of megahertz. Note that, the loss of the electro-optic phase shifter is larger than that of the thermo-optic phase shifter due to the carrier injection in the waveguide region.

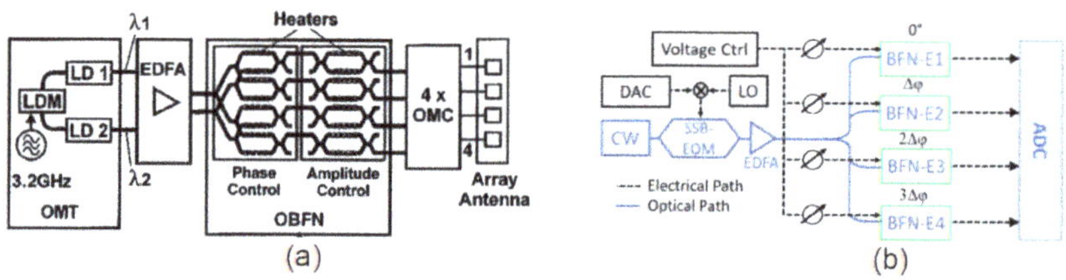

Figure 19. Schematic diagrams of OBFNs based on phase shifter array with (**a**) thermo-optic phase shifters [22] (Copyright 2002, IEEE, #5491100730358), and (**b**) electro-optic phase shifters [28] (Copyright 2020, IEEE, #5491100991991).

4.4. Matrix Architectures

Matrix architectures, including the Butler matrix, Blass matrix and Nolen matrix are popularly adopted in electric beamforming networks [103–106]. A distinguished advantage of these architectures is that multi-beam beamforming can be straightforwardly realized when injecting signals into different input ports, in addition to the fact that these architectures have compact structures. There are few works that have adopted these architectures in OBFNs [29,91,95,96,107–110], in which the optical Butler matrix architecture has the superiorities of footprint and optical loss. Figure 20a shows a 4 × 4 optical Butler matrix which consists of four 3 dB couplers, two phase shifters and a waveguide crossing [111], which is similar to the electrical counterpart. With this structure, each optical signal from one of the input ports will be split into four output ports with a linear phase relationship and an even power. For example, the phases of four optical outputs, when launching from the first input port (In_1), are φ_1, $\pi/4 + \varphi_1$, $\pi/2 + \varphi_1$, and $3\pi/4 + \varphi_1$, separately. Using the similar Butler matrix, several implementations for OBFN transmitters and receivers were established [91,107–110]. For example, a Butler matrix architecture for an OBFN receiver has been demonstrated with an 8 × 8 Butler matrix fabricated on a lithium niobate ($LiNbO_3$) platform with a footprint of ~32 mm × 0.9 mm [91]. Employing this Butler matrix in an OBFN receiver has advantages over using other OBFN receivers since each output of the matrix can combine a high-power LO for increasing the receiving power of the system.

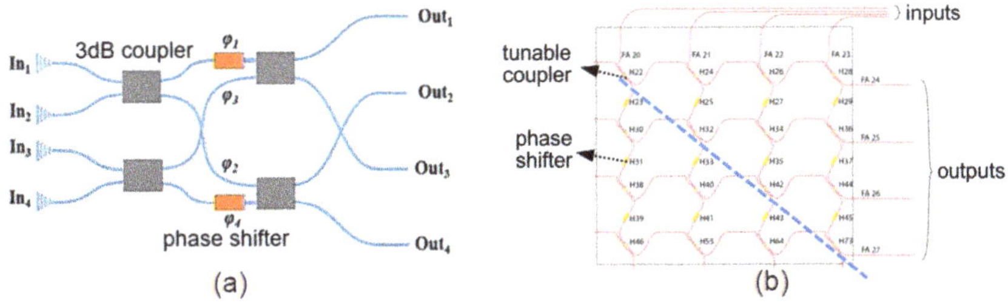

Figure 20. Schematic diagrams of (**a**) a 4 × 4 Butler matrix (reprinted with permission from [111] © The Optical Society), and (**b**) 4 × 4 Blass matrix [29] (Copyright 2021, IEEE, #5491101159355).

The typical Butler matrix shown above cannot tune the amplitude of the output ports, in order to control the side lobe and grating lobe of the beam pattern formed by the PAA. This issue can be solved by the Blass matrix and Nolen matrix, although these architectures have a larger footprint than that of the Butler matrix. Figure 20b presents a 4 × 4 Blass matrix, in which a tunable coupler together with a phase shifter is served as one node [29]. The phase and amplitude of lights at four outputs can be tuned simultaneously, and each output has the ability to export four optical signals from the four inputs. Based on this principle, an $M \times N$ Blass matrix is designed for feeding an N-element PAA [96]. This architecture adopts the self-heterodyne detection technique to produce RF signals. The main operation is as follows. M parts of the optical carrier are modulated by M independent RF signals; then, the upper sidebands are selected by the optical sideband filter and tuned by the $M \times N$ Blass matrix, and are coupled with the optical carrier at the couplers and mixed back to the RF signals at PDs. Finally, RF signals are fed to N antennas. In this architecture, $M \times N$ phase shifters and tunable couplers are utilized, which occupy a relatively large footprint and need a great number of electric control units. One approach to remediate this issue is to use a Nolen matrix, which reduces the number of nodes to half of that in the Blass matrix. The Nolen matrix has a triangular structure which can also be seen in Figure 20b by ignoring the section below the dashed line. A 144 × 36 Nolen matrix has been reported, and its beamforming pattern has been theoretically investigated [95]. It is worthy to note that the Nolen matrix architecture for OBFN has a footprint that is three orders of magnitude smaller than that of its electric counterpart.

PAAs using phase shifter architectures addressed above are not squint-free, since the phase shift provided by a phase shifter does not depend on the RF frequency, causing a change in the beam angle under different RF frequencies. However, there is a great number of applications with limited bandwidth, while exploiting phase shifters in a TTD architecture can reduce its complexity and cost [18,20,112,113]. For instance, one architecture combined with the phase shifters and TTDs has been proposed in [92,114]. In these two architectures, phase shifters and TTD can realize fine tuning and coarse tuning, separately. This feature is beneficial for scanning the beam in a broad coverage, and achieving a balance between bandwidth and system complexity. Therefore, OBFN architectures with a phase shifter array and TTDs are conducive to PAAs with a limited bandwidth.

5. Scalability and Application Potentials

Up to now, the OBFN architectures implemented have been tested for PAAs with few antenna elements, which may have resulted from the limited optical power of lasers and the large energy dissipation at several components, such as modulators, phase shifters, delay lines and PD. The scalability of OBFNs, including the performances of size, weight, power consumption, loss, bandwidth, multi-beam beamforming and linearity, should be kept in line with the increase in the scale of PAAs. Here, the scalability represents the capability to catch up the development of a wireless communication system with the explosive growth of capacity. In this section, a scalability comparison of different OBFN architectures, scalable techniques and the application potential of OBFNs is discussed.

5.1. Scalability Comparison of Different OBFN Architectures

Table 1 summarizes six performances related to the scalability of OBFN architectures which are elaborated above. The size and weight of OBFNs are intuitive performances corresponding to the number of free-space components and fiber components used in architectures [14]. The power consumptions of OBFNs are mainly induced by lasers, except for the electric and photonic amplifiers [65]. Here, the power consumptions of OBFNs are simply evaluated using the number of lasers applied in these architectures. OBFNs' loss originates from the insertion loss of components and propagation loss of waveguides. OBFNs' bandwidth is determined by the mechanisms of time delay and the bandwidth of optical devices. TTD architectures have a wider bandwidth than that of phase shifter array architectures, while MRR and MZI-TTD have a smaller bandwidth than other TTD architectures do since time delays of MRR and MZI-TTD remain constant within a limited frequency range [23,45]. Multi-beam beamforming is a significant ability of OBFNs for massive PAAs which have a narrow beam and should meet the demand of multiusers [9]. The capability of the multi-beam beamforming is assessed via considering the feasibility of producing multiple beams with current architectures. It is shown in Table 1 that, architectures of MRR group delay, MZI delay, photonic crystal, time delay selection with integrated devices, and integrated phase shifter array are of a small size, are light weight and have relatively low power consumption, which are amendable for aerospace applications and pole-mounted base stations. The fiber dispersion delay architectures and time delay selection architectures of SLMs have the advantage of low loss and wide bandwidth. Meanwhile, matrix architectures have the inherently ability of multi-beam beamforming with a relatively small size, weight and power consumption. Therefore, the scalability of these architectures may be evaluated according to the application scenario, while there is not an architecture that can meet the requirements of all applications. For example, the time delay selection architectures of SLMs can easily establish a massive OBFN for massive PAAs in a scenario without considering its size and weight; however, integrated architectures are more appropriate for aerospace antennas and pole-mounted antennas.

Table 1. Comparison of scalability for various OBFN architectures.

Schemes \ Performance	Size	Weight	Power Consumption *	Loss	Bandwidth	Multi-Beam Beamforming
I. Fiber dispersion delay architectures	+	+	+	+++	+++	+
II. MRR group delay architectures	+++	+++	++	++	++	++
III. MZI delay architectures	++	++	++	++	++	++
IV. PC delay architectures with integrated PC	+++	+++	+	+	++	++
V. Time delay selection architectures with SLMs and PBSs	+	+	+	+++	+++	++
VI. Time delay selection architectures with integrated optical switches and delay lines	++	++	++	++	+++	++
VII. Time delay selection architectures with integrated wavelength de-multiplexer	++	++	++	++	+++	+
VIII. Polarization-modulated phase shifter architectures	+	+	+	++	+	++
IX. Integrated phase shifter array architectures	+++	+++	++	++	+	++
X. Matrix architectures	++	++	++	++	+	+++

The more "+" appears, the better the corresponding performance is. * power consumption is evaluated by the number of lasers and the integration level in various architectures.

5.2. Scalable Techniques for OBFNs

Among the above architectures, MRR group delay architectures, MZI delay architectures, PC delay architectures, integrated time–time delay selection architectures and integrated phase shifter array architectures and matrix architectures are promising for small-scale integration. Nevertheless, these architectures are still of a relatively large size and demand a large number of delay units. For example, a 16×1 MRR group delay architecture is adopted to form and steer the beam received by the 4 × 4 2D PAA, which needs 20 MRRs to provide the accurate delays for 16 antennas (Figure 9). It is difficult to scale up this architecture for massive PAAs, since a thermal compensation circuit and a delay control circuit are required for each MRR except for the complexity of delays. One approach is to use the wavelength multiplexing technique as addressed earlier, dramatically decreasing the total number of MRRs from 20 to 8 for the 4 × 4 2D PAA. However, a large amount of lasers, including an integrated laser array, are required for the wavelength multiplexing technique which greatly increases the footprint and cost. One solution is to use micro-comb source (Figure 21a [115]) which can largely reduce the size and has the flexibility to meet the demands of different 2D PAAs by changing number of comb lines [56,116]. Furthermore, combining wavelength multiplexing with mode multiplexing, the channel number of integrating OBFNs can be further increased by several times [117], relaxing the requirement of a microcomb source. In addition, photonic field-programmable gate arrays based on microdisk (Figure 21b) and MRR (Figure 21c) can achieve a small footprint, time delay selection, wavelength filtering and reconfiguration simultaneously [118,119]. These structures have potential to be applied in large-scale OBFNs. To sum this up, microcomb sources and programmable structures may provide high scalability for OBFNs used in PAAs with massive antenna elements.

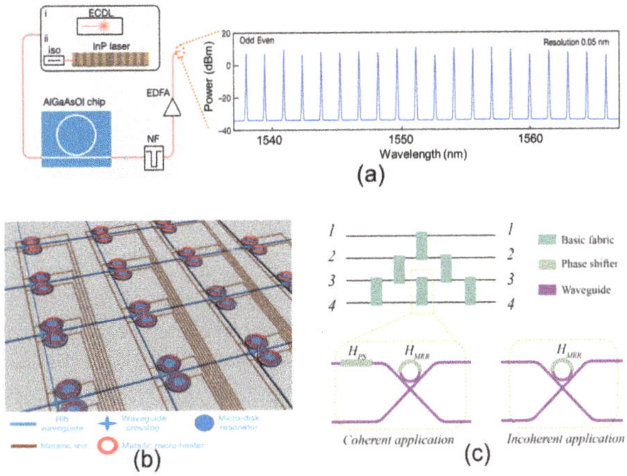

Figure 21. (**a**) A microcomb source and its spectrum [115] (figure licensed under a Creative Commons Attribution 4.0 License), (**b**) microdisk-based photonic field-programmable gate arrays [118] (figure licensed under a Creative Commons Attribution 4.0 License), and (**c**) MRR-based photonic field-programmable gate arrays [119] (Copyright 2021, AIP Publishing, #5443540906967).

For small-scale integration, a material platform with very low loss is vital for large-scale OBFNs. Currently, photonic integration circuits are predominantly implemented on platforms of silicon (Si), silicon nitride (Si_3N_4) and indium phosphide (InP). Table 2 lists the propagation losses of these material platforms [120–125], in which it is exhibited that the Si_3N_4 deposited via low-pressure chemical vapor deposition (LPCVD) has the smallest propagation loss, of less than 0.1 dB/cm. The propagation losses of Si and InP fabricated by generic foundries is about 1 dB/cm and 2 dB/cm, respectively, while Si_3N_4 deposited via plasma-enhanced chemical vapor deposition (PECVD), inductively coupled plasma chemical vapor deposition (ICP-CVD) and reactive sputtering (RS) can also achieve a small loss of ~1 dB/cm. As a result, Si_3N_4 material, especially the Si_3N_4 deposited via LPCVD, is promising for the integration of large-scale OBFNs. More recently, a wideband erbium waveguide amplifier was realized based on Si_3N_4 deposited via LPCVD [126], which may have greatly promoted the photonic integration on the Si_3N_4 platform. However, the lack of a modulator and PD hinders the full integration of photonic circuits on a Si_3N_4

platform. One probable approach is to develop a Si-Si_3N_4 monolithic integration platform as presented in Figure 22, which combines the Si active devices (such as modulator and PD) and Si_3N_4 passive devices [120,127]. Thus, this Si-Si_3N_4 photonic platform can capitalize on the advantages of Si_3N_4 passive devices, Si_3N_4 waveguide amplifiers and Si active devices.

Table 2. Typical propagation losses of Si, Si_3N_4 and InP waveguides.

Performance \ Material	Si	Si_3N_4 (LPCVD)	Si_3N_4 (PECVD)	Si_3N_4 (ICP-CVD)	Si_3N_4 (RS)	InP
Propagation loss (dB/cm)	~1.0 [120]	<0.1 [121]	~2.0 [122]	~0.8 [123]	~0.8 [124]	~2.0 [125]

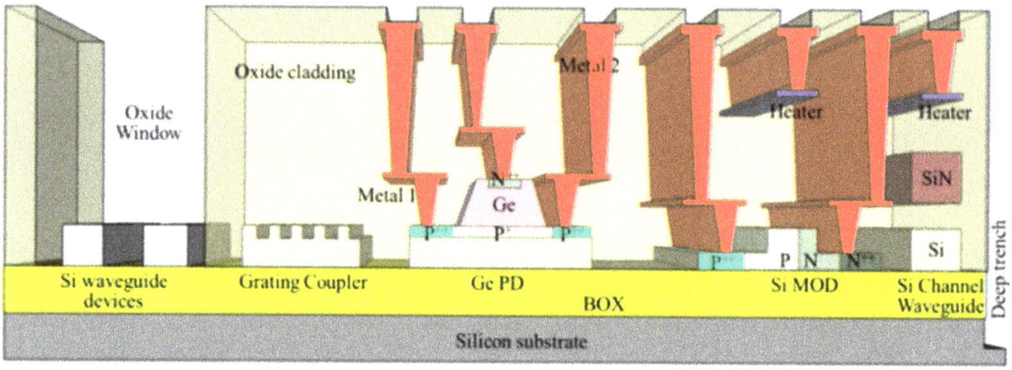

Figure 22. A Si-Si3N4 monolithic integration platform provided by Advanced Micro Foundry (AMF) [120]. Figure licensed under a Creative Commons Attribution 4.0 License.

The heterogeneous integration and monolithic integration of photonic circuits and electrical circuits (electro-photonic systems) are prominent for the compactness, energy efficiency and stability of beamforming systems [11]. The heterogeneous integration of an electro-photonic system can present the advantages of Si_3N_4 deposition via LPCVD, thanks to the separate fabrication of electric circuits and photonic circuits on different wafers. Meanwhile, the monolithic integration of an electro-photonic system can accelerate the innovation of an electro-photonic system on one substrate [128]. This electro-photonic monolithic integration can be implemented on a silicon-on-insulator (SOI) platform and bulk silicon platform, the latter being more CMOS-compatible [128,129]. To integrate Si_3N_4 devices on these platforms, Si_3N_4 material should be fabricated via low-temperature processes such as PECVD, ICP-CVD and RS, instead of LPCVD. Therefore, more efforts should be devoted to manufacturing low-loss photonic circuits and waveguide amplifiers via a low-temperature fabrication process for exploiting the advantages of a Si_3N_4 platform and the monolithic integration of electro-photonic systems.

In addition, the linearity of OBFNs is also a key factor for scalability, which is mainly determined by the linearity of modulators and PDs [130–132]. For microwave photonic systems, such as OBFNs, the linearity of the modulator is critical for the performance of whole system [132]. The linearity of a modulator can be characterized by a spurious free dynamic range (SFDR) defined as the ratio of the maximum RF power which produces third-order intermodulation distortions to noise power. Typically, the SFDR of a silicon-based MZM is smaller than 100 dB·$Hz^{2/3}$ at a 1 GHz modulation frequency, exhibiting worse linearity than do $LiNbO_3$ modulators [133]. The linearity of a modulator can be improved using response compensation techniques of MRR and the Kerr effect in a MZM [132–134]. Especially, the SFDR of a heterogeneously integrated III–V/Si MZM has been increased to ~117 dB·$Hz^{2/3}$ at 10 GHz with the assistance of MRR, as shown in Figure 23a. For mmWave wireless communication networks, high-speed and high-power

PDs are significant devices as well. The linearity of PDs can be evaluated via a consideration of the maximum output power that approaches the 1 dB compression point [130,135]. One alternative to using high-speed and high-power PDs is to use uni-traveling carrier (UTC) PDs. These PDs have superiorities of low bias voltage, high operation speed and high output power compared to common PDs, since electrons are the majority carriers as shown in Figure 23b [136–138]. Currently, UTC-PDs are mainly fabricated on Si/Germanium, InP and InGaAs material platforms [139–141]. The maximum output power of Si/Germanium UTC-PDs has reached ~0 dBm at 20 GHz [140–142]. Compared to those with a Si/Germanium platform, InP/InGaAs-based UTC-PDs have a wider bandwidth and higher output power, which has resulted in a maximum output power larger than 20 dBm at a low-frequency band of mmWave spectrum such as at 28 GHz, 40 GHz and 48 GHz [130,143–146]. Therefore, heterogeneous integrations of InP-based UTC-PDs on Si and Si_3N_4 can be employed [147–149], to obtain high-speed and high-power PDs which are comparable to their InP-based counterparts. In summary, it is promising to integrate high-linearity modulators and PDs on Si/ Si_3N_4 platforms, achieving compact and high-linearity OBFNs.

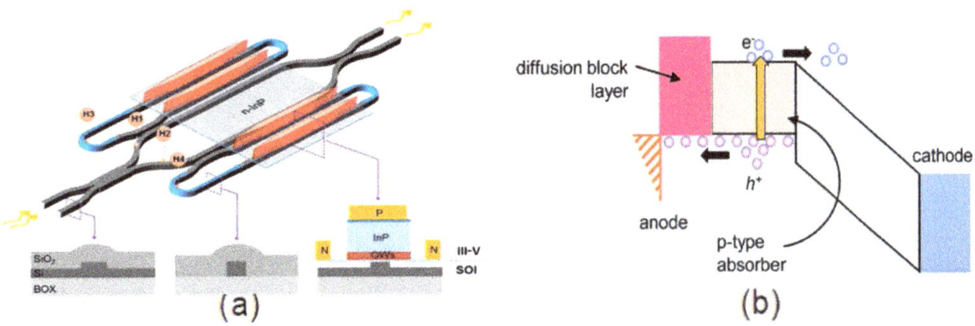

Figure 23. (**a**) A MRR-assisted MZM (reprinted with permission from [134] © The Optical Society), and (**b**) band diagram of UTC-PD [136] (Copyright 2017, AIP Publishing, #5443550099755).

5.3. Application Potential in Wireless Communication Systems

The application of OBFNs in practical wireless communication system may not be realized in a short time, suffering from the relatively low energy efficiency of electro-optic conversion at the modulator/laser and optic-electric conversion at the PD. Nevertheless, one approach that combines the advantages of analog radio over fiber (A-RoF) and OBFNs has demonstrated application potential in high-frequency wireless communications such as mmWave wireless communication [32–35,38,150]. This approach implements A-RoF in the mobile fronthaul and OBFNs as beamformers for PAAs. A-RoF technology is a promising alternative for mobile fronthaul, thanks to its high bandwidth efficiency, carrying RF signals directly on the optical signals [7,151]. Recently, A-RoF fronthaul linked with RF signals of high-level modulation formats such as 16 QAM, 32 QAM and 64 QAM have experimentally succeeded and achieved a data rate larger than 1 Gb/s for each beam [152–154]. One issue that may hinder the application of the A-RoF with OBFN is its nonlinearity, which results from the laser nonlinear effect, four-wave mixing in optical amplifier and fibers, the nonlinear transfer function of modulators, and the nonlinearity of PDs, as well as the power amplifier [153,155]. OBFNs with high-linearity modulators and high-power PDs provide a solution for this issue. The convergence of an A-RoF fronthaul and OBFN-based PAAs can not only eliminate mixers and digital–analog convertors/analog–digital convertors [32,35], but also remove the lasers and modulators required for OBFNs in antenna sites. This is because OBFNs can be deployed at the central office, which simplifies the antenna units and improves the cost effectiveness and installation flexibility [38,114,152].

6. Conclusions and Outlook

Beamforming and beamsteering through OBFNs provide promising enabling technologies for mmWave wireless communications. In this review, we analyzed typical OBFNs with TTD architectures and phase shifter architectures, introduced their principles and basic features, and conducted a performance comparison of different architectures. Furthermore, several technologies that can scale-up OBFNs were recommended, which include wavelength multiplexing using a microcomb source, MRR/microdisk-based photonic field-programmable gate arrays, the use of low-loss material platforms such as Si_3N_4, and the heterogeneous and monolithic integration of electro-photonic systems. In addition, integrated devices such as high-linearity modulators and UTC-PDs on Si/Si_3N_4 platforms are also key components in OBFNs for meeting the demands of mmWave PAAs with massive antenna elements and high excitation power. These two research topics may receive much more concern in the future. For practical applications, the convergence of an OBFN-based PAA and A-RoF is a competitive candidate technology for mmWave wireless communications.

Author Contributions: Conceptualization, F.D.; investigation, F.D., Y.G., Z.G. and Y.Y.; writing—original draft preparation, F.D.; writing—review and editing, F.D., Y.Y., Y.W. and T.C.; funding acquisition, T.C. All authors have read and agreed to the published version of the manuscript.

Funding: This work was supported by Huawei Technologies Co., Ltd.

Data Availability Statement: No new data were created or analyzed in this study. Data sharing is not applicable to this article.

Acknowledgments: The authors would like to thank Xiongbin Yu and Zhipeng Luo from the Huawei Technologies Co., Ltd., for their help with academic writing of this paper.

Conflicts of Interest: The authors declare no conflict of interest.

References

1. Li, X.; Yu, J.; Chang, G.-K. Photonics-Assisted Technologies for Extreme Broadband 5G Wireless Communications. *J. Light. Technol.* **2019**, *37*, 2851–2865. [CrossRef]
2. Chen, Y.W.; Zhang, R.; Hsu, C.W.; Chang, G.K. Key Enabling Technologies for the Post-5G Era: Fully Adaptive, All-Spectra Coordinated Radio Access Network with Function Decoupling. *IEEE Commun. Mag.* **2020**, *58*, 60–66. [CrossRef]
3. Tataria, H.; Shafi, M.; Molisch, A.F.; Dohler, M.; Sjöland, H.; Tufvesson, F. 6G Wireless Systems: Vision, Requirements, Challenges, Insights, and Opportunities. *Proc. IEEE* **2021**, *109*, 1166–1199. [CrossRef]
4. Hong, W.; Jiang, Z.H.; Yu, C.; Hou, D.; Wang, H.; Guo, C.; Hu, Y.; Kuai, L.; Yu, Y.; Jiang, Z.; et al. The Role of Millimeter-Wave Technologies in 5G/6G Wireless Communications. *IEEE J. Microw.* **2021**, *1*, 101–122. [CrossRef]
5. Rappaport, T.S.; Sun, S.; Mayzus, R.; Zhao, H.; Azar, Y.; Wang, K.; Wong, G.N.; Schulz, J.K.; Samimi, M.; Gutierrez, F. Millimeter Wave Mobile Communications for 5G Cellular: It Will Work! *IEEE Access* **2013**, *1*, 335–349. [CrossRef]
6. Rappaport, T.S.; Xing, Y.; Kanhere, O.; Ju, S.; Madanayake, A.; Mandal, S.; Alkhateeb, A.; Trichopoulos, G.C. Wireless Communications and Applications Above 100 GHz: Opportunities and Challenges for 6G and beyond. *IEEE Access* **2019**, *7*, 78729–78757. [CrossRef]
7. Lim, C.; Tian, Y.; Ranaweera, C.; Nirmalathas, T.A.; Wong, E.; Lee, K.L. Evolution of Radio-Over-Fiber Technology. *J. Light. Technol.* **2019**, *37*, 1647–1656. [CrossRef]
8. Paul, B.; Sertel, K.; Nahar, N.K. Photonic Beamforming for 5G and Beyond: A Review of True Time Delay Devices Enabling Ultra-Wideband Beamforming for mmWave Communications. *IEEE Access* **2022**, *10*, 75513–75526. [CrossRef]
9. Hong, W.; Jiang, Z.H.; Yu, C.; Zhou, J.; Chen, P.; Yu, Z.; Zhang, H.; Yang, B.; Pang, X.; Jiang, M.; et al. Multibeam Antenna Technologies for 5G Wireless Communications. *IEEE Trans. Antennas Propag.* **2017**, *65*, 6231–6249. [CrossRef]
10. Robert, M. *Phased Array Antenna Handbook*, 3rd ed.; Artech House: Norwood, MA, USA, 2017; p. 1.
11. Cao, Z.; Ma, Q.; Smolders, A.B.; Jiao, Y.; Wale, M.J.; Oh, C.W.; Wu, H.; Koonen, A.M.J. Advanced Integration Techniques on Broadband Millimeter-Wave Beam Steering for 5G Wireless Networks and Beyond. *IEEE J. Quantum Electron.* **2016**, *52*, 0600620. [CrossRef]
12. Aldaya, I.; Campuzano, G.; Castañón, G.; Aragón-Zavala, A. A Tutorial on Optical Feeding of Millimeter-Wave Phased Array Antennas for Communication Applications. *Int. J. Antennas Propag.* **2015**, *2015*, 264812. [CrossRef]
13. Pan, S.; Ye, X.; Zhang, Y.; Zhang, F. Microwave Photonic Array Radars. *IEEE J. Microw.* **2021**, *1*, 176–190. [CrossRef]

14. Anzalchi, J.; Perrott, R.; Latunde-Dada, K.; Oldenbeuving, R.; Roeloffzen, C.G.; Van Dijk, P.W.; Hoekman, M.; Leeuwis, H.; Leinse, A. Optical beamforming based on microwave photonic signal processing. In Proceedings of the International Conference on Space Optics (ICSO), Biarritz, France, 18–21 October 2016.
15. Pan, S.; Zhang, Y. Microwave Photonic Radars. *J. Light. Technol.* **2020**, *38*, 5450–5484. [CrossRef]
16. Cumming, R.C.; Matt, L.; Wright, M.L. Optically Operated Microwave Phased-Array Antenna System. U.S. Patent 3,878,520, 15 April 1975.
17. Levine, A.M. Fiber Optic Phased Array Antenna System for RF Transmission. U.S. Patent 4,028,702, 7 June 1977.
18. Goutzoulis, A.; Davies, K.; Zomp, J.; Hrycak, P.; Johnson, A. Development and field demonstration of a hardware-compressive fiber-optic true-time-delay steering system for phased-array antennas. *Appl. Opt.* **1994**, *33*, 8173–8185. [CrossRef]
19. Dolfi, D.; Michel-Gabriel, F.; Bann, S.; Huignard, J.P. Two-dimensional optical architecture for time-delay beam forming in a phased-array antenna. *Opt. Lett.* **1991**, *16*, 255–257. [CrossRef] [PubMed]
20. Frigyes, I.; Seeds, A.J. Optically generated true-time delay in phased-array antennas. *IEEE Trans. Microw. Theory Tech.* **1995**, *43*, 2378–2386. [CrossRef]
21. Capmany, J.; Novak, D. Microwave photonics combines two worlds. *Nat. Photonics* **2007**, *1*, 319–330. [CrossRef]
22. Grosskopf, G.; Eggemann, R.; Zinal, S.; Kuhlow, B.; Przyrembel, G.; Rohde, D.; Kortke, A.; Ehlers, H. Photonic 60-GHz maximum directivity beam former for smart antennas in mobile broad-band communications. *IEEE Photonics Technol. Lett.* **2002**, *14*, 1169–1171. [CrossRef]
23. Meijerink, A.; Roeloffzen, C.G.H.; Meijerink, R.; Zhuang, L.; Marpaung, D.A.I.; Bentum, M.J.; Burla, M.; Verpoorte, J.; Jorna, P.; Hulzinga, A.; et al. Novel Ring Resonator-Based Integrated Photonic Beamformer for Broadband Phased Array Receive Antennas—Part I: Design and Performance Analysis. *J. Light. Technol.* **2010**, *28*, 3–18. [CrossRef]
24. Zhuang, L.; Roeloffzen, C.G.H.; Meijerink, A.; Burla, M.; Marpaung, D.A.I.; Leinse, A.; Hoekman, M.; Heideman, R.G.; Etten, W.v. Novel Ring Resonator-Based Integrated Photonic Beamformer for Broadband Phased Array Receive Antennas—Part II: Experimental Prototype. *J. Light. Technol.* **2010**, *28*, 19–31. [CrossRef]
25. Pan, S.; Zhang, Y. Tunable and wideband microwave photonic phase shifter based on a single-sideband polarization modulator and a polarizer. *Opt. Lett.* **2012**, *37*, 4483–4485. [CrossRef] [PubMed]
26. Liu, Y.; Wichman, A.R.; Isaac, B.; Kalkavage, J.; Adles, E.J.; Clark, T.R.; Klamkin, J. Ultra-Low-Loss Silicon Nitride Optical Beamforming Network for Wideband Wireless Applications. *IEEE J. Sel. Top. Quantum Electron.* **2018**, *24*, 8300410. [CrossRef]
27. Duarte, V.C.; Prata, J.G.; Ribeiro, C.F.; Nogueira, R.N.; Winzer, G.; Zimmermann, L.; Walker, R.; Clements, S.; Filipowicz, M.; Napierała, M.; et al. Modular coherent photonic-aided payload receiver for communications satellites. *Nat. Commun.* **2019**, *10*, 1984. [CrossRef] [PubMed]
28. Serafino, G.; Porzi, C.; Hussain, B.; Scotti, F.; Falconi, F.; Chiesa, M.; Toccafondo, V.; Bogoni, A.; Ghelfi, P. High-Performance Beamforming Network Based on Si-Photonics Phase Shifters for Wideband Communications and Radar Applications. *IEEE J. Sel. Top. Quantum Electron.* **2020**, *26*, 6101011. [CrossRef]
29. Muñoz, R.; Rommel, S.; Dijk, P.v.; Brenes, J.; Grivas, E.; Manso, C.; Roeloffzen, C.; Vilalta, R.; Fabrega, J.M.; Landi, G.; et al. Experimental Demonstration of Dynamic Optical Beamforming for Beyond 5G Spatially Multiplexed Fronthaul Networks. *IEEE J. Sel. Top. Quantum Electron.* **2021**, *27*, 8600216. [CrossRef]
30. Santacruz, J.P.; Rommel, S.; Roeloffzen, C.G.H.; Timens, R.B.; Dijk, P.W.L.; Jurado-Navas, A.; Monroy, I.T. Incoherent Optical Beamformer for ARoF Fronthaul in Mm-Wave 5G/6G Networks. *J. Light. Technol.* **2022**, *41*, 1325–1334. [CrossRef]
31. Sun, H.; Lu, L.; Liu, Y.; Ni, Z.; Chen, J.; Zhou, L. Broadband 1×8 Optical Beamforming Network Based on Anti-resonant Microring Delay Lines. *J. Light. Technol.* **2022**, *40*, 6919–6928. [CrossRef]
32. Cao, Z.; Zhao, X.; Soares, F.M.; Tessema, N.; Koonen, A.M.J. 38-GHz Millimeter Wave Beam Steered Fiber Wireless Systems for 5G Indoor Coverage: Architectures, Devices, and Links. *IEEE J. Quantum Electron.* **2017**, *53*, 8000109. [CrossRef]
33. Zhang, X.; Zhao, M.; Jiao, Y.; Cao, Z.; Koonen, A.M.J. Integrated Wavelength-Tuned Optical mm-Wave Beamformer with Doubled Delay Resolution. *J. Light. Technol.* **2020**, *38*, 2353–2359. [CrossRef]
34. Moerman, A.; Kerrebrouck, J.V.; Caytan, O.; Paula, I.L.d.; Bogaert, L.; Torfs, G.; Demeester, P.; Rogier, H.; Lemey, S. Beyond 5G without Obstacles: mmWave-over-Fiber Distributed Antenna Systems. *IEEE Commun. Mag.* **2022**, *60*, 27–33. [CrossRef]
35. Paula, I.L.d.; Bogaert, L.; Caytan, O.; Kerrebrouck, J.V.; Moerman, A.; Muneeb, M.; Brande, Q.V.d.; Torfs, G.; Bauwelinck, J.; Rogier, H.; et al. Air-Filled SIW Remote Antenna Unit with True Time Delay Optical Beamforming for mmWave-Over-Fiber Systems. *J. Light. Technol.* **2022**, *40*, 6961–6975. [CrossRef]
36. Lu, F.; Xu, M.; Shen, S.; Alfadhli, Y.M.; Cho, H.J.; Chang, G.K. Demonstration of Inter-Dimensional Adaptive Diversity Combining and Repetition Coding in Converged MMW/FSO Links for 5G and beyond Mobile Fronthaul. In Proceedings of the Optical Fiber Communications Conference and Exposition, San Diego, CA, USA, 11 March 2018.
37. Zhang, R.; Lu, F.; Xu, M.; Liu, S.; Peng, P.C.; Shen, S.; He, J.; Cho, H.J.; Zhou, Q.; Yao, S.; et al. An Ultra-Reliable MMW/FSO A-RoF System Based on Coordinated Mapping and Combining Technique for 5G and Beyond Mobile Fronthaul. *J. Light. Technol.* **2018**, *36*, 4952–4959. [CrossRef]
38. Morant, M.; Trinidad, A.; Tangdiongga, E.; Koonen, T.; Llorente, R. Experimental Demonstration of mm-Wave 5G NR Photonic Beamforming Based on ORRs and Multicore Fiber. *IEEE Trans. Microw. Theory Tech.* **2019**, *67*, 2928–2935. [CrossRef]
39. Rotman, R.; Tur, M.; Yaron, L. True Time Delay in Phased Arrays. *Proc. IEEE* **2016**, *104*, 504–518. [CrossRef]
40. John, L.V. *Antenna Engineering Handbook*, 4th ed.; McGraw-Hill Education: New York, NY, USA, 2007; pp. 3-4–3-13.

41. Zhou, L.; Wang, X.; Lu, L.; Chen, J. I Integrated optical delay lines: A review and perspective nvited. *Chin. Opt. Lett.* **2018**, *16*, 101301. [CrossRef]
42. Esman, R.D.; Frankel, M.Y.; Dexter, J.L.; Goldberg, L.; Parent, M.G.; Stilwell, D.; Cooper, D.G. Fiber-optic prism true time-delay antenna feed. *IEEE Photonics Technol. Lett.* **1993**, *5*, 1347–1349. [CrossRef]
43. Frankel, M.Y.; Matthews, P.J.; Esman, R.D. Two-dimensional fiber-optic control of a true time-steered array transmitter. *IEEE Trans. Microw. Theory Tech.* **1996**, *44*, 2696–2702. [CrossRef]
44. Riza, N.A. Transmit/receive time-delay beam-forming optical architecture for phased-array antennas. *Appl. Opt.* **1991**, *30*, 4594–4595. [CrossRef]
45. Duarte, V.C.; Drummond, M.V.; Nogueira, R.N. Photonic True-Time-Delay Beamformer for a Phased Array Antenna Receiver based on Self-Heterodyne Detection. *J. Light. Technol.* **2016**, *34*, 5566–5575. [CrossRef]
46. Lin, C.-Y.; Subbaraman, H.; Hosseini, A.; Wang, A.X.; Zhu, L.; Chen, R.T. Silicon nanomembrane based photonic crystal waveguide array for wavelength-tunable true-time-delay lines. *Appl. Phys. Lett.* **2012**, *101*, 051101. [CrossRef]
47. Burla, M.; Roeloffzen, C.G.H.; Zhuang, L.; Marpaung, D.; Khan, M.R.; Maat, P.; Dijkstra, K.; Leinse, A.; Hoekman, M.; Heideman, R. System integration and radiation pattern measurements of a phased array antenna employing an integrated photonic beamformer for radio astronomy applications. *Appl. Opt.* **2012**, *51*, 789–802. [CrossRef] [PubMed]
48. Jung, B.M.; Shin, J.D.; Kim, B.G. Optical True Time-Delay for Two-Dimensional X-Band Phased Array Antennas. *IEEE Photonics Technol. Lett.* **2007**, *19*, 877–879. [CrossRef]
49. Piqueras, M.A.; Grosskopf, G.; Vidal, B.; Herrera, J.; Martinez, J.M.; Sanchis, P.; Polo, V.; Corral, J.L.; Marceaux, A.; Galiere, J.; et al. Optically beamformed beam-switched adaptive antennas for fixed and mobile broad-band wireless access networks. *IEEE Trans. Microw. Theory Tech.* **2006**, *54*, 887–899. [CrossRef]
50. Rasras, M.S.; Madsen, C.K.; Cappuzzo, M.A.; Chen, E.; Gomez, L.T.; Laskowski, E.J.; Griffin, A.; Wong-Foy, A.; Gasparyan, A.; Kasper, A.; et al. Integrated resonance-enhanced variable optical delay lines. *IEEE Photonics Technol. Lett.* **2005**, *17*, 834–836. [CrossRef]
51. Ye, X.; Zhang, F.; Pan, S. Optical true time delay unit for multi-beamforming. *Opt. Express* **2015**, *23*, 10002–10008. [CrossRef]
52. Furuya, K.; Hirasawa, T.; Oishi, M.; Akiba, S.; Hirokawa, J.; Ando, M. 60 GHz-Band Photonic-Integrated Array-Antenna and Module for Radio-over-Fiber-Based Beam Forming. *IEICE Trans. Commun.* **2017**, *100-B*, 1717–1725. [CrossRef]
53. Esman, R.D.; Monsma, M.J.; Dexter, J.L.; Cooper, D.G. Microwave true time-delay modulator using fibre-optic dispersion. *Electron. Lett.* **1992**, *28*, 1905–1908. [CrossRef]
54. Gustavsson, U.; Frenger, P.; Fager, C.; Eriksson, T.; Zirath, H.; Dielacher, F.; Studer, C.; Pärssinen, A.; Correia, R.; Matos, J.N.; et al. Implementation Challenges and Opportunities in Beyond-5G and 6G Communication. *IEEE J. Microw.* **2021**, *1*, 86–100. [CrossRef]
55. Hu, Y.; Zhan, J.; Jiang, Z.H.; Yu, C.; Hong, W. An Orthogonal Hybrid Analog–Digital Multibeam Antenna Array for Millimeter-Wave Massive MIMO Systems. *IEEE Trans. Antennas Propag.* **2021**, *69*, 1393–1403. [CrossRef]
56. Zhang, C.; Lei, P.; Liu, R.; He, B.; Chen, Z.; Xie, X.; Hu, W. Large-scale true-time-delay remote beamforming with EO frequency combs and multicore fiber. *Opt. Lett.* **2021**, *46*, 3793–3796. [CrossRef]
57. Burla, M.; Khan, M.R.H.; Marpaung, D.A.I.; Roeloffzen, C.G.H.; Maat, P.; Dijkstra, K.; Leinse, A.; Hoekman, M.; Heideman, R. Squint-free beamsteering demonstration using a photonic integrated beamformer based on optical ring resonators. In Proceedings of the IEEE International Topical Meeting on Microwave Photonics, Montreal, QC, Canada, 5–9 October 2010.
58. Liu, Y.; Wichman, A.; Isaac, B.; Kalkavage, J.; Adles, E.J.; Clark, T.R.; Klamkin, J. Tuning Optimization of Ring Resonator Delays for Integrated Optical Beam Forming Networks. *J. Light. Technol.* **2017**, *35*, 4954–4960. [CrossRef]
59. Abbas, G.; Chan, V.; Ting, Y. A dual-detector optical heterodyne receiver for local oscillator noise suppression. *J. Light. Technol.* **1985**, *3*, 1110–1122. [CrossRef]
60. Meijerink, A.; Roeloffzen, C.G.H.; Zhuang, L.; Marpaung, D.A.I.; Heideman, R.G.; Borreman, A.; Etten, W.v. Phased Array Antenna Steering Using a Ring Resonator-Based Optical Beam Forming Network. In Proceedings of the Symposium on Communications and Vehicular Technology, Liege, Belgium, 23 November 2006.
61. Zhuang, L.; Roeloffzen, C.G.H.; Heideman, R.G.; Borreman, A.; Meijerink, A.; Etten, W.v. Single-Chip Ring Resonator-Based 1×8 Optical Beam Forming Network in CMOS-Compatible Waveguide Technology. *IEEE Photonics Technol. Lett.* **2007**, *19*, 1130–1132. [CrossRef]
62. Schippers, H.; Verpoorte, J.; Jorna, P.; Hulzinga, A.; Zhuang, L.; Meijerink, A.; Roeloffzen, C.G.H.; Marpaung, D.A.I.; Etten, W.v.; Heideman, R.G.; et al. Broadband optical beam forming for airborne phased array antenna. In Proceedings of the IEEE Aerospace conference, Big Sky, MT, USA, 7–14 March 2009.
63. Burla, M.; Khan, R.; Zhuang, L.; Roeloffzen, C. Multiwavelength optical beam forming network with ring resonator-based binary-tree architecture for broadband phased array antenna systems. In Proceedings of the 13th Annual Symposium of the IEEE/LEOS Benelux Chapter, Enschede, The Netherlands, 27–28 November 2008.
64. Burla, M.; Marpaung, D.A.I.; Zhuang, L.; Khan, M.R.; Leinse, A.; Beeker, W.; Hoekman, M.; Heideman, R.G.; Roeloffzen, C.G.H. Multiwavelength-Integrated Optical Beamformer Based on Wavelength Division Multiplexing for 2-D Phased Array Antennas. *J. Light. Technol.* **2014**, *32*, 3509–3520. [CrossRef]
65. Duarte, V.; Prata, J.; Nogueira, R.; Winzer, G.; Zimmermann, L.; Walker, R.; Clements, S.; Filipowicz, M.; Napierala, M.; Nasilowski, T.; et al. Modular and smooth introduction of photonics in high-throughput communication satellites—Perspective of project BEACON. In Proceedings of the International Conference on Space Optics (ICSO), Chania, Greece, 9–12 October 2018.

66. Drummond, M.V.; Monteiro, P.P.; Nogueira, R.N. Photonic True-Time Delay Beamforming Based on Polarization-Domain Interferometers. *J. Light. Technol.* **2010**, *28*, 2492–2498. [CrossRef]
67. Baba, T. Slow light in photonic crystals. *Nat. Photonics* **2008**, *2*, 465–473. [CrossRef]
68. Yongqiang, J.; Howley, B.; Zhong, S.; Qingjun, Z.; Chen, R.T.; Chen, M.Y.; Brost, G.; Lee, C. Dispersion-enhanced photonic crystal fiber array for a true time-delay structured X-band phased array antenna. *IEEE Photonics Technol. Lett.* **2005**, *17*, 187–189. [CrossRef]
69. Ishikura, N.; Hosoi, R.; Hayakawa, R.; Tamanuki, T.; Shinkawa, M.; Baba, T. Photonic crystal tunable slow light device integrated with multi-heaters. *Appl. Phys. Lett.* **2012**, *100*, 221110. [CrossRef]
70. Takeuchi, G.; Terada, Y.; Takeuchi, M.; Abe, H.; Ito, H.; Baba, T. Thermally controlled Si photonic crystal slow light waveguide beam steering device. *Optics Express* **2018**, *26*, 11529–11537. [CrossRef]
71. Ito, H.; Kusunoki, Y.; Maeda, J.; Akiyama, D.; Kodama, N.; Abe, H.; Tetsuya, R.; Baba, T. Wide beam steering by slow-light waveguide gratings and a prism lens. *Optica* **2020**, *7*, 47–52. [CrossRef]
72. Tamanuki, T.; Ito, H.; Baba, T. Thermo-Optic Beam Scanner Employing Silicon Photonic Crystal Slow-Light Waveguides. *J. Light. Technol.* **2021**, *39*, 904–911. [CrossRef]
73. Fathpour, S.; Riza, N. Silicon-photonics-based wideband radar beamforming: Basic design. *Opt. Eng.* **2010**, *49*, 018201. [CrossRef]
74. Wang, X.; Zhou, L.; Li, R.; Xie, J.; Lu, L.; Wu, K.; Chen, J. Continuously tunable ultra-thin silicon waveguide optical delay line. *Optica* **2017**, *4*, 507–515. [CrossRef]
75. Liu, Y.; Isaac, B.; Kalkavage, J.; Adles, E.; Clark, T.; Klamkin, J. 93-GHz Signal Beam Steering with True Time Delayed Integrated Optical Beamforming Network. In Proceedings of the Optical Fiber Communication Conference, San Diego, CA, USA, 3 March 2019.
76. Trinidad, A.M.; Cao, Z.; van Zantvoort, J.H.C.; Tangdiongga, E.; Koonen, A.M.J. Broadband and continuous beamformer based on switched delay lines cascaded by optical ring resonator. In Proceedings of the Optical Fiber Communication Conference, San Diego, CA, USA, 3 March 2019.
77. Zhu, C.; Lu, L.; Shan, W.; Xu, W.; Zhou, G.; Zhou, L.; Chen, J. Silicon integrated microwave photonic beamformer. *Optica* **2020**, *7*, 1162–1170. [CrossRef]
78. Yaron, L.; Rotman, R.; Zach, S.; Tur, M. Photonic Beamformer Receiver with Multiple Beam Capabilities. *IEEE Photonics Technol. Lett.* **2010**, *22*, 1723–1725. [CrossRef]
79. Tessema, N.; Yan, F.; Cao, Z.; Tangdiongga, E.; Koonen, A.M.J. Compact and tunable AWG-based true-time delays for multi-Gbps radio beamformer. In Proceedings of the European Conference on Optical Communication (ECOC), Valencia, Spain, 27 September 2015.
80. Vidal, B.; Mengual, T.; Martí, J. Fast Optical Beamforming Architectures for Satellite-Based Applications. *Adv. Opt. Technol.* **2012**, *2012*, 385409. [CrossRef]
81. Riza, N.A.; Khan, S.A.; Arain, M.A. Flexible beamforming for optically controlled phased array antennas. *Opt. Commun.* **2003**, *227*, 301–310. [CrossRef]
82. Beeckman, J.; Neyts, K.; Vanbrabant, P. Liquid-crystal photonic applications. *Opt. Eng.* **2011**, *50*, 081202. [CrossRef]
83. Bleha, W.; Lei, L.A. Advances in Liquid Crystal on Silicon (LCOS) spatial light modulator technology. In Proceedings of the SPIE Defense, Security, and Sensing, Baltimore, MA, USA, 4 June 2013.
84. Jianping, Y.; Jianliang, Y.; Yunqi, L. Continuous true-time-delay beamforming employing a multiwavelength tunable fiber laser source. *IEEE Photonics Technol. Lett.* **2002**, *14*, 687–689. [CrossRef]
85. Zhang, J.; Yao, J. Photonic True-Time Delay Beamforming Using a Switch-Controlled Wavelength-Dependent Recirculating Loop. *J. Light. Technol.* **2016**, *34*, 3923–3929. [CrossRef]
86. Jung, B.M.; Yao, J. A Two-Dimensional Optical True Time-Delay Beamformer Consisting of a Fiber Bragg Grating Prism and Switch-Based Fiber-Optic Delay Lines. *IEEE Photonics Technol. Lett.* **2009**, *21*, 627–629. [CrossRef]
87. Ye, X.; Zhang, F.; Pan, S. Compact optical true time delay beamformer for a 2D phased array antenna using tunable dispersive elements. *Opt. Lett.* **2016**, *41*, 3956–3959. [CrossRef]
88. Painchaud, Y.; Paquet, C.; Guy, M. Optical Tunable Dispersion Compensators based on Thermally Tuned Fiber Bragg Gratings. *Opt. Photon. News* **2007**, *18*, 48–53. [CrossRef]
89. Spasojevic, M.; Chen, L.R. Discretely tunable optical delay lines using serial and step-chirped sidewall Bragg gratings in SOI. *Electron. Lett.* **2013**, *49*, 608–610. [CrossRef]
90. Sun, H.; Wang, Y.; Chen, L.R. Integrated Discretely Tunable Optical Delay Line Based on Step-Chirped Subwavelength Grating Waveguide Bragg Gratings. *J. Light. Technol.* **2020**, *38*, 5551–5560. [CrossRef]
91. Charczenko, W.; Surette, M.; Matthews, P.; Klotz, H.; Mickelson, A. Integrated optical Butler matrix for beam forming in phased-array antennas. In Proceedings of the Optoelectronic Signal Processing for Phase-Array Antennas II, Los Angeles, CA, USA, 1 June 1990.
92. Jofre, L.; Stoltidou, C.; Blanch, S.; Mengual, T.; Vidal, B.; Marti, J.; McKenzie, I.; Cura, J.M.d. Optically Beamformed Wideband Array Performance. *IEEE Trans. Antennas Propag.* **2008**, *56*, 1594–1604. [CrossRef]
93. Zhang, Y.; Wu, H.; Zhu, D.; Pan, S. An optically controlled phased array antenna based on single sideband polarization modulation. *Optics Express* **2014**, *22*, 3761–3765. [CrossRef]

94. Shi, S.; Bai, J.; Schneider, G.J.; Zhang, Y.; Nelson, R.; Wilson, J.; Schuetz, C.; Grund, D.W.; Prather, D.W. Conformal Wideband Optically Addressed Transmitting Phased Array with Photonic Receiver. *J. Light. Technol.* **2014**, *32*, 3468–3477. [CrossRef]
95. Roeloffzen, C.G.H.; Oldenbeuving, R.M.; Timens, R.B.; van Dijk, P.W.L.; Taddei, C.; Leinse, A.; Hoekman, M.; Heideman, R.G.; Zhuang, L.; Marpaung, D.A.I.; et al. Integrated Optical Beamformers. In Proceedings of the Optical Fiber Communication Conference, Los Angeles, CA, USA, 22 March 2015.
96. Tsokos, C.; Mylonas, E.; Groumas, P.; Katopodis, V.; Gounaridis, L.; Timens, R.B.; Oldenbeuving, R.M.; Roeloffzen, C.G.H.; Avramopoulos, H.; Kouloumentas, C. Analysis of a Multibeam Optical Beamforming Network Based on Blass Matrix Architecture. *J. Light. Technol.* **2018**, *36*, 3354–3372. [CrossRef]
97. Gao, Y.; Wen, A.; Tu, Z.; Zhang, W.; Lin, L. Simultaneously photonic frequency downconversion, multichannel phase shifting, and IQ demodulation for wideband microwave signals. *Opt. Lett.* **2016**, *41*, 4484–4487. [CrossRef]
98. Gao, Y.; Wen, A.; Liu, L.; Tian, S.; Xiang, S.; Wang, Y. Compensation of the Dispersion-Induced Power Fading in an Analog Photonic Link Based on PM–IM Conversion in a Sagnac Loop. *J. Light. Technol.* **2015**, *33*, 2899–2904. [CrossRef]
99. Zhang, W.; Yao, J. Ultrawideband RF Photonic Phase Shifter Using Two Cascaded Polarization Modulators. *IEEE Photonics Technol. Lett.* **2014**, *26*, 911–914. [CrossRef]
100. Jiang, T.; Yu, S.; Wu, R.; Wang, D.; Gu, W. Photonic downconversion with tunable wideband phase shift. *Opt. Lett.* **2016**, *41*, 2640–2643. [CrossRef] [PubMed]
101. Dubovitsky, S.; Steier, W.H.; Yegnanarayanan, S. Analysis and Improvement of Mach-Zehnder Modulator Linearity Performance for Chirped and Tunable Optical Carriers. *J. Light. Technol.* **2002**, *20*, 858. [CrossRef]
102. Porzi, C.; Serafino, G.; Sans, M.; Falconi, F.; Sorianello, V.; Pinna, S.; Mitchell, J.E.; Romagnoli, M.; Bogoni, A.; Ghelfi, P. Photonic Integrated Microwave Phase Shifter up to the mm-Wave Band with Fast Response Time in Silicon-on-Insulator Technology. *J. Light. Technol.* **2018**, *36*, 4494–4500. [CrossRef]
103. Lin, T.H.; Hsu, S.K.; Wu, T.L. Bandwidth Enhancement of 4 × 4 Butler Matrix Using Broadband Forward-Wave Directional Coupler and Phase Difference Compensation. *IEEE Trans. Microw. Theory Tech.* **2013**, *61*, 4099–4109. [CrossRef]
104. Zhong, L.H.; Ban, Y.L.; Lian, J.W.; Yang, Q.L.; Guo, J.; Yu, Z.F. Miniaturized SIW Multibeam Antenna Array Fed by Dual-Layer 8 × 8 Butler Matrix. *IEEE Antennas Wirel. Propag. Lett.* **2017**, *16*, 3018–3021. [CrossRef]
105. Lian, J.W.; Ban, Y.L.; Xiao, C.; Yu, Z.F. Compact Substrate-Integrated 4 × 8 Butler Matrix with Sidelobe Suppression for Millimeter-Wave Multibeam Application. *IEEE Antennas Wirel. Propag. Lett.* **2018**, *17*, 928–932. [CrossRef]
106. Ren, H.; Arigong, B.; Zhou, M.; Ding, J.; Zhang, H. A Novel Design of 4 ×4 Butler Matrix with Relatively Flexible Phase Differences. *IEEE Antennas Wirel. Propag. Lett.* **2016**, *15*, 1277–1280. [CrossRef]
107. Madrid, D.; Vidal, B.; Martinez, A.; Polo, V.; Corral, J.L.; Marti, J. A novel 2N beams heterodyne optical beamforming architecture based on N×N optical Butler matrices. In Proceedings of the International Microwave Symposium Digest, Seattle, WA, USA, 2–7 June 2002.
108. Piqueras, M.A.; Cuesta-Soto, F.; Villalba, P.; Martí, A.; Hakansson, A.; Perdigués, J.; Caille, G. Photonic beamforming network for multibeam satellite-on-board phased-array antennas. In Proceedings of the International Conference on Space Optics, Toulouse, France, 14–17 October 2008.
109. Piqueras, M.A.; Mengual, T.; Navasquillo, O.; Sotom, M.; Caille, G. Opto-microwave, Butler matrixes based front-end for a multi-beam large direct radiating array antenna. In Proceedings of the International Conference on Space Optics Tenerife, Canary Islands, Spain, 6–10 October 2014.
110. Belkin, M.E.; Fofanov, D.A.; Sigov, A.S. Computer-Aided Design of an Integrated-Photonic Butler Matrix for a True-Time Delay Millimeter-Wave Antenna Feeder Network. In Proceedings of the Radiation and Scattering of Electromagnetic Waves (RSEMW), Divnomorskoe, Russia, 28 June–2 July 2021.
111. Lu, P.; Xu, W.; Zhu, C.; Liu, C.; Lu, L.; Zhou, L.; Chen, J. Integrated multi-beam optical phased array based on a 4 × 4 Butler matrix. *Opt. Lett.* **2021**, *46*, 1566–1569. [CrossRef]
112. Vidal, B.; Mengual, T.; Ibanez-Lopez, C.; Marti, J. Optical Beamforming Network Based on Fiber-Optical Delay Lines and Spatial Light Modulators for Large Antenna Arrays. *IEEE Photonics Technol. Lett.* **2006**, *18*, 2590–2592. [CrossRef]
113. Mengual, T.; Vidal, B.; Stoltidou, C.; Blanch, S.; Martí, J.; Jofre, L.; McKenzie, I.; del Cura, J.M. Optical phase-based beamformer using MZM SSB modulation combined with crystal polarization optics and a spatial light modulator. *Opt. Commun.* **2008**, *281*, 217–224. [CrossRef]
114. Ito, K.; Suga, M.; Shirato, Y.; Kita, N.; Onizawa, T. Remote Beamforming Scheme with Fixed Wavelength Allocation for Radio-Over-Fiber Systems Employing Single-Mode Fiber. *J. Light. Technol.* **2022**, *40*, 997–1006. [CrossRef]
115. Shu, H.; Chang, L.; Tao, Y.; Shen, B.; Xie, W.; Jin, M.; Netherton, A.; Tao, Z.; Zhang, X.; Chen, R.; et al. Microcomb-driven silicon photonic systems. *Nature* **2022**, *605*, 457–463. [CrossRef] [PubMed]
116. Xue, X.; Xuan, Y.; Bao, C.; Li, S.; Zheng, X.; Zhou, B.; Qi, M.; Weiner, A.M. Microcomb-Based True-Time-Delay Network for Microwave Beamforming with Arbitrary Beam Pattern Control. *J. Light. Technol.* **2018**, *36*, 2312–2321. [CrossRef]
117. Xu, H.; Liu, C.; Dai, D.; Shi, Y. Direct-access mode-division multiplexing switch for scalable on-chip multi-mode networks. *Nanophotonics* **2021**, *10*, 4551–4566. [CrossRef]
118. Zhang, W.; Yao, J. Photonic integrated field-programmable disk array signal processor. *Nat. Commun.* **2020**, *11*, 406. [CrossRef]
119. Yi, D.; Wang, Y.; Tsang, H.K. Multi-functional photonic processors using coherent network of micro-ring resonators. *APL Photonics* **2021**, *6*, 100801. [CrossRef]

120. Siew, S.Y.; Li, B.; Gao, F.; Zheng, H.Y.; Zhang, W.; Guo, P.; Xie, S.W.; Song, A.; Dong, B.; Luo, L.W.; et al. Review of Silicon Photonics Technology and Platform Development. *J. Light. Technol.* **2021**, *39*, 4374–4389. [CrossRef]
121. Roeloffzen, C.G.H.; Hoekman, M.; Klein, E.J.; Wevers, L.S.; Timens, R.B.; Marchenko, D.; Geskus, D.; Dekker, R.; Alippi, A.; Grootjans, R.; et al. Low-Loss Si3N4 TriPleX Optical Waveguides: Technology and Applications Overview. *IEEE J. Sel. Top. Quantum Electron.* **2018**, *24*, 4400321. [CrossRef]
122. Mao, S.C.; Tao, S.H.; Xu, Y.L.; Sun, X.W.; Yu, M.B.; Lo, G.Q.; Kwong, D.L. Low propagation loss SiN optical waveguide prepared by optimal low-hydrogen module. *Optics Express* **2008**, *16*, 20809–20816. [CrossRef] [PubMed]
123. Shao, Z.; Chen, Y.; Chen, H.; Zhang, Y.; Zhang, F.; Jian, J.; Fan, Z.; Liu, L.; Yang, C.; Zhou, L.; et al. Ultra-low temperature silicon nitride photonic integration platform. *Optics Express* **2016**, *24*, 1865–1872. [CrossRef]
124. Frigg, A.; Boes, A.; Ren, G.; Abdo, I.; Choi, D.-Y.; Gees, S.; Mitchell, A. Low loss CMOS-compatible silicon nitride photonics utilizing reactive sputtered thin films. *Optics Express* **2019**, *27*, 37795–37805. [CrossRef] [PubMed]
125. Augustin, L.M.; Santos, R.; Haan, E.d.; Kleijn, S.; Thijs, P.J.A.; Latkowski, S.; Zhao, D.; Yao, W.; Bolk, J.; Ambrosius, H.; et al. InP-Based Generic Foundry Platform for Photonic Integrated Circuits. *IEEE J. Sel. Top. Quantum Electron.* **2018**, *24*, 6100210. [CrossRef]
126. Liu, Y.; Qiu, Z.; Ji, X.; Lukashchuk, A.; He, J.; Riemensberger, J.; Hafermann, M.; Wang, R.N.; Liu, J.; Ronning, C.; et al. A photonic integrated circuit-based erbium-doped amplifier. *Science* **2022**, *376*, 1309–1313. [CrossRef] [PubMed]
127. Sacher, W.D.; Mikkelsen, J.C.; Huang, Y.; Mak, J.C.C.; Yong, Z.; Luo, X.; Li, Y.; Dumais, P.; Jiang, J.; Goodwill, D.; et al. Monolithically Integrated Multilayer Silicon Nitride-on-Silicon Waveguide Platforms for 3-D Photonic Circuits and Devices. *Proc. IEEE* **2018**, *106*, 2232–2245. [CrossRef]
128. Atabaki, A.H.; Moazeni, S.; Pavanello, F.; Gevorgyan, H.; Notaros, J.; Alloatti, L.; Wade, M.T.; Sun, C.; Kruger, S.A.; Meng, H.; et al. Integrating photonics with silicon nanoelectronics for the next generation of systems on a chip. *Nature* **2018**, *556*, 349–354. [CrossRef]
129. Sun, C.; Wade, M.T.; Lee, Y.; Orcutt, J.S.; Alloatti, L.; Georgas, M.S.; Waterman, A.S.; Shainline, J.M.; Avizienis, R.R.; Lin, S.; et al. Single-chip microprocessor that communicates directly using light. *Nature* **2015**, *528*, 534–538. [CrossRef]
130. Beling, A.; Xie, X.; Campbell, J.C. High-power, high-linearity photodiodes. *Optica* **2016**, *3*, 328–338. [CrossRef]
131. Bass, J.; Tran, H.; Du, W.; Soref, R.; Yu, S.-Q. Impact of nonlinear effects in Si towards integrated microwave-photonic applications. *Optics Express* **2021**, *29*, 30844–30856. [CrossRef]
132. Feng, H.; Zhang, K.; Sun, W.; Ren, Y.; Zhang, Y.; Zhang, W.; Wang, C. Ultra-high-linearity integrated lithium niobate electro-optic modulators. *Photonics Res.* **2022**, *10*, 2366–2373. [CrossRef]
133. Bottenfield, C.G.; Thomas, V.A.; Ralph, S.E. Silicon Photonic Modulator Linearity and Optimization for Microwave Photonic Links. *IEEE J. Sel. Top. Quantum Electron.* **2019**, *25*, 3400110. [CrossRef]
134. Zhang, C.; Morton, P.A.; Khurgin, J.B.; Peters, J.D.; Bowers, J.E. Ultralinear heterogeneously integrated ring-assisted Mach–Zehnder interferometer modulator on silicon. *Optica* **2016**, *3*, 1483–1488. [CrossRef]
135. Yang, Z.; Yu, Q.; Zang, J.; Campbell, J.C.; Beling, A. Phase-Modulated Analog Photonic Link with a High-Power High-Linearity Photodiode. *J. Light. Technol.* **2018**, *36*, 3805–3814. [CrossRef]
136. Ishibashi, T.; Ito, H. Uni-traveling-carrier photodiodes. *J. Appl. Phys.* **2020**, *127*, 031101. [CrossRef]
137. Umezawa, T.; Kanno, A.; Kashima, K.; Matsumoto, A.; Akahane, K.; Yamamoto, N.; Kawanishi, T. Bias-Free Operational UTC-PD above 110 GHz and Its Application to High Baud Rate Fixed-Fiber Communication and W-Band Photonic Wireless Communication. *J. Light. Technol.* **2016**, *34*, 3138–3147. [CrossRef]
138. Muramoto, Y.; Yoshimatsu, T.; Nada, M.; Ishibashi, T. High-speed photodetector technologies. *NTT Tech. Rev.* **2012**, *10*, 1–5.
139. Ito, H.; Kodama, S.; Muramoto, Y.; Furuta, T.; Nagatsuma, T.; Ishibashi, T. High-speed and high-output InP-InGaAs unitraveling-carrier photodiodes. *IEEE J. Sel. Top. Quantum Electron.* **2004**, *10*, 709–727. [CrossRef]
140. Piels, M.; Bowers, J.E. Si/Ge uni-traveling carrier photodetector. *Optics Express* **2012**, *20*, 7488–7495. [CrossRef]
141. Piels, M.; Bowers, J.E. 40 GHz Si/Ge Uni-Traveling Carrier Waveguide Photodiode. *J. Light. Technol.* **2014**, *32*, 3502–3508. [CrossRef]
142. Fu, Z.; Yu, H.; Wei, Z.; Xia, P.; Zhang, Q.; Wang, X.; Huang, Q.; Wang, Y.; Yang, J. High-Power and High-Speed Ge/Si Traveling-Wave Photodetector Optimized by Genetic Algorithm. *J. Light. Technol.* **2022**, *41*, 240–248. [CrossRef]
143. Carey, V.A.; Konkol, M.R.; Harrity, C.E.; Shahid, E.L.; Schuetz, C.A.; Yao, P.; Prather, D.W. W-Band Pulse Generation Using Phase-Locked Lasers and High-Power Photodiode. *IEEE Photonics Technol. Lett.* **2022**, *34*, 645–648. [CrossRef]
144. Peng, Y.; Sun, K.; Shen, Y.; Beling, A.; Campbell, J.C. High-Power and High-Linearity Photodiodes at 1064 nm. *J. Light. Technol.* **2020**, *38*, 4850–4856. [CrossRef]
145. Cross, A.S.; Zhou, Q.; Beling, A.; Fu, Y.; Campbell, J.C. High-power flip-chip mounted photodiode array. *Optics Express* **2013**, *21*, 9967–9973. [CrossRef]
146. Xie, X.; Zhou, Q.; Li, K.; Shen, Y.; Li, Q.; Yang, Z.; Beling, A.; Campbell, J.C. Improved power conversion efficiency in high-performance photodiodes by flip-chip bonding on diamond. *Optica* **2014**, *1*, 429–435. [CrossRef]
147. Maes, D.; Reis, L.; Poelman, S.; Vissers, E.; Avramovic, V.; Zaknoune, M.; Roelkens, G.; Lemey, S.; Peytavit, E.; Kuyken, B. High-Speed Photodiodes on Silicon Nitride with a Bandwidth beyond 100 GHz. In Proceedings of the Conference on Lasers and Electro-Optics (CLEO), San Jose, CA, USA, 15–20 May 2022.

148. Xie, X.; Qiugui, Z.; Norberg, E.; Jacob-Mitos, M.; Yaojia, C.; Ramaswamy, A.; Fish, G.; Bowers, J.E.; Campbell, J.; Beling, A. Heterogeneously integrated waveguide-coupled photodiodes on SOI with 12 dBm output power at 40 GHz. In Proceedings of the Optical Fiber Communications Conference and Exhibition (OFC), Los Angeles, CA, USA, 22–26 March 2015.
149. Wang, Y.; Wang, Z.; Yu, Q.; Xie, X.; Posavitz, T.; Jacob-Mitos, M.; Ramaswamy, A.; Norberg, E.J.; Fish, G.A.; Beling, A. High-Power Photodiodes With 65 GHz Bandwidth Heterogeneously Integrated onto Silicon-on-Insulator Nano-Waveguides. *IEEE J. Sel. Top. Quantum Electron.* **2018**, *24*, 6000206. [CrossRef]
150. Cao, Z.; Lu, R.; Wang, Q.; Tessema, N.; Jiao, Y.; van den Boom, H.P.; Tangdiongga, E.; Koonen, A.M. Cyclic additional optical true time delay for microwave beam steering with spectral filtering. *Opt. Lett.* **2014**, *39*, 3402–3405. [CrossRef]
151. Ruggeri, E.; Tsakyridis, A.; Vagionas, C.; Leiba, Y.; Kalfas, G.; Pleros, N.; Miliou, A. Multi-User V-Band Uplink Using a Massive MIMO Antenna and a Fiber-Wireless IFoF Fronthaul for 5G mmWave Small-Cells. *J. Light. Technol.* **2020**, *38*, 5368–5374. [CrossRef]
152. Morant, M.; Trinidad, A.; Tangdiongga, E.; Koonen, T.; Llorente, R. Multi-Beamforming Provided by Dual-Wavelength True Time Delay PIC and Multicore Fiber. *J. Light. Technol.* **2020**, *38*, 5311–5317. [CrossRef]
153. Vagionas, C.; Ruggeri, E.; Tsakyridis, A.; Kalfas, G.; Leiba, Y.; Miliou, A.; Pleros, N. Linearity Measurements on a 5G mmWave Fiber Wireless IFoF Fronthaul Link with Analog RF Beamforming and 120° Degrees Steering. *IEEE Commun. Lett.* **2020**, *24*, 2839–2843. [CrossRef]
154. Tsakyridis, A.; Ruggeri, E.; Kalfas, G.; Oldenbeuving, R.M.; Dijk, P.W.L.v.; Roeloffzen, C.G.H.; Leiba, Y.; Miliou, A.; Pleros, N.; Vagionas, C. Reconfigurable Fiber Wireless IFoF Fronthaul with 60 GHz Phased Array Antenna and Silicon Photonic ROADM for 5G mmWave C-RANs. *IEEE J. Sel. Areas Commun.* **2021**, *39*, 2816–2826. [CrossRef]
155. Liu, S.; Xu, M.; Wang, J.; Lu, F.; Zhang, W.; Tian, H.; Chang, G.K. A Multilevel Artificial Neural Network Nonlinear Equalizer for Millimeter-Wave Mobile Fronthaul Systems. *J. Light. Technol.* **2017**, *35*, 4406–4417. [CrossRef]

MDPI AG
Grosspeteranlage 5
4052 Basel
Switzerland
Tel.: +41 61 683 77 34

Applied Sciences Editorial Office
E-mail: applsci@mdpi.com
www.mdpi.com/journal/applsci

www.ingramcontent.com/pod-product-compliance
Lightning Source LLC
Chambersburg PA
CBHW061506210926
51850CB00008B/1837